Representation Type of Commutative Noetherian Rings III: Global Wildness and Tameness

MEMOIRS
of the
American Mathematical Society

Number 832

Representation Type of Commutative Noetherian Rings III: Global Wildness and Tameness

Lee Klingler
Lawrence S. Levy

July 2005 • Volume 176 • Number 832 (end of volume) • ISSN 0065-9266

American Mathematical Society
Providence, Rhode Island

2000 *Mathematics Subject Classification.*
Primary 13E05, 16G60.

Library of Congress Cataloging-in-Publication Data

Klingler, Lee, 1955–
Representation type of commutative Noetherian rings III : global wildness and tameness / Lee Klingler, Lawrence S. Levy.
p. cm. — (Memoirs of the American Mathematical Society, ISSN 0065-9266 ; no. 832)
"Volume 176, number 832 (end of volume)."
Includes bibliographical references and index.
ISBN 0-8218-3738-9 (alk. paper)
1. Noetherian rings. 2. Representations of rings (Algebra). I. Levy, Lawrence S., 1933– II. Title. III Series.

QA3.A57 no. 832
[QA251.3]
510 s—dc22
[512′.4] 2005044092

Memoirs of the American Mathematical Society

This journal is devoted entirely to research in pure and applied mathematics.

Subscription information. The 2005 subscription begins with volume 173 and consists of six mailings, each containing one or more numbers. Subscription prices for 2005 are $606 list, $485 institutional member. A late charge of 10% of the subscription price will be imposed on orders received from nonmembers after January 1 of the subscription year. Subscribers outside the United States and India must pay a postage surcharge of $31; subscribers in India must pay a postage surcharge of $43. Expedited delivery to destinations in North America $35; elsewhere $130. Each number may be ordered separately; *please specify number* when ordering an individual number. For prices and titles of recently released numbers, see the New Publications sections of the *Notices of the American Mathematical Society.*

Back number information. For back issues see the *AMS Catalog of Publications.*

Subscriptions and orders should be addressed to the American Mathematical Society, P. O. Box 845904, Boston, MA 02284-5904, USA. *All orders must be accompanied by payment.* Other correspondence should be addressed to 201 Charles Street, Providence, RI 02904-2294, USA.

Memoirs of the American Mathematical Society is published bimonthly (each volume consisting usually of more than one number) by the American Mathematical Society at 201 Charles Street, Providence, RI 02904-2294, USA. Periodicals postage paid at Providence, RI. Postmaster: Send address changes to Memoirs, American Mathematical Society, 201 Charles Street, Providence, RI 02904-2294, USA.

This publication is indexed in *Science Citation Index®, SciSearch®, Research Alert®, CompuMath Citation Index®, Current Contents®/Physical, Chemical & Earth Sciences.*
Printed in the United States of America.

∞ The paper used in this book is acid-free and falls within the guidelines established to ensure permanence and durability.
Visit the AMS home page at `http://www.ams.org/`

10 9 8 7 6 5 4 3 2 1 10 09 08 07 06 05

Contents

1. Abstract

This memoir completes the series of papers beginning with [**KL1, KL2**], showing that, for a commutative noetherian ring Λ, either the category of Λ-modules of finite length has wild representation type or else we can describe the category of finitely generated Λ-modules, including their direct-sum relations and local-global relations. (There is a possible exception to our results, involving characteristic 2.)

Levy wishes to express his appreciation to the NSA for continuing to support this project after he retired from teaching so that he could finish it, to the University of Nebraska for its willingness to administer this support, and to Professor Roger Wiegand for his help in setting this up, as well as for his mathematical help.

We are indebted to Jan Trlifaj for several insiteful observations that allowed Section 18 to take a much more definitive form than it originally had.

1991 *Mathematics Subject Classification.* 13E05, 16G60

Key words and phrases. Tame representation type, wild representation type, commutative noetherian ring, Dedekind-like ring.

2. Introduction

In 1911 Steinitz extended the structure theorem for finitely generated abelian groups by proving his classification of isomorphism classes in fingen(Ω), the category of finitely generated modules over (what are now called) Dedekind domains Ω [**S**]. One of the most interesting parts of Steinitz's results is that he was also able to describe the direct-sum relations in fingen(Ω). Steinitz's theorem has been particularly resistant to generalization. In fact, the only noetherian integral domains Ω (other than Dedekind domains) for which such a module classification for fingen(Ω) was known prior to completion of this project seem to be the Dedekind-like rings studied in [**L2**] (a subclass of the rings called "Dedekind-like" in the present memoir). However, for commutative noetherian rings that are not domains, some other results exist [**NR, NRSB**].

On the other hand, during the past 30 years or so, the concept of wild repsentation type has arisen as an insurmountable obstacle to getting a classification theorem for the isomorphism classes in fingen(A) when A is a finite dimensional algebra over a field. For an early result that provided part of the motivation for this series of papers, see [**Ri**]. The results there and after it show that the vast majority of finite dimensional algebras over algebraically closed fields have wild representation type. Moreover, no wild algebra A has a complete classification theorem for the isomorphism classes in fingen(A). (However, we know of no theorem in mathematical logic saying that such a classification cannot be proved.)

The objective of this memoir is to describe the category fingen(Ω) where Ω is any commutative noetherian ring such that fingen(Ω) does not have wild representation type. This completes the project begun in [**KL1, KL2**] where Ω was restrictred to be a complete local ring. We say that Ω is fingen-*tame* — more completely, fingen(Ω) *has tame representation type* — if we can describe the isomorphism classes, direct-sum relations, and local-global relations in this category, and give some information about the homomorphisms in this category.

Let $\mathfrak{m}$ be a maximal ideal of a commutative noetherian ring Ω. Informally, we say that Ω is finlen-*wild* (wrt $\mathfrak{m}$) — more completely, the category finlen(Ω) of Ω-modules of finite length has *wild representation type (with respect to* $\mathfrak{m}$) — if any description of all isomorphism classes in finlen(Ω) would have to contain a description of all isomorphism classes in finlen(A) for *all* finite-dimensional (noncommutative!) algebras A over the field $k = \Omega/\mathfrak{m}$. See Definition 14.5 for our formal definition of finlen-wild, and [**KL1**, Remarks 2.3] for a brief introduction to the notion of wildness from the standpoint of commutative noetherian rings.

We define a class of commutative noetherian rings Λ called "Dedekind-like" [Definition 10.1], and a (very small) class of commutative artinian local rings of length 4 called "Klein rings" [Definitions 14.2]. The reason for these definitions is the following result (whose "indecomposability" hypothesis involves no loss of generality, since Ω is noetherian).

THEOREM 2.1 (see Theorem 14.5). *If an indecomposable a commutative noetherian ring* Ω *is such that* finlen(Ω) *fails to be wild (wrt* $\mathfrak{m}$) *for all maximal ideals* $\mathfrak{m}$, *then* Ω *is a homomorphic image of a Dedekind-like ring, or else is a Klein ring.*

[1]Received by the editor February 20, 2003, and in revised form June 30, 2004. Corrections February 2, 2005.

After proving this we spend the rest of the memoir proving tameness of fingen(Λ) when Λ is Dedekind-like — that is, describing its isomorphism classes, direct-sum relations, and local-golobal relations — and giving examples. The fingen-tameness of all homomorphic images of Dedekind-like rings is an obvious consequence. A possible slight exception, involving characteristic 2, to our tameness results is stated in Additional Hypothesis 10.2. The complete local case of all this was the subject of [**KL1, KL2**].

The much simpler case of fingen-tameness of Klein rings is contained in [**KL2**, §11, especially Theorem 11.3], because all commutative artinian local rings are complete local rings.

Dedekind-like rings are a small class of reduced noetherian rings of Krull dimension 1 whose ideals are all generated by two elements [Corollary 10.9]. Thus, as in the case of finite dimensional algebras, the overwhelming majority of noetherian rings are finlen-wild (wrt some $\mathfrak{m}$). In particular, all noetherian rings of Krull dimension ≥ 2 are finlen-wild. This paucity of tame rings immediately raises the question of whether there are interesting examples of Dedekind-like rings other than direct sums of Dedekind domains.

EXAMPLES 2.2 (Natural Examples of Dedekind-like rings).

(i) $\mathbb{Z}[\sqrt{n}\,]$ when n is squarefree.
(ii) $\mathbb{Z}G_n$ (integral group ring of any cyclic group of order n) when n squarefree.
(iii) All subrings of squarefree index in $\mathbb{Z} \oplus \ldots \oplus \mathbb{Z}$.
(iv) (Over any algebraically closed field:) The coordinate ring of any affine curve whose singularities are all simple nodes.
(v) $k[x,y]/(xy)$ and $k[[x,y]]/(xy)$ for any field k.
(vi) $\mathbb{R}+x\mathbb{C}[x]$ and $\mathbb{R}+x\mathbb{C}[[x]]$ (polynomials and formal power series rings over the complex numbers, with real constant term).

For proofs, see the following. (i): [Example 36.3]. (ii): [**L3**] (where the more restrictive definition of "Dedekind-like" is phrased somewhat differently than the definition in this memoir). (iii): By Corollary 10.7 it suffices to show that all localizations of this ring at maximal ideals are Dedekind-like. This is done in [**KL2**, 12.5]. (v) and (vi): The power series rings are Dedekind-like by [**KL1**, 2.17 and 2.18]. For the polynomial version, use this and the fact that noetherian rings are Dedekind-like if and only if all completions at maximal ideals are Dedekind-like [Theorem 11.9] (and note that discrete valuation rings are Dedekind-like). (iv) A purely algebraic statement of a property, from which (iv) follows, is: *Suppose that the completion of a noetherian ring Ω at each maximal ideal is either isomorphic to $k[[x,y]]/(xy)$ or $k[[x]]$, for some field k. Then Ω is Dedekind-like.* The proof is the same as that of statement (v), with the polynomial ring replaced by Ω.

We included rings (i) in our list because rings of algebraic integers are the rings that interested Steinitz. Dedekind-like rings of algebraic integers may be the only non-integrally closed rings of algebraic integers whose finitely generated module category has been described.

We included examples (vi) because they seem to be the simplest fingen-tame rings that are infinite dimensional algebras over a non-algebraically-closed field and have no counterpart over algebraically closed fields.

We now describe the main contents of this memoir — omitting most technical details and definitions (but giving references to them). From now on, unless otherwise specified, *ring* means "commutative ring", and *local ring* means "noetherian local ring". Moreover Λ always denotes Dedekind-like ring with normalization Γ, in this introduction..

Chapter 1: Dedekind-like Rings. We define these rings and establish some of their basic properties. For example, a noetherian ring Λ is Dedekind-like if and only if all localizations $\Lambda_{\mathfrak{m}}$ at maximal ideals are Dedekind-like [Corollary 10.7]; equivalently, if and only if all completions $\hat{\Lambda}_{\mathfrak{m}}$ are Dedekind-like [Theorem 11.9].

We also state the possible characteristic-2 exception to our tameness results [Additional Hypothesis 10.2]. In Examples 10.3 we note that this exception never arises if Λ is a ring of algebraic integers or an algebra over a field of characteristic $\neq 2$, or an algebra over any algebraically closed field. Our wildness results (e.g. Theorem 2.1) are not affected by this possible exception. We do not comment further on this possible exception.

Chapter 2: Wildness. We prove Theorem 2.1 (above). The interesting feature of the proof is that it is pure commutative algebra: "wildness" never explicitly arises here. But it does arise in [**KL1**], where the theorem is proved for complete local rings. If Ω is not a Klein ring and fails to be finlen-wild (wrt every $\mathfrak{m}$) we conclude from [**KL1**, Theorem 2.10] that every completion $\hat{\Lambda}_{\mathfrak{m}}$ is a homomorphic image of a Dedekind-like ring, and then prove that Λ itself is a homomorphic image of a Dedekind-like ring. This is done by a series of manipulations of surjective pullback squares, the subject of Section 12 [in Chapter 1].

An interesting "loose end" is that Theorem 2.1 does *not* state that no Ω can be both tame and wild. We were unable to prove this.

Chapter 3: Structure of a Genus. This begins our study of tameness. Recall that Λ denotes a Dedekind-like ring, and recall the definition of the *genus* of $M \in \text{fingen}(\Lambda)$:

$$\text{(2.2.1)} \qquad \begin{aligned} \text{genus}(M) &= \{N \in \text{fingen}(\Lambda) \mid \hat{N}_{\mathfrak{m}} \cong \hat{M}_{\mathfrak{m}} \quad \forall \mathfrak{m} \in \text{maxspec}(\Lambda)\} \\ &= \{N \in \text{fingen}(\Lambda) \mid N_{\mathfrak{m}} \cong M_{\mathfrak{m}} \quad \forall \mathfrak{m} \in \text{maxspec}(\Lambda)\} \end{aligned}$$

For every $M \in \text{fingen}(\Lambda)$ and $(\forall \mathfrak{m} \in \text{maxspec}(\Lambda))$ [**KL2**] describes the detailed structure of the $\mathfrak{m}$-adic completion $\hat{M}_{\mathfrak{m}}$. (Remarkably little of this detail is needed for what we do here; and we review this when we need it.) This chapter answers the following two question.

(i) The *package deal* question: For which families $\{M(\mathfrak{m})\}$, where each $M(\mathfrak{m}) \in \text{fingen}(\hat{\Lambda}_{\mathfrak{m}})$, does there exist $M \in \text{fingen}(\Lambda)$ such that $\hat{M}_{\mathfrak{m}} \cong M(\mathfrak{m})$ for all $\mathfrak{m}$? The question is complicated by the fact that, unlike the rings studied in integral representation theory, it is not necessarily true that $\Lambda_{\mathfrak{m}}$ is a DVR (discrete valuation ring) for all but finitely many $\mathfrak{m}$. Examples should eventually appear in [**HL**]. However, the problem is made tractable by the fact that — for every $M \in \text{fingen}(\Lambda)$ — $M_{\mathfrak{m}}$ is a free $\Lambda_{\mathfrak{m}}$-module for all but finitely many $\mathfrak{m}$. The difference between this and the classical situation is that the set of nontrivial (= nonfree) localizations changes from module to module. There is one more obviously necessary condition for the existence of M, namely suitable consistency of the (torsionfree) ranks of the $M(\mathfrak{m})$. These two conditions are proved to be sufficient in Package Deal Theorem 15.6. They are actually sufficient for rings much more general than Dedekind-like rings, a topic dealt with in detail in [**LO**].

(ii) *Indecomposability.* Given the collection of completions $\hat{M}_{\mathfrak{m}}$ of $M \in \text{fingen}(\Lambda)$, when is M indecomposable? Part of the answer is that a certain diagram — the "connections graph" of M — must be connected and all ranks of M must be at most 2 [Theorem 16.8].

Two interesting consequences are: (a) If M is indecomposable, then each $\hat{M}_{\mathfrak{m}}$ is the direct sum of at most 4 indecomposable modules [Theorem 16.10]. (b) If some rank — always meaning "torsionfree rank" — equals 2, then the indecomposable module M cannot be torsionfree [Corollary 16.9].

Chapter 4: Substitute for Conductor Squares. Let Γ be the normalization Λ. Unlike the torsionfree modules studied in integral representation theory, it is not true that every $M \in \text{fingen}(\Lambda)$ is contained in a Γ-module. We call M a *Γ-separated Λ-module* if it is contained in some Γ-module [Definitions 17.1]. Moreover, it can be that there is no conductor ideal for Λ and Γ (because ${}_{\Lambda}\Gamma$ need not be finitely generated).

However, $\big(\forall \mathfrak{m} \in \text{maxspec}(\Lambda)\big)$ $\mathfrak{m}_{\mathfrak{m}}$ is a conductor ideal for $\Lambda_{\mathfrak{m}}$ and $\Gamma_{\mathfrak{m}}$. We call the inclusion $(\Lambda/\mathfrak{m} \subseteq \Gamma/\Gamma\mathfrak{m}) = (\Lambda_{\mathfrak{m}}/\mathfrak{m}_{\mathfrak{m}} \subseteq \Gamma_{\mathfrak{m}}/\mathfrak{m}_{\mathfrak{m}})$ the *$\mathfrak{m}$-residue inclusion* of the inclusion $(\Lambda \subseteq \Gamma)$ [Definition 11.2], and show that the inclusion $(M \subseteq X = \Gamma M)$ — for every Γ-separated Λ-module M — has a family of residue inclusions that forms a satisfactory substitute for conductor squares. See Section 17.

The next step is to deal with Λ-modules that are not Γ-separated. We define a "best approximation" to any such module by a Λ-submodule of a Γ-module, and call this a "separated cover" of M. This is a fundamental tool in both this memoir and in the complete local case studied in [**KL2**]. See Section 18.

Chapter 5: Isomorphism Classes in a Genus, Idèle Group Action. Given $M \in \text{fingen}(\Lambda)$, what do we need to know beyond the collection of completions $\hat{M}_{\mathfrak{m}}$, in order to determine the isomorphism class of M? For this purpose, we can ignore Λ-modules of finite length, because: (i) Each indecomposable such module U equals one of its $\mathfrak{m}$-adic completions, and hence is described in detail in [**KL2**]; and (ii) If U appears as a direct summand of some other module M, then the complement of U in M is unique up to isomorphism. Therefore there is no loss of generality in restricting our attention to modules in $\text{fingen}_{\infty}(\Lambda)$ the category of finitely generated Λ-modules without nonzero direct summands of finite length. See Section 7 [Chapter 0] for more about this reduction. Note that we do allow submodules of finite length that are not direct summands.

We consider the Γ-isomorphism class of $\Gamma \otimes_{\Lambda} M$ to be a "first approximation" to the Λ-isomorphism classe of M, for $M \in \text{fingen}_{\infty}(\Lambda)$. The object of this chapter is to define a group action $M \to M^{\boldsymbol{u}}$ such that the set of all resulting Λ-isomorphism classes $M^{\boldsymbol{u}}$ is the *restricted genus* of M, that is, the collection of all Λ-isomorphism classes $N \in \text{genus}(M)$ such that $\Gamma \otimes_{\Lambda} N \cong \Gamma \otimes_{\Lambda} M$ as Γ-modules.

The exponent $\boldsymbol{u}$ is an element of a group that we call the group of "residue unit idèles", a group formed from the groups of units of the residue fields of Λ and Γ. Such group actions have been studied elsewhere, but always under the assumption of the existence of a conductor ideal for Λ and Γ. We make use of the (possibly infinite family) of nontrivial residue inclusions of $(\Lambda \subseteq \Gamma)$, thus taking advantage of the existence of local conductor ideals for Dedekind-like rings. The term "idèle" indicates that we have to deal with infinitely many such inclusions $(\Lambda/\mathfrak{m} \subseteq \Gamma/\Gamma\mathfrak{m})$; but for any individual idèle, almost all coordinates are trivial (units of $\Lambda/\mathfrak{m}$ not just of the larger $\Gamma/\Gamma\mathfrak{m}$).

This chapter is unfortunately very complicated. Moreover, the process of determining the stabilizer of this action is the only place in this memoir that requires detailed matrix-theoretic results from [**KL2**]. This is not surprising because, even in classically-studied situations where conductor ideals exist, there is a paucity of results about stabilizers of such actions.

But we do get a clear answer to the question of what nonlocal information is needed to determine the isomorphism class of M, given the genus of M.

We can be more explicit about this, by using the definition of "genus class group" reviewed in (2.2.3) below: Since Γ is a direct sum of Dedekind domains, say Γ_h, the group $\mathcal{G}_\Gamma(\Gamma)$ can be identified with the direct sum of the genus class groups $\mathcal{G}_{\Gamma_h}(\Gamma_h)$ of the various Γ_h, more classically known as their *ideal class groups.* Thus the direct sum of these ideal class groups forms the first approximation to every $\mathcal{G}(M)$ when ${}_\Lambda M$ is faithful. In slightly more detail, we get a "Mayer-Vietoris" short exact sequence for every faithful ${}_\Lambda M$

$$0 \to K \to \mathcal{G}(M) \to \mathcal{G}_\Gamma(\Gamma) \to 0 \tag{2.2.2}$$

where K is formed from the group of residue unit idèles. Readers familiar with classical Mayer-Vietoris sequences might be interested to know that our sequences (2.2.2) are valid even when there is no conductor ideal for Λ and Γ, the situation whenever ${}_\Lambda\Gamma$ is not finitely generated. To get this result we take advantage of the fact that local conductors always exist for Dedekind-like rings. See Theorem 25.14.

Chapter 6: Web of Class Groups; and Chapter 7: Direct Sums. These chapters are the core of our study of direct-sum behavior in fingen(Λ). As before, we may restrict our attention to $\text{fingen}_\infty(\Lambda)$. The first step is to recall that, for any $M \in \text{fingen}_\infty(\Lambda)$ the set of isomorphism classes $[M'] \in \text{genus}(M)$ can be made into an abelian group $\mathcal{G}(M)$ — the *genus class group of* M — in such a way that $[M]$ is the zero element of the group and:

$$\begin{gathered} \oplus_{i=1}^m [M_i] \cong \oplus_{i=1}^n [N_i] \qquad (M_i, N_i \in \text{genus}(M)) \\ \Longleftrightarrow \\ m = n \quad \text{and} \quad \sum_i [M_i] = \sum_i [N_i] \quad \text{in } \mathcal{G}(M) \end{gathered} \tag{2.2.3}$$

In fact, this can be done in exactly one way [Definition 25.1]. This (known fact) actually holds for all $M \in \text{fingen}(\Omega)$ where Ω is any reduced noetherian ring of Krull dimension 1.

Remark on the zero element $[M]$ *of* $\mathcal{G}(M)$. Note that, in the above definition, $[M]$ was an arbitrarily chosen isomorphism class in its genus. We call it the *base point* of genus(M), to emphasize this arbitrariness.

Very little seems to be known about the relation of genus class groups to each other, or about extending the notion of genus class group to deal with direct sums when the terms are not all in a single genus. Informally, we deal with the latter problem by making sense of the notion of adding elements in different genus class groups. Formally, we need the *web of genus class groups.* This consists of one genus class group $\mathcal{G}(M)$, together with its arbitrarily selected base point, and natural maps $\xi = \xi^{M,N}\colon \mathcal{G}(M) \to \mathcal{G}(N)$ for certain pairs M, N. For the precise definition of these ξ-maps, and the basic situations in which they exist, see Section 25. We give a brief indication here, of how they are used to study direct sums. (Without this overview, it might be difficult to understand the long technical development of this subject.)

First note [Corollary 25.9] that $\xi^{M,P}$ and $\xi^{N,P}$ are always defined when $P \in \text{genus}(M \oplus N)$. Now let $M', N' \in \text{fingen}_\infty(\Lambda)$ be given, and let M, N be the base points in their respective genera. Also, let P be the base point in genus$(M \oplus N)$. The three boxes in diagram (2.2.4) show the three corresponding genus class groups, the given $[M'], [N']$, their natural images in $\mathcal{G}(P)$, and the element $[M' \oplus N']$ of $\mathcal{G}(P)$.

(2.2.4)

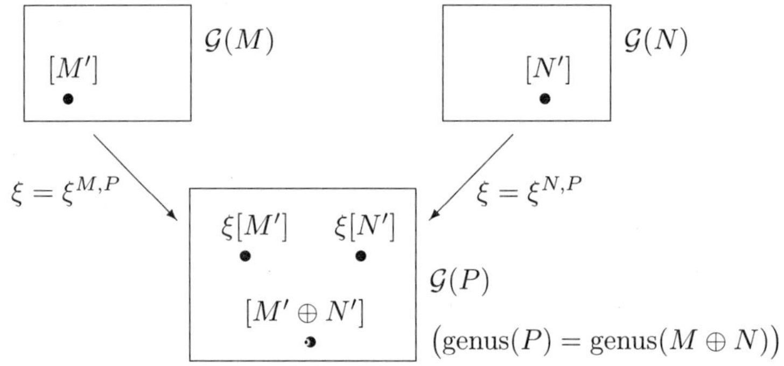

THEOREM 2.3 (see Theorem 29.1). *Keep the above notation. Then:*

(2.3.1) $$[M' \oplus N'] = \xi[M'] + \xi[N'] + [P_0] \quad \text{in } \mathcal{G}(P)$$

where the "correction term" $[P_0]$ depends only on the (arbitrarily selected) base points M, N, P, and not on M' or N'.

Informally, the theorem says that the "sum" $[M'] + [N']$ is obtained by adding the natural images $\xi[M']$ and $\xi[N']$ in $\mathcal{G}(P)$, and then adding an (annoying) correction term whose presence is due to the completely arbitrary choice of base points M, N, P in the three genera. This correction term seems unavoidable. For example, if we happen to choose $P = M \oplus N$, then the correction term becomes zero, and we get the more pleasant formula

(2.3.2) $$[M' \oplus N'] = \xi[M'] + \xi[N'] \quad \text{in } \mathcal{G}(P)$$

An open problem is whether one can choose a base point in each $\mathcal{G}(M)$ in such a way that the set of base points is closed under direct sums. If so, the more pleasant formula (2.3.2) would hold throughout the web of class groups. Nevertheless, Theorem 2.3 shows that *all direct-sum behavior is determined by two types of invariants:* (i) "counting" invariants, due to the fact that the Krull-Schmidt theorem holds in each $\mathfrak{m}$-adic completion; and (ii) abelian groups and group homomorphisms, determined by genus class groups and ξ-maps.

Using these ideas we get a definitive answer to how direct-sum cancellation can fail, in terms of group-theoretic invariants [Theorem 26.2], and we apply this to a number of examples in these chapters and the final chapter of this memoir.

Our other major application of these ideas is to the way that the set of all genus class groups is organized. The set of all genus class groups of faithful $M \in \text{fingen}_\infty(\Lambda)$ has an inverse limit s$\mathcal{G}$ (with respect to the ξ-maps), and this *super genus class group* maps onto all actual genus class groups of modules, faithful or not, in a natural way [Sections 27 and 28]. Moreover, if the normalization Γ is a finitely generated Λ-module, then there is an identification s$\mathcal{G} = \mathcal{G}(P)$ for some faithful P. This P can often be chosen to be Λ itself, and can always be chosen to have all ranks equal to 2 [Theorem 27.10, Corollary 27.11, and Theorem 27.12].

Moreover, if we must choose P to have its (torsionfree) ranks equal to 2, then P cannot be torsionfree! [Theorem 27.12] This is a counterexample to the generally held belief that torsionfree modules can hold more "global information" than any other types of modules.

Chapter 8: Finite Normalization. As might be expected, we can get stronger results in the classical situation of finite normalization (i.e. ${}_\Lambda\Gamma$ is finitely generated) than in general. For example:

• Γ serves as a test-module for direct-sum cancellation [Theorem 35.2].

• When does "power-isomorphism" hold in every genus? [i.e. $\text{genus}(A) = \text{genus}(B) \implies A^e \cong B^e$ for some e]? And when it holds, what is the smallest e? [Theorem 35.1] The answer is determined by the exponent of the super genus class group $s\mathcal{G}$.

• When does power-cancellation hold? (i.e. $A \oplus X \cong B \oplus X \implies A^e \cong B^e$ for some e) and when it holds, what is the smallest e? [Theorem 35.3]. The answer here is determined by the exponent of a certain subgroup of $s\mathcal{G}$.

• When does direct-sum cancellation hold in $\Lambda = \mathbb{Z}[\sqrt{n}]$ (n squarefree)? Depending upon the value of n we either answer the question or else reduce the answer to the question of whether the units group of the normalization Γ of Λ maps onto the units group of $\Gamma/\Gamma\mathfrak{m}$, where $\mathfrak{m}$ is the maximal ideal of Λ that lies over 2. [Example 36.3].

CHAPTER 0

Preliminaries

This chapter assembles some facts that apply to rings more general than the Dedekind-like rings that occupy the bulk of this memoir. Many of these facts are known, but are included so that readers not familiar with them can find clear statements in the form in which we use them. We include short proofs of known results when these proofs are no longer than a precise series of references, or when we do not know good references

3. Standard Notation and Terminology

This section collects some terminology and notation that occurs often and retains its significance throught this memoir. The more complete Terminology Index and Notation Index at the end of the memoir make it easy to find other definitions when they are encountered outside of the section in which they are defined.

NOTATION 3.1. Throughout this memoir, *ring* means "commutative ring" and *local ring* means "noetherian local ring" unless otherwise specified. A ring is *reduced* if it has no nonzero nilpotent elements.

Λ always denotes a Dedekind-like ring [defined in §10] unless otherwise specified. Essentially the only "otherwise specified" situation is that we regularly use it for rings that we are proving to be Dedekind-like.

A *regular element* of a ring is a non-zero-divisor in that ring. The *total quotient ring* of a ring is the localization that inverts all regular elements of the ring. Maps denoted by $\hookrightarrow$ and $\twoheadrightarrow$ denote injections and surjections, respectively.

Γ always denotes the *normalization* of a reduced noetherian ring (integral closure in its total quotient ring), and denotes the normalization of Λ if no other ring is explicitly specified. There is a unique decomposition

$$\Gamma = \oplus_{h \in \mathcal{H}} \Gamma_h \tag{3.1.1}$$

where each Γ_h is an integral domain. If Γ is the normalization of Λ, then each Γ_h is a Dedekind domain [Definition 10.1]. (Note that we do not consider fields to be Dedekind domains.) We call the rings Γ_h the *coordinate rings* of Γ.

$(\dots)_Q = (\dots)_{Q(\Omega)}$ denotes localization at the complement of the finite set of minimal prime ideals of a noetherian ring Ω. When Ω is reduced, this is the same as the localization that inverts all regular elements of Λ, that is, Ω_Q is the total quotient ring of Ω. See Section 4 for the many equivalent interpretations of $(\dots)_Q$, which we make extensive use of. In particular, if Γ is the normalization of a reduced noetherian ring Ω and $Q = Q(\Omega)$, then $\Gamma_Q = \Omega_Q$ and — in the notation of (3.1.1) — each $(\Gamma_h)_Q$ is the field of quotients of Γ_h [Corollary 4.4].

As in the earlier papers of this series, we write functions on the left, with one exception: functions that will be represented by matrix multiplication are written on the right.

fingen(Ω) denotes the category of all finitely generated modules over a ring Ω.

finlen(Ω) denotes the category of all finite-length Ω-modules.

$\text{fingen}_\infty(\Omega)$ denotes the category of all modules in fingen(Ω) that have no nonzero direct summands in finlen(Ω). In particular, $\{0\} \in \text{fingen}_\infty(\Omega)$. Section 7 discusses the unique decomposition of $M \in \text{fingen}(\Omega)$ into a direct sum $M = M_\infty \oplus M_0$ where $M_\infty \in \text{fingen}_\infty(\Omega)$ and $M_0 \in \text{finlen}(\Omega)$.

maxspec(Ω) and minspec(Ω) denote, respectively, the sets of maximal ideals and minimal prime ideals of a ring Ω.

singspec(Ω) — when Ω is reduced and noetherian of Krull dimension 1, with normalization Γ — denotes the set of *singular* maximal ideals of Ω, that is, the set of $\mathfrak{m} \in \text{maxspec}(\Omega)$ such that $\Omega_\mathfrak{m} \subset \Gamma_\mathfrak{m}$ (proper inclusion). If $\Omega_\mathfrak{m} = \Gamma_\mathfrak{m}$ we call $\mathfrak{m}$ *nonsingular.*

$\text{maxsupp}_\Lambda(H)$ denotes the set of maximal ideals $\mathfrak{m}$ of Λ in the *support* of the Λ-module H, that is, such that $H_\mathfrak{m} \neq 0$.

$\bar{X}(\mathfrak{m})$ denotes $X/\mathfrak{m}X$ where X is any Λ-module and $\mathfrak{m} \in \text{maxspec}(\Lambda)$. If X is a Γ-module, so is $\bar{X}(\mathfrak{m})$. When Λ is a local ring, we often shorten $\bar{X}(\mathfrak{m})$ to $\bar{X}$.

$\rho^\mathfrak{m}$ denotes the natural homomorphism $X \twoheadrightarrow \bar{X}(\mathfrak{m})$.

$\hat{\Omega}_\mathfrak{m}$ and $\hat{M}_\mathfrak{m}$ denote the $\mathfrak{m}$-adic completions of a ring Ω and a module $M \in \text{fingen}(\Omega)$ respectively. See Section 5 for a discussion of this. See Remarks 5.3 for a discussion of $\hat{\Gamma}_\mathfrak{m}$ $\big(\mathfrak{m} \in \text{maxspec}(\Omega)\big)$. When Ω is local we often shorten $\hat{\Omega}_\mathfrak{m}$ to $\hat{\Omega}$.

$(\dots)\hat{}_\mathfrak{m}$ denotes the $\mathfrak{m}$-adic completion of an expression consisting of more than a single letter.

(a$\forall\mathfrak{m}$) denotes *for almost all* $\mathfrak{m}$, that is, "for all except finitely many $\mathfrak{m}$."

$(\dots)^\times$ denotes the group of units of $(\dots)$.

4. Q-Localization and Minimal Primes

LEMMA 4.1. *Let $\Gamma = \oplus_{h\in\mathcal{H}}\Gamma_h$ be the normalization of a reduced noetherian ring Ω, where each Γ_h is an integral domain. Then:*

(i) *The finite set of minimal prime ideals $\mathfrak{p}_h$ of Ω is in one-to-one correspondence with the set of coordinate rings Γ_h of Γ, via $\mathfrak{p}_h = \ker(\Omega \to \Gamma_h)$.*
(ii) *The set of zero-divisors of Ω is the union of the minimal prime ideals of Ω.*
(iii) *Each $\ker(\Gamma \to \Gamma_h)$ is the unique minimal prime ideal of Γ lying over $\mathfrak{p}_h$.*
(iv) *Every element of Ω that is a zero-divisor in Γ is already a zero-divisor in Ω*

PROOF. (i) We claim that there are no inclusion relations among the ideals $\mathfrak{p}_h$. To keep the notation simple, select two indices h, calling them 1 and 2, and choose a nonzero element of Γ_2. Since Γ is contained in the total quotient ring of Λ, this element has the form x/d where $x, d \in \Lambda$ and d is a regular element of Λ. Moreover, coordinate 2 of x is nonzero, while $x_1 = 0$. Therefore $x \in \mathfrak{p}_1 - \mathfrak{p}_2$. Thus the claim is proved.

Since the product of the prime ideals $\mathfrak{p}_h$ is zero, every minimal prime ideal is among them. Since, in addition, there are no inclusion relations between these kernels, each is a minimal prime.

(iv) This holds since every element of Γ has the form x/d with $x, d \in \Omega$ and d regular in Ω.

(ii) Write an arbitrary element $\omega \in \Omega$ as a tuple $\omega = (\omega_h)_{h \in \mathcal{H}}$. Every zero divisor ω of Ω is also a zero divisor of Γ. Since each Γ_h is an integral domain, it follows that at least one coordinate ω_h of ω must be zero. Therefore the set of zero divisors of Λ is contained in the union of the ideals $\mathfrak{p}_h$ in (i). The converse is a special case of the general fact that every minimal prime ideal of a ring consists of zero divisors, and is also obvious in view of statement (iv).

(iii) By the case $\Lambda = \Gamma$ of statement (i) the minimal prime ideals of Γ are the ideals $\mathfrak{p}'_h = \ker(\Gamma \to \Gamma_h)$. Obviously, each $\mathfrak{p}'_h \cap \Lambda = \mathfrak{p}_h$. Thus, if we had $\mathfrak{p}'_h \cap \Lambda = \mathfrak{p}'_k \cap \Lambda$ for some $h \neq k$, we would have $\mathfrak{p}_h = \mathfrak{p}_k$, contrary to (i). □

DEFINITION 4.2 (Ω_Q, M_Q). Let Ω be a noetherian ring, $Q = Q(\Omega)$ the finite set of minimal prime ideals of Ω. We denote by Ω_Q the localization of Ω that inverts the elememts of $\Omega - \cup Q$. When Ω is a reduced ring — the situation in which we almost always encounter it in this memoir — Ω_Q is the total quotient ring of Ω, that is, the localization that inverts the regular elements of Ω [Lemma 4.1(ii)]. We make use of both of these versions of Ω_Q.

If M is an Ω-module, M_Q denotes the localization of M obtained by using the denomimator set $\Omega - \cup Q$.

If Ω is also an algebra over a ring R, then the notation M_Q sometimes needs more precision, which we obtain by writing $M_{Q(R)}$ or $M_{Q(\Omega)}$. In fact, *less* precision is often sufficient, as we show in Lemma 4.3 and Corollary 4.4 below.

Before proceeding, we note:

(4.2.1) The ring Ω_Q is artinian.

This holds since every prime ideal of the noetherian ring Ω_Q is a minimal prime ideal.

LEMMA AND DEFINITION 4.3 (The many Q-localizations). *Let Ω be a reduced noetherian ring, Γ its normalization, and $\mathfrak{m} \in \operatorname{maxspec}(\Omega)$. Then we have the following natural isomorphisms (which we write as equality).*

(i) $\Gamma_{Q(\Gamma)} = \Gamma_{Q(\Omega)} = \Omega_{Q(\Omega)}$.
(ii) $(\Omega_\mathfrak{m})_{Q(\Omega_\mathfrak{m})} = (\Omega_\mathfrak{m})_{Q(\Omega)}$.
(iii) *If $\Omega_\mathfrak{m}$ has Krull dimension 1 and $\Gamma_\mathfrak{m}$ is module-finite over $\Omega_\mathfrak{m}$, then $(\hat{\Omega}_\mathfrak{m})_{Q(\hat{\Omega}_\mathfrak{m})} = (\hat{\Omega}_\mathfrak{m})_{Q(\Omega)}$. ($\hat{\Omega}_\mathfrak{m}$ denotes the $\mathfrak{m}$-adic completion of Ω.)*

We usually write simply $(\ldots)_Q$ for all of the above localizations. In particular:

(4.3.1) *The localizations $(\ldots)_Q$ in* (i)–(iii) *are direct sums of fields.*

PROOF. The rings Ω and Γ are reduced, and hence so are all of their localizations. By the finite generation hypothesis in (iii), $\hat{\Omega}_\mathfrak{m}$ is also reduced. [See Remarks 5.3(iii) and (iv).] Therefore the Q-localizations $\Omega_{Q(\Omega)}$, $\Gamma_{Q(\Gamma)}$, $(\Omega_\mathfrak{m})_{Q(\Omega_\mathfrak{m})}$, and $(\hat{\Omega}_\mathfrak{m})_{Q(\Omega)}$ are again reduced.

Thus the remaining content of the lemma is that all of the denominator sets can be replaced by $\Omega - \cup Q(\Omega)$ without altering the ring.

(i) Obviously $\Gamma_{Q(\Gamma)} \supseteq \Gamma_{Q(\Omega)} \supseteq \Omega_{Q(\Omega)}$. Since Γ and Λ both have the same total quotient ring, namely the left and right sides of this chain of inclusions, the inclusions can be replaced by equality.

(ii) The identification $(\Omega_{Q(\Omega)})_\mathfrak{m} = (\Omega_\mathfrak{m})_{Q(\Omega_\mathfrak{m})}$ is proved in [**LO**, 2.3(i)]. (The proof there does not actually use the standing Krull dimension 1 hypothesis of that paper.) Reversing the order of localization on the left yields (ii).

Before proving (iii), we recall a familiar fact (see [**AM**, Proposition 1.11], for example).

(4.3.2) If a maximal ideal $\mathfrak{m}$ of a noetherian ring Ω is a not a minimal prime, then $\mathfrak{m}$ is not contained in the union of the minimal primes.

(iii) In view of (ii), it suffices to prove that $(\hat{\Omega}_{\mathfrak{m}})_{Q(\hat{\Omega}_{\mathfrak{m}})} = (\hat{\Omega}_{\mathfrak{m}})_{Q(\Omega_{\mathfrak{m}})}$. Thus, we may assume that Ω is local with maximal ideal $\mathfrak{m}$. Moreover, if $\mathfrak{m}$ is also a minimal prime, then $\Omega = \Omega_{\mathfrak{m}}$ is a field, whence the proof becomes trivial; so we may also assume that $\mathfrak{m}$ is not minimal.

It suffices to show that the same prime ideals survive in both localizations (since any localization is naturally isomorphic to the localization at the complement of the union of the prime ideals that survive). Since the rings we are working with are reduced, the denominator set in each localization can be taken to be the set of regular elements of the ring in the subscript. Since regular elements of Ω remain regular in $\hat{\Omega}_{\mathfrak{m}}$, no minimal prime of $\hat{\Omega}_{\mathfrak{m}}$ is destroyed by either localization. On the other hand, the maximal ideals of both local rings Ω and $\hat{\Omega}_{\mathfrak{m}}$ contain regular elements. For Ω, this holds because its maximal ideal is not contained in the union of the minimal primes [(4.3.2) and Lemma 4.1(ii)]. For $\hat{\Omega}_{\mathfrak{m}}$, this holds because regular elements of Ω remain regular in $\hat{\Omega}_{\mathfrak{m}}$. Therefore, the maximal ideal of $\hat{\Omega}_{\mathfrak{m}}$ is destroyed by both localizations. Since $\hat{\Omega}_{\mathfrak{m}}$ has Krull dimension 1, the statement follows.

In view of statements (i)–(iii) and (4.2.1), all of the rings in (i)–(iii) are reduced artinian rings, and hence are direct sums of fields. This proves (4.3.1). □

COROLLARY 4.4. *Let $\Gamma = \oplus_h \Gamma_h$, where each Γ_h is an integral domain, be the normalization of a reduced noetherian ring Ω. Then each $(\Gamma_h)_Q$, where $Q = Q(\Omega)$, is the field of quotients of Γ_h.*

PROOF. The total quotient ring of Γ is $\Gamma_Q = \oplus_h (\Gamma_h)_Q$. Therefore each $(\Gamma_h)_Q$ is a field or zero. The latter possibility never occurs because — since Ω is reduced — the natural map $\Gamma \to \Gamma_Q$ is an injection. Since the localization $(\Gamma_h)_Q$ of Γ_h is a field containing Γ_h, it follows that $(\Gamma_h)_Q$ is the field of quotients of Γ_h. □

LEMMA 4.5. *Let Ω be a noetherian ring of Krull dimension 1 such that no maximal ideal is a minimal prime ideal, and $M \in$ fingen(Ω). Then $M_Q = 0$ if and only if M has finite length.*

PROOF. First, suppose that M has finite length. To prove that $M_Q = 0$, we may assume that M is a simple module and therefore is annihilated by some maximal ideal $\mathfrak{m}$. Since $\mathfrak{m}$ is not a minimal prime, (4.3.2) shows that $\mathfrak{m}$ contains an element c that is not in any minimal prime ideal of Ω. Thus, c acts invertibly on M_Q but also annihilates it, so that $M_Q = 0$.

Conversely, suppose that $M_Q = 0$. Since M is finitely generated, there is some element $c \in \Omega - \cup Q$ such that $cM = 0$. Thus, since Ω has Krull dimension 1, M is a finitely generated module over the artinian ring $\Omega/\Omega c$, and hence M has finite length. □

5. Ideal-adic Completions

NOTATION AND QUICK REVIEW 5.1 (Completion). Let Ω be a noetherian ring, $\mathfrak{m}$ a maximal ideal of Ω, and $M \in$ fingen(Ω). Then we can identify the $\mathfrak{m}$-adic completion $\hat{M}_{\mathfrak{m}}$ of M with the $\hat{\Omega}_{\mathfrak{m}}$-module $\hat{\Omega}_{\mathfrak{m}} \otimes_{\Omega} M$ [**M**, Theorem 8.7], and write $\hat{\nu}_{\mathfrak{m}}\colon M \to \hat{M}_{\mathfrak{m}}$ for the *natural map* given by $x \to 1 \otimes x$. Moreover, $\hat{\Omega}_{\mathfrak{m}}$ is a flat Ω-module [**M**, Theorem 8.8]. Some additional facts that we shall need:

(i) The $\mathfrak{m}$-adic completion of M can be identified with the $\mathfrak{m}_{\mathfrak{m}}$-adic completion of $M_{\mathfrak{m}}$ (because $\hat{\Omega}_{\mathfrak{m}} \otimes_{\Omega} M = \hat{\Omega}_{\mathfrak{m}} \otimes_{\Omega_{\mathfrak{m}}} \Omega_{\mathfrak{m}} \otimes_{\Omega} M$).

(ii) If Ω is a local ring, then $\hat{\nu}_{\mathfrak{m}}$ is an embedding [**AM**, Theorem 10.17 and Corollary 10.19]. We almost always write this as actual inclusion.

(iii) If Ω is a local ring, then $\hat{\Omega}_{\mathfrak{m}}$ is a faithfully flat Ω-module [**M**, Theorem 8.14].

(iv) Whether or not Ω is local, $\hat{\Omega}_{\mathfrak{m}} \cdot \hat{\nu}_{\mathfrak{m}}(M) = \hat{M}_{\mathfrak{m}}$, by the definition of $\hat{\nu}_{\mathfrak{m}}$.

Recall that the completion of a ring Ω with respect to a filter of ideals of Ω can be viewed as equivalence classes of Cauchy sequences. If $\{G_n\}_{n\geq 1}$ and $\{H_n\}_{n\geq 1}$ are filters of ideals of Ω, we call them *equivalent* if, for each positive integer i, there are positive integers j and k such that $G_j \subseteq H_i$ and $H_k \subseteq G_i$. Equivalent filters yield completions that are canonically isomorphic rings.

We denote the Jacobson radical of any ring Ω by $\mathrm{rad}(\Omega)$.

LEMMA 5.2. *Let Ω be a noetherian semilocal reduced ring of Krull dimension one, and Υ a ring, $\Omega \subseteq \Upsilon \subseteq \Omega_Q$, such that Υ is a finitely generated Ω-module. Then the* $\mathrm{rad}(\Omega)$*-adic completion of Υ as a Ω-algebra — i.e. the completion with respect to the filter of ideals $\{\Upsilon\cdot \mathrm{rad}(\Omega)^n\}_{n\geq 1}$ of Υ — can be canonically identified with the* $\mathrm{rad}(\Upsilon)$*-adic completion of Υ — i.e. the completion with respect to the filter of ideals $\{\mathrm{rad}(\Upsilon)^n\}_{n\geq 1}$.*

PROOF. We may assume that Ω is indecomposable. Moreover, Ω is not artinian, because it is assumed to be of Krull dimension one. Then, since ${}_{\Omega}\Upsilon$ is finitely generated, the two rings contain a common ideal $\mathfrak{c}$ ("conductor") which is contained in both $\mathrm{rad}(\Omega)$ and $\mathrm{rad}(\Upsilon)$, and which contains a unit of Ω_Q [**LO**, Lemma 1.4].

We claim that the filter $\{\Upsilon\cdot \mathrm{rad}(\Omega)^n\}_{n\geq 1}$ is equivalent to the filter $\{\mathfrak{c}^n\}_{n\geq 1}$. Since the ideal $\mathfrak{c}$ of Ω contains a regular element and Ω has Krull dimension 1, the ring $\Omega/\mathfrak{c}$ is artinian. And, since $\mathfrak{c}$ is contained in $\mathrm{rad}(\Omega)$, the radical $\mathrm{rad}(\Omega)/\mathfrak{c}$ of the artinian ring $\Omega/\mathfrak{c}$ is nilpotent. This is equivalent to saying that $\mathfrak{c}$ contains some power of $\mathrm{rad}(\Omega)$. The rest of the proof of the claim is easy.

Similarly the filter $\{\mathfrak{c}^n\}_{n\geq 1}$ is equivalent to the filter $\{\mathrm{rad}(\Upsilon)^n\}_{n\geq 1}$. These two equivalences, along with the remarks in 5.1, prove the lemma. □

REMARKS 5.3 (Completion of normalization). Let Ω be a reduced noetherian ring of Krull dimension 1, with normalization Γ, and let $\mathfrak{m} \in \mathrm{maxspec}(\Omega)$. We summarize the basic facts about the $\mathfrak{m}$-adic completion of Γ in the case where Ω is not necessarily local and Γ is not necessarily module-finite over Ω. The assertions needed in this memoir are (i)–(iii). We reduce (ii) and (iii) to standard properties of completions of finitely generated modules over semilocal rings. Fortunately, the peculiarities listed in (iv) do not occur in this memoir, because $\Gamma_{\mathfrak{m}}$ is always module-finite over $\Lambda_{\mathfrak{m}}$ when Λ is Dedekind-like [Definition 10.1].

(i) *The localization $\Gamma_{\mathfrak{m}}$ is the normalization of $\Omega_{\mathfrak{m}}$ and is a direct sum of semilocal principal ideal domains and fields.*

The first assertion is a special case of [**LO**, 4.3]. The second assertion is a property of all integrally closed reduced noetherian rings of Krull dimension 1.

(ii) Let $M \in \text{fingen}(\Gamma)$. *Whether or not* Γ *is a finitely generated* Ω*-module, there is a natural isomorphism between the* $\mathfrak{m}$*-adic completion of* M *(viewed as an* Ω*-module) and the* $\mathfrak{m}_\mathfrak{m}$*-adic completion of* $M_\mathfrak{m}$ *(viewed as an* $\Omega_\mathfrak{m}$*-module). Furthermore, both of these can be identified with the* $\Gamma\mathfrak{m}$*-adic completion of* M.

Therefore we write all of these completions simply as $\hat{M}_\mathfrak{m}$.

The $\mathfrak{m}$-adic completion of M is the inverse limit of the modules $M/(\mathfrak{m}^n M)$, while the $\mathfrak{m}_\mathfrak{m}$-adic completion of $M_\mathfrak{m}$ is the inverse limit of the modules $M_\mathfrak{m}/(\mathfrak{m}_\mathfrak{m}^n M_\mathfrak{m}) = (M/\mathfrak{m}^n M)_\mathfrak{m} = M/(\mathfrak{m}^n M)$. (The last equality holds because every element of $\Omega - \mathfrak{m}$ is invertible modulo all powers of $\mathfrak{m}$.) Therefore we can identify the first two completions mentioned in statement (ii).

Consider the $\Gamma\mathfrak{m}$-adic completion. Since $\Gamma M = M$, the $\Gamma\mathfrak{m}$-adic completion is the inverse limit of the modules $M/\Gamma\mathfrak{m}M = M/\mathfrak{m}M$, and therefore equals the previous two completions.

(iii) *Suppose that* $\Gamma_\mathfrak{m}$ *is a finitely generated* $\Omega_\mathfrak{m}$*-module. Then for every* $M \in \text{fingen}(\Gamma)$, *we can identify* $\hat{M}_\mathfrak{m}$ *with* $\hat{\Omega}_\mathfrak{m} \otimes_\Omega M$. *In particular, we have the identification* $\hat{\Omega}_\mathfrak{m} \otimes_\Omega \Gamma = \hat{\Gamma}_\mathfrak{m}$.

Furthermore, $\hat{\Omega}_\mathfrak{m}$ *is reduced,* $\hat{\Gamma}_\mathfrak{m}$ *is its normalization and* $\hat{\Gamma}_\mathfrak{m}$ *is a direct sum of DVRs and fields, each being the completion of* $\Gamma_\mathfrak{m}$ *with respect to one of its maximal ideals.*

If every maximal ideal of Ω *has height 1, then no fields occur in the decomposition in the previous paragraph.*

We have that $\hat{\Omega}_\mathfrak{m} \otimes_\Omega \Gamma = \hat{\Omega}_\mathfrak{m} \otimes_{\Omega_\mathfrak{m}} \Omega_\mathfrak{m} \otimes_\Omega \Gamma = \hat{\Omega}_\mathfrak{m} \otimes_{\Omega_\mathfrak{m}} \Gamma_\mathfrak{m}$. Since $\Gamma_\mathfrak{m}$ is module-finite over $\Omega_\mathfrak{m}$, by hypothesis, this last tensor product is the $\mathfrak{m}_\mathfrak{m}$-adic completion of $\Gamma_\mathfrak{m}$, which by (ii) is the $\mathfrak{m}$-adic completion of Γ. That is, $\hat{\Omega}_\mathfrak{m} \otimes_\Omega \Gamma = \hat{\Gamma}_\mathfrak{m}$, as desired. The first assertion of (iii) then follows by tensoring with M over Γ, and using assertion (ii).

Consider the second paragraph of (iii). By the final assertion in the first paragraph we have $\hat{\Gamma}_\mathfrak{m} = \hat{\Omega}_\mathfrak{m} \otimes_\Omega \Gamma$ and this can be identified with $\hat{\Omega}_\mathfrak{m} \otimes_{\Omega_\mathfrak{m}} \Gamma_\mathfrak{m}$, the $\mathfrak{m}_\mathfrak{m}$-adic completion of $\Gamma_\mathfrak{m}$ because of our finite generation hypothesis.

By (i) $\Gamma_\mathfrak{m}$ is the normalization of $\Omega_\mathfrak{m}$, and is a direct sum of rings Υ, each of which is a semilocal principal ideal domain or a field. Moreover, by (ii), the $\mathfrak{m}_\mathfrak{m}$-adic completion of $\Gamma_\mathfrak{m}$ can be identified with the $\Gamma_\mathfrak{m}\,\mathfrak{m}_\mathfrak{m}$-adic completion of $\Gamma_\mathfrak{m}$.

The proof of the decomposition of $\hat{\Gamma}_\mathfrak{m}$ is finished by recalling that the completion of a semilocal ring, with respect to its Jacobson radical, is naturally isomorphic to the direct sum of its completions at its maximal ideals. $\hat{\Gamma}_\mathfrak{m}$ is the normalization of $\hat{\Omega}_\mathfrak{m}$ because it is a direct sum of normal rings and is module-finite over $\hat{\Omega}_\mathfrak{m}$.

Third paragraph of (iii). Since every maximal ideal of Ω has height 1, the same is true of $\Omega_\mathfrak{m}$ and hence of its normalization $\Gamma_\mathfrak{m}$ [Lemma 10.5]. Therefore no fields can appear in the decomposition of $\Gamma_\mathfrak{m}$, and hence in the decomposition of $\hat{\Gamma}_\mathfrak{m}$.

(iv) *Suppose that* $(\Omega, \mathfrak{m})$ *is local, but* Γ *is not module-finite over* Ω. *Then* $\hat{\Omega}_\mathfrak{m}$ *has nonzero nilpotent elements* [**M**, page 264], *and* $\hat{\Omega}_\mathfrak{m} \otimes_\Omega \Gamma \not\cong \hat{\Gamma}_\mathfrak{m}$. *In fact, the ring* $\hat{\Omega}_\mathfrak{m} \otimes_\Omega \Gamma$ *is not noetherian* [**LO**, Lemma 6.7(i)].

When is a module X over some $\hat{\Omega}_\mathfrak{m}$ the $\mathfrak{m}$-adic completion of an Ω-module? Informally, the next result states that we can answer this by looking only at the torsionfree structure of X.

LEMMA 5.4. *Let* $(\Omega, \mathfrak{m})$ *be a local (noetherian) ring of Krull dimension 1, and let* $X \in \text{fingen}(\hat{\Omega}_\mathfrak{m})$. *Then the following are equivalent.*

(i) $X \cong \hat{A}_{\mathfrak{m}}$ *for some* $A \in \text{fingen}(\Omega)$.
(ii) $X_Q \cong \hat{\Omega}_{\mathfrak{m}} \otimes_\Omega V$ *for some* $V \in \text{fingen}(\Omega_Q)$.

PROOF. This is part of [**LO**, Theorem 3.4]. *Caution.* Readers consulting this reference should note that $\hat{\Omega}_{\mathfrak{m}} \otimes_\Omega V$ is denoted by $\hat{V}$ and called the "rad(Ω)-induced completion of V" [**LO**, Definitions 3.1]. This is a type of completion but definitely not an ideal-adic completion: For example, if $\hat{\Omega}_{\mathfrak{m}}$ is an integral domain, then its Q-localization $(\hat{\Omega}_{\mathfrak{m}})_Q$ is a field, and hence $\hat{\mathfrak{m}}^n_{\mathfrak{m}} V = V$ for all n. The rad(Ω)-induced completion is an extension of the notion of the completion of a field with respect to a valuation. □

REMARK 5.5. Lemma 5.4 takes a particularly simple form when Ω is reduced and its normalization Γ is a finitely generated Ω-module. In this situation, the completion $\hat{\Omega}_{\mathfrak{m}}$ is reduced, and hence $(\hat{\Omega})_Q$ is a direct sum of fields. Therefore verifying the isomorphism in Lemma 5.4(ii) is just a matter of comparing dimensions of vector spaces over these fields. We refer to these dimensions as "(torsionfree) ranks" in many parts of this memoir.

6. Primary Components (terminology from abelian groups)

We often use the following terminology, which is borrowed from the theory of abelian groups and, unfortunately, conflicts with some other very commonly-used terminology in commutative ring theory.

DEFINITION 6.1 ($\mathfrak{m}$-primary component). Let M be a module over a commutative ring Ω and $\mathfrak{m} \in \text{maxspec}(\Omega)$. We define the $\mathfrak{m}$-*primary component* $P(\mathfrak{m})$ of M by

$$P(\mathfrak{m}) = \{x \in M \mid (\exists n)\ \mathfrak{m}^n x = 0\} \tag{6.1.1}$$

LEMMA 6.2. *Let M be a finitely generated module over a noetherian ring Ω, and $\mathfrak{m}$ a maximal ideal of Ω. Then:*

(i) *$P(\mathfrak{m}) = P(\mathfrak{m})_{\mathfrak{m}}$ (canonical Ω-isomorphism via $p \to p/1$). Moreover, every Ω-submodule of $P(\mathfrak{m})$ is a $\Omega_{\mathfrak{m}}$-submodule.*
(ii) *When viewed as a submodule of $M_{\mathfrak{m}}$, $P(\mathfrak{m})_{\mathfrak{m}}$ is the largest finite-length $\Omega_{\mathfrak{m}}$-submodule (equivalently, Ω-submodule) of $M_{\mathfrak{m}}$.*
(iii) *If we identify $P(\mathfrak{m})$ with its image in $M_{\mathfrak{m}}$ under the isomorphism in statement (i), then every finite-length direct summand of $M_{\mathfrak{m}}$ becomes a direct summand of M.*
(iv) *If $\mathfrak{n}$ is a maximal ideal distinct from $\mathfrak{m}$, then $P(\mathfrak{m})_{\mathfrak{n}} = 0$.*
(v) *If $M_{\mathfrak{m}} \to U$ is a surjective homomorphism of $\Omega_{\mathfrak{m}}$-modules, and U has finite length, then the composition $M \to M_{\mathfrak{m}} \to U$ is again surjective.*

PROOF. (i) Since M is noetherian, its submodule $P(\mathfrak{m})$ is finitely generated, and therefore there exists s such that $\mathfrak{m}^s P(\mathfrak{m}) = 0$. Therefore $P(\mathfrak{m})$ is a finitely generated module over the ring $\Omega/\mathfrak{m}^s$, which is artinian because it is noetherian of Krull dimension 0. We conclude:

(6.2.1) The Ω-module (equivalently, $\Omega/\mathfrak{m}^s$-module) $P(\mathfrak{m})$ has finite length.

We have the canonical ring isomorphism $\Omega/\mathfrak{m}^s = \Omega_{\mathfrak{m}}/\mathfrak{m}^s_{\mathfrak{m}}$ because elements of $\Omega - \mathfrak{m}$ become invertible modulo $\mathfrak{m}^s$. Therefore (the details of) the proof of the first statement in (i) follow by viewing $P(\mathfrak{m})$ as a $\Omega/\mathfrak{m}^s$-module.

It suffices to prove the second statement in (i) for cyclic submodules of $P(\mathfrak{m})$; and this again follows from the canonical isomorphism $\Omega/\mathfrak{m}^s = \Omega_\mathfrak{m}/\mathfrak{m}^s_\mathfrak{m}$.

(ii) We have $P(\mathfrak{m}) = P(\mathfrak{m})_\mathfrak{m}$ by (i). The left-hand side has finite length by (6.2.1). Therefore the right-hand side has finite Ω-length; and therefore it has finite $\Omega_\mathfrak{m}$-length by the second statement in (i).

Now we need to show that every finite-length $\Omega_\mathfrak{m}$-submodule of $M_\mathfrak{m}$ is contained in $P(\mathfrak{m})_\mathfrak{m}$. It suffices to do this for cyclic submodules, say for $\Omega_\mathfrak{m}\cdot(x/1)$. Since this submodule has finite length it is annihilated by some power of the maximal ideal $\mathfrak{m}_\mathfrak{m}$ of $\Omega_\mathfrak{m}$, and hence $\mathfrak{m}^n(x/1) = 0$ for some n. Finite generation of the ideal $\mathfrak{m}$ of Ω then yields an element $c \in \Omega - \mathfrak{m}$ such that $c\mathfrak{m}^n x = 0$, and hence $cx \in P(\mathfrak{m})$. Since c becomes invertible in $\Omega_\mathfrak{m}$, we now have $x/1 \in P(\mathfrak{m})_\mathfrak{m}$ as desired.

(iii) Let X be the given direct summand of $M_\mathfrak{m}$. By (ii) and then (i) we have $X \subseteq P(\mathfrak{m})_\mathfrak{m} = P(\mathfrak{m})$. Therefore, for a projection of M to X we can use the composition of the natural map $M \to M_\mathfrak{m}$ with any projection map $M_\mathfrak{m} \to X$.

(iv) Given $x \in P(\mathfrak{m})$, choose n as in (6.1.1) and $t \in \mathfrak{m} - \mathfrak{n}$. Then $t^n x = 0$ and therefore $x/1 = 0$ in $P(\mathfrak{m})_\mathfrak{n}$.

(v) Let M' and M'' be the images of M in $M_\mathfrak{m}$ and U respectively. Then $\Omega_\mathfrak{m} M' = M_\mathfrak{m}$. Since $M_\mathfrak{m} \to U$ is surjective, we therefore have $\Omega_\mathfrak{m} M'' = U$. But since U has finite length, every Ω-submodule of U is an $\Omega_\mathfrak{m}$-submodule, by statement (i). Therefore $M'' = U$, as desired. □

LEMMA 6.3. *Let M be a finitely generated module over a noetherian ring Ω, and for each maximal ideal $\mathfrak{m}$ let $P(\mathfrak{m})$ be the $\mathfrak{m}$-primary component of M. Then:*

(i) *There are only finitely many maximal ideals $\mathfrak{m}$ such that $M_\mathfrak{m}$ has a nonzero submodule of finite length (equivalently, such that $P(\mathfrak{m}) \neq 0$).*

(ii) *The sum $\sum_\mathfrak{m} P(\mathfrak{m})$ is direct and is the largest finite-length submodule of M.*

PROOF. (i) If $\mathfrak{m}$ and $\mathfrak{n}$ are distinct maximal ideals of Ω, then $\hom_\Omega\big(P(\mathfrak{m}), P(\mathfrak{n})\big) = 0$. Therefore, the sum $\sum_\mathfrak{m} P(\mathfrak{m})$ is direct. Since M is noetherian, it follows that only finitely many $P(\mathfrak{m})$ can be nonzero. The desired conclusions now follow from Lemma 6.2(i) and (ii).

(ii) Let $S = \sum_\mathfrak{m} P(\mathfrak{m})$. Directness and finiteness of the number of terms were proved above. Therefore finite length of S follows from finite length of each of the terms [Lemma 6.2(i) and (ii).].

It now suffices to show that every finite-length submodule X of M is contained in S. To show that $X \subseteq S$, it suffices to show that $X_\mathfrak{m} \subseteq S_\mathfrak{m}$ in $M_\mathfrak{m}$ for every maximal ideal $\mathfrak{m}$. Since X has Ω-finite length, $X_\mathfrak{m}$ has $\Omega_\mathfrak{m}$-finite length.

On the other hand, $S_\mathfrak{m} = P(\mathfrak{m})_\mathfrak{m}$ because $P(\mathfrak{n})_\mathfrak{m} = 0$ whenever $\mathfrak{n} \neq \mathfrak{m}$ [Lemma 6.2(iv)]. Since $P(\mathfrak{m})_\mathfrak{m}$ is the largest finite-length submodule of $M_\mathfrak{m}$ [Lemma 6.2(ii)], the proof is complete. □

COROLLARY 6.4. *Let $M \in \text{fingen}(\Omega)$, with Ω noetherian.*

(i) *If M has finite length then $M = \oplus_{\mathfrak{m}\in\text{maxspec}(\Omega)} M_\mathfrak{m}$ (canonical isomorphism).*

(ii) *(Whether or not M has finite length:)*

(6.4.1) $$M_Q = \oplus_{\mathfrak{p}\in\text{minspec}(\Omega)} M_\mathfrak{p} \qquad \text{(canonical isomorphism)}$$

PROOF. To be explicit: The canonical isomorphism in (i) is the map that sends each element m of M to the indexed family whose $\mathfrak{m}$-coordinate is the element $m/1$

of $M_{\mathfrak{m}}$. Of course, one only needs to use the finite number of coordinates such that $M_{\mathfrak{m}} \neq 0$. Similarly, in (ii) each fraction m/d in M_Q is sent to the tuple whose $\mathfrak{p}$-coordinate is the fraction m/d in $M_{\mathfrak{p}}$.

(i) We have $M = \oplus_{\mathfrak{m}\in \mathrm{maxspec}(\Omega)} P(\mathfrak{m})$ since M has finite length [Lemma 6.3]. Again, since M has finite length, we have $P(\mathfrak{m}) = M_{\mathfrak{m}}$ [Lemma 6.2(i)].

(ii) Recall that the ring Ω_Q is artinian [see (4.2.1)]. Therefore the finitely generated Ω_Q-module M_Q has finite length. The desired result therefore follows by applying statement (i) to the Ω_Q-module M_Q and realizing that the Q-localizations of the minimal prime ideals of Ω are the maximal ideals of Ω_Q. □

LEMMA 6.5. *Let $M \in \mathrm{fingen}(\Omega)$, with Ω noetherian, and let $\mathfrak{m}$ be a maximal ideal of Ω.*

(i) *If U is an $\hat{\Omega}_{\mathfrak{m}}$-module of finite length, then every Ω-submodule of U is a $\hat{\Omega}_{\mathfrak{m}}$-submodule.*
(ii) *If $\hat{M}_{\mathfrak{m}} \to U$ is any surjective $\hat{\Omega}_{\mathfrak{m}}$-homomorphism and U has finite length, then the natural composition $M \to \hat{M}_{\mathfrak{m}} \to U$ is again surjective.*
(iii) *If Ω is local and the $\hat{\Omega}_{\mathfrak{m}}$-module $\hat{M}_{\mathfrak{m}}$ has finite length, then the natural inclusion $M \subseteq \hat{M}_{\mathfrak{m}}$ is equality, and the Ω-module M has finite length.*

PROOF. (i) Since U has finite length as a $\hat{\Omega}_{\mathfrak{m}}$-module, we have $\hat{\mathfrak{m}}_{\mathfrak{m}}^n U = 0$ for some n, and hence $\mathfrak{m}^n U = 0$. Therefore U is a module over each of the isomorphic artinian rings $\Omega/\mathfrak{m}^n \cong \hat{\Omega}_{\mathfrak{m}}/\hat{\mathfrak{m}}_{\mathfrak{m}}^n$ [artinian because they are noetherian of Krull dimension zero]. Naturality of this isomorphism shows that every cyclic Ω-submodule of U is a $\hat{\Omega}$-submodule, and this is enough to complete the proof.

(ii) In view of (i), the proof of Lemma 6.2(v) works here, too.

(iii) Since Ω is local, the natural map $M \to \hat{M}_{\mathfrak{m}}$ is injective [see 5.1(ii)]. For surjectivity, apply statement (ii) to the case $U = \hat{M}_{\mathfrak{m}}$ and the identity map $\hat{M}_{\mathfrak{m}} \to U$. Finite length of M follows from (i). □

LEMMA 6.6. *Let Ω be an indecomposable noetherian ring, and suppose that, for some maximal ideal $\mathfrak{m}$, either $\Omega_{\mathfrak{m}}$ or $\hat{\Omega}_{\mathfrak{m}}$ is an artinian ring. Then $\Omega = \Omega_{\mathfrak{m}} = \hat{\Omega}_{\mathfrak{m}}$ (canonical isomorphisms).*

PROOF. If some $\hat{\Omega}_{\mathfrak{m}}$ is artinian, then the natural inclusion $\Omega_{\mathfrak{m}} \subseteq \hat{\Omega}_{\mathfrak{m}}$ is equality [Lemma 6.5(iii)]. Conversely, if $\Omega_{\mathfrak{m}}$ is artinian, then $\mathfrak{m}_{\mathfrak{m}}$-adic completion leaves $\Omega_{\mathfrak{m}}$ unchanged.

For the rest of the proof we may assume that $\Omega_{\mathfrak{m}}$ is an artinian ring. Then Lemma 6.2 (i) and (iii), applied to the case $M = \Omega$, show that $\Omega_{\mathfrak{m}}$ is a Ω-direct summand of Ω. Then indecomposability of Ω shows that $\Omega = \Omega_{\mathfrak{m}}$, as claimed. □

7. Elimination of finite-length summands

Suppose that we want to describe the local-global and direct-sum relations in the category of finitely generated modules over a noetherian ring Ω. The purpose of this section is to explain why *it suffices to study modules in* $\mathrm{fingen}_\infty(\Omega)$, *the category of finitely generated Ω-modules without nonzero direct summands of finite length.* Much of the material in this section (or at least its crux) is well-known or scattered throughout [**L2**], and some of it requires Ω to have Krull dimension 1.

First consider the subcategory finlen(Ω) of Ω-modules of finite length. This is a Krull-Schmidt category, and hence has no nontrivial direct-sum relations. Moreover, locally isomorphic modules (at maximal ideals of Ω) are actually isomorphic. Thus there are no nontrivial local-global relations.

Now consider a general $M \in \text{fingen}(\Omega)$. Such an M is uniquely the direct sum of a module in finlen(Ω) and a module in $\text{fingen}_\infty(\Omega)$ [Lemma 7.1(i)]. Moreover, the property of not having direct summands of finite length is preserved by taking direct sums [Lemma 7.1(ii)], localization at maximal ideals [Lemma 7.1(iii)], and — in Krull dimension 1 — $\mathfrak{m}$-adic completion [Proposition 7.3].

LEMMA 7.1. *Let M be a noetherian module over any ring Ω (possibly noncommutative). Then:*

(i) *There is a decomposition $M = F \oplus M'$ where $F \in \text{finlen}(\Omega)$ and $M' \in \text{fingen}_\infty(\Omega)$. Moreover, F and M' are unique up to isomorphism.*
(ii) *If $M = M_1 \oplus \ldots \oplus M_n$ and each $M_i \in \text{fingen}_\infty(\Omega)$, then the same is true of M.*
(iii) *Suppose that Ω is commutative and noetherian. Then $M \in \text{fingen}_\infty(\Omega)$ if and only if, for every $\mathfrak{m} \in \text{maxspec}(\Omega)$, the localization $M_\mathfrak{m} \in \text{fingen}_\infty(\Omega_\mathfrak{m})$.*

PROOF. The proofs of statements (i) and (ii) use the well-known fact that indecomposable modules of finite length have local endomorphism rings.

(ii) Given a decomposition $M = M_1 \oplus \ldots \oplus M_n$, it suffices to show that, if C is a direct summand of M such that the endomorphism ring of C is local, then C is isomorphic to a direct summand of some M_i. This follows from the direct-summand "exchange" property [**F**, 2.7], a property posessed by modules with local endomorphism rings [**F**, 2.8].

(i) The existence of such a decomposition $M = F \oplus M'$ is an immediate consequence of the fact that M is noetherian. To obtain the uniqueness assertion, suppose that $M = F_1 \oplus M_1'$ is another such decomposition of M, and let C be any indecomposable direct summand of F. Then C is isomorphic to a direct summand of F_1 by the proof of statement (ii) above and our hypothesis that M_1' has no direct summands of finite length. Since modules with local endomorphism rings cancel from direct sums [**F**, 4.5], we now have that $F' \oplus M' \cong F_1' \oplus M_1'$, where F' and F_1' are the complements of C and its copy in F and F_1, respectively. Repeating this procedure sufficiently often completes the proof of the uniqueness of F and M'.

(iii) Suppose that M has a nonzero finite-length summand, and hence an indecomposable such summand C. Say $M \cong C \oplus X$. Since C is indecomposable, it is contained in some primary component $P(\mathfrak{m})$ of M [Lemma 6.3], and hence $C \cong C_\mathfrak{m}$ [Lemma 6.2(i)]. Then $M_\mathfrak{m} \cong C_\mathfrak{m} \oplus X_\mathfrak{m} \cong C \oplus X_\mathfrak{m}$ shows that $M_\mathfrak{m}$ has a nonzero direct summand of finite length.

Conversely, if C is any finite-length direct summand of some $M_\mathfrak{m}$, then C is isomorphic to a direct summand of M itself [Lemma 6.2(iii)], and has finite length as an Ω-module [Lemma 6.2(i)]. □

It is obvious that submodules of finite length are unchanged by the completion process (over local rings). We prove one partial converse to this in Proposition 7.2: completion never introduces new direct summands of finite length, in dimension 1. We prove another partial converse in Proposition 7.4 (a stronger conclusion under more restrictive hypotheses).

PROPOSITION 7.2. *Let* $(\Omega, \mathfrak{m}, k)$ *be a local (noetherian) ring of Krull dimension 1, and let* $M \in \text{fingen}(\Omega)$. *Then every finite-length* $\hat{\Omega}_\mathfrak{m}$*-direct summand* X *of* $\hat{M}_\mathfrak{m}$ *is isomorphic to a (finite-length)* Ω*-direct summand of* M.

PROOF. Let $\hat{M}_\mathfrak{m} = W \oplus X$ for some $\hat{\Omega}_\mathfrak{m}$-module X. Since X has finite length and the unique maximal ideal of the local ring $\hat{\Omega}_\mathfrak{m}$ of Krull dimension 1 is not a minimal prime ideal, we have $X_Q = 0$ [Lemma 4.5]. Therefore $(\hat{M}_\mathfrak{m})_Q \cong W_Q$. Since $\hat{M}_\mathfrak{m}$ is the completion of a finitely generated Ω-module, so therefore is W [Lemma 5.4]. Say $W = \hat{A}_\mathfrak{m}$.

Since $(A \oplus X)\hat{}_\mathfrak{m} \cong W \oplus X \cong \hat{M}_\mathfrak{m}$, we therefore have $A \oplus X \cong M$ [Lemma 9.1]. □

PROPOSITION 7.3. *Let* $M \in \text{fingen}(\Omega)$, *with* Ω *noetherian of Krull dimension 1. Then* $M \in \text{fingen}_\infty(\Omega)$ *if and only if* $\hat{M}_\mathfrak{m} \in \text{fingen}_\infty(\hat{\Omega}_\mathfrak{m})$ $(\forall \mathfrak{m} \in \text{maxspec}(\Omega))$.

PROOF. First suppose that some $\hat{M}_\mathfrak{m}$ has a direct summand $X \neq 0$ of finite length. If $\Omega_\mathfrak{m}$ has Krull dimension 1, then $M_\mathfrak{m}$ has a direct summand that is isomorphic to X [Proposition 7.2], and hence the same it true of M [Lemma 6.2(iii)].

Otherwise $\Omega_\mathfrak{m}$ has Krull dimension 0, that is, is artinian. In this case $M_\mathfrak{m}$ has finite length, and hence $M_\mathfrak{m} = \hat{M}_\mathfrak{m}$. Therefore $M_\mathfrak{m}$ has a nonzero direct summand of finite length. Passage from $M_\mathfrak{m}$ to M itself is the same as before.

Conversely, suppose that $M = N \oplus X$ with $X \neq 0$ of finite length. We may assume that X is indecomposable. Since the direct sum of the primary components of M is the largest finite-length submodule of M [Lemma 6.3(ii)], indecomposability of X implies that $X \subseteq P(\mathfrak{m})$ for some primary component $P(\mathfrak{m})$. But then $M_\mathfrak{m} \cong N_\mathfrak{m} \oplus X$ [Lemma 6.2(i)]. Since $\mathfrak{m}$-adic completion leaves finite-length $\Omega_\mathfrak{m}$-modules unchanged, we get $\hat{M}_\mathfrak{m} \cong \hat{N}_\mathfrak{m} \oplus X$, as desired. □

PROPOSITION 7.4. *Let* $\mathfrak{m}$ *be a maximal ideal of height 1 of a reduced noetherian ring* Ω, *let* $M \in \text{fingen}(\Omega)$, *and suppose that* $\Omega_\mathfrak{m}$ *has finite normalization (i.e. its normalization is a finitely generated* $\Omega_\mathfrak{m}$*-module).*

If the $\mathfrak{m}$*-adic completion* $\hat{M}_\mathfrak{m}$ *has a nonzero submodule of finite length, then* M *has a submodule of finite length with a composition factor isomorphic to* $\Omega/\mathfrak{m}$.

PROOF. If we can prove that $M_\mathfrak{m}$ has a nonzero submodule of finite length, then it will follow that M has a submodule of finite length with $\Omega/\mathfrak{m}$ as a composition factor [Lemma 6.2(ii)]. For the rest of this proof, we therefore assume that *Ω is a reduced local ring whose maximal ideal $\mathfrak{m}$ has height 1, and Ω has finite normalization.* In particular, Ω has Krull dimension 1.

Let M be an arbitrary module in $\text{fingen}(\Omega)$. We claim:

(7.4.1) M has no nonzero submodules of finite length if and only if M is a submodule of a free Ω-module of finite rank.

Suppose that M has no nonzero submodules of finite length, and let $K = \ker(M \to M_Q)$. Then $K_Q = 0$. Since $\mathfrak{m}$ is not a minimal prime ideal, this implies that K has finite length [Lemma 4.5], and hence $K = 0$. Thus we may write $M \subseteq M_Q$. Since the ring Ω is reduced, the ring Ω_Q is a direct sum of fields, and therefore M_Q, like every finitely generated Ω_Q-module is contained in a free Ω_Q-module of finite rank. Say $M \subseteq M_Q \subseteq \oplus_{i=1}^n \Omega_Q b_i$, where the b's are a free basis. Writing some finite generating set of M in terms of these basis elements, and choosing a common denominator d for the coefficients that arise then shows that $M \subseteq \oplus_i \Omega(b_i/d)$, a free Ω-module of finite rank.

Conversely, suppose that $M \subseteq F$, a free Ω-module, and M contains a nonzero submodule of finite length. Then some nonzero submodule of M, and hence of F, is annihilated by the maximal ideal $\mathfrak{m}$ of Ω. Hence some nonzero element of Ω is annihilated by $\mathfrak{m}$. But since $\mathfrak{m}$ is not a minimal prime ideal and Ω is reduced, $\mathfrak{m}$ contains a regular element of Ω [Lemma 4.1], and hence cannot annihilate any nonzero element of Ω. This completes the proof of (7.4.1).

Now we prove the proposition itself by showing that if M has no nonzero submodules of finite length, then the same is true of $\hat{M}_{\mathfrak{m}}$.

Since the local ring Ω is reduced and has finite normalization, its $\mathfrak{m}$-adic completion $\hat{\Omega}_{\mathfrak{m}}$ is again a reduced ring [Remarks 5.3(iii)]. Moreover, since M is contained in a free Ω-module of finite rank, $\hat{M}_{\mathfrak{m}}$ is contained in a free $\hat{\Omega}_{\mathfrak{m}}$-module of finite rank. Therefore, by (7.4.1), $\hat{M}_{\mathfrak{m}}$ has no nonzero submodules of finite length. □

8. Basic Finiteness Conditions

Let Ω be a reduced noetherian ring of Krull dimension 1, with normalization Γ, and recall $\mathfrak{m} \in \operatorname{maxspec}(\Lambda)$ is called *singular* if $\Omega_{\mathfrak{m}} \neq \Gamma_{\mathfrak{m}}$. This section describes the basic finiteness conditions satisfied by finitely generated Ω-modules. Almost everything in this section is well-known and easy to prove in the classical situation that Ω has only finitely many singular maximal ideals. Extending these well-known results to allow for infinitely many singular maximal ideals relies mainly on the results of [**GL2, LO**].

Recall [Notation 3.1] that we use the notation (a$\forall\mathfrak{m}$) to mean "for *almost all* $\mathfrak{m}$," that is, "for all except finitely many $\mathfrak{m}$", and Recall Q denotes the finite set of minimal prime ideals of Ω..

Essentially all special finiteness properties in Krull dimension 1 are consequences of the following well-known fact.

LEMMA 8.1. *Let Ω be a noetherian ring of Krull dimension 1, and let $d \in \Omega - \cup Q$. (Thus, if Ω is reduced, d is an arbitrary regular element of Λ.) Then d is contained in only finitely many maximal ideals of Ω and the ring $\Omega/\Omega d$ is artinan.*

PROOF. For $d \in \Omega - \cup Q$, the ring $\Omega/\Omega d$ has Krull dimension 0, so that $\Omega/\Omega d$ is artinian and therefore has only finitely many maximal ideals. The statement in parentheses is proved in Lemma 4.1(ii). □

Decomposition (6.4.1) shows that one can construct a module $M \in \operatorname{fingen}(\Omega)$ with arbitrarily prescribed localizations $M_{\mathfrak{p}}$ at the (finitely many) minimal primes $\mathfrak{p}$ of Ω. On the other hand, once this is done, almost all of the localizations $M_{\mathfrak{m}}$ at maximal ideals are determined. The next lemma makes this more precise.

LEMMA 8.2. *Let $L, M \in \operatorname{fingen}(\Omega)$, where Ω is noetherian of Krull dimension 1.*

(i) *If $L_Q \cong M_Q$, then $L_{\mathfrak{m}} \cong M_{\mathfrak{m}}$* $\big(\text{a}\forall\mathfrak{m} \in \operatorname{maxspec}(\Omega)\big)$.
(ii) *If Ω is reduced, then $L_{\mathfrak{m}}$ is $\Omega_{\mathfrak{m}}$-free* $\big(\text{a}\forall\mathfrak{m} \in \operatorname{maxspec}(\Omega)\big)$.
(iii) *If L and M are submodules of an Ω-module X (not necessarily finitely generated) and $L_Q \subseteq M_Q$ in X_Q, then $L_{\mathfrak{m}} \subseteq M_{\mathfrak{m}}$ in $X_{\mathfrak{m}}$* $\big(\text{a}\forall\mathfrak{m} \in \operatorname{maxspec}(\Omega)\big)$.

PROOF. (i) See [**GL2**, 2.1].

(ii) Temporarily ignore the fact that Ω is reduced. Part of the assertion of [**GL2**, 2.2] is that, if L_Q is Ω_Q-projective, then $L_\mathfrak{m}$ is $\Omega_\mathfrak{m}$-free $\big(\text{a}\forall\mathfrak{m} \in \text{maxspec}(\Omega)\big)$. But since Ω is reduced, Ω_Q is a direct sum of fields [see (4.3.1)]. Therefore every Ω_Q-module — in particular, L_Q — is projective.

(iii) This statement is proved in [**GL2**, 2.4] under the additional hypothesis that X be finitely generated. To reduce to that situation, let $X' = L + M$, a finitely generated Ω-submodule of X, and apply [**GL2**, 2.4] with X' in place of X. □

COROLLARY 8.3. *Let Ω be a noetherian ring of Krull dimension 1. If $\alpha \in \Omega$ and $\alpha/1$ is a unit in Ω_Q, then $\alpha/1$ is a unit in $\Omega_\mathfrak{m}$* $\big(\text{a}\forall\mathfrak{m} \in \text{maxspec}(\Omega)\big)$.

PROOF. Ω and $\Omega\alpha$ are finitely generated submodules of the Ω-module Ω and, by hypothesis, $\Omega_Q = (\Omega\alpha)_Q$ in $(\Omega)_Q$. Therefore $\Omega_\mathfrak{m} \subseteq (\Omega\alpha)_\mathfrak{m}$ in $\Omega_\mathfrak{m}$ $\big(\text{a}\forall\mathfrak{m} \in \text{maxspec}(\Omega)\big)$ [Lemma 8.2(iii)]. Since the oppposite inclusion is obvious, we have $\Omega_\mathfrak{m} = (\Omega\alpha)_\mathfrak{m}$ in $\Omega_\mathfrak{m}$ $\big(\text{a}\forall\mathfrak{m} \in \text{maxspec}(\Omega)\big)$. In other words, $\alpha/1$ is a unit in $\Omega_\mathfrak{m}$ $\big(\text{a}\forall\mathfrak{m} \in \text{maxspec}(\Omega)\big)$. □

Our next result shows that Lemma 8.2(i) gives the full set restrictions for changing the localizations $M_\mathfrak{m}$ without changing M_Q.

LEMMA 8.4. *Let $L \in \text{fingen}(\Omega)$, where Ω is noetherian of Krull dimension 1. For each maximal ideal $\mathfrak{m}$, let $M(\mathfrak{m}) \in \text{fingen}(\Omega_\mathfrak{m})$, and suppose that*

(i) *$M(\mathfrak{m}) \cong L_\mathfrak{m}$ $\big(\text{a}\forall\mathfrak{m} \in \text{maxspec}(\Omega)\big)$, and*

(ii) *$M(\mathfrak{m})_Q \cong (L_\mathfrak{m})_Q$ $\big(\forall\mathfrak{m} \in \text{maxspec}(\Omega)\big)$.*

Then there exists $M \in \text{fingen}(\Omega)$ such that $M_\mathfrak{m} \cong M(\mathfrak{m})$ for all maximal ideals $\mathfrak{m}$.

PROOF. This is just a paraphrase of part of [**LO**, 2.9]. □

The next result shows that equality of submodules can be specified locally. (Compare Lemma 8.2(iii).)

THEOREM 8.5. *Let Ω be a noetherian ring of Krull dimension 1. Let $L \subseteq X$ be Ω-modules, with L finitely generated and $L_Q = X_Q$ in X_Q. For each maximal ideal $\mathfrak{m}$, let $M(\mathfrak{m})$ be a finitely generated $\Omega_\mathfrak{m}$-submodule of $X_\mathfrak{m}$, and suppose that*

(i) *$M(\mathfrak{m}) = L_\mathfrak{m}$ in $X_\mathfrak{m}$ $\big(\text{a}\forall\mathfrak{m} \in \text{maxspec}(\Omega)\big)$, and*

(ii) *$M(\mathfrak{m})_Q = (L_\mathfrak{m})_Q$ in X_Q $\big(\forall\mathfrak{m} \in \text{maxspec}(\Omega)\big)$.*

Then there is exactly one finitely generated Ω-submodule M of X such that $M_\mathfrak{m} = M(\mathfrak{m})$ for all maximal ideals $\mathfrak{m}$. Moreover, $M_Q = X_Q$.

PROOF. Uniqueness of M is clear, since every Ω-submodule of X is determined by its localizations at maximal ideals of Ω. (See, for example, [**AM**, Proposition 3.8].) If ${}_\Omega X$ is finitely generated, the existence of M becomes part of [**LO**, 2.6]. We reduce to this situation.

Each of the $\Omega_\mathfrak{m}$-modules $M(\mathfrak{m})$ can be generated by finitely many elements of the form $x/1$, with $x \in X$. Therefore, there is a finite subset $\{x_1, \ldots, x_n\}$ of X such that the subset $\{x_1/1, \ldots, x_n/1\}$ of $X_\mathfrak{m}$ contains an $\Omega_\mathfrak{m}$-generating set for $M(\mathfrak{m})$, *for each of the finitely many maximal ideals $\mathfrak{m}$ such that $M(\mathfrak{m}) \neq L_\mathfrak{m}$*. Let $X' = L + \sum_i \Omega x_i$, a finitely generated Ω-submodule of X. Then we can apply [**LO**, 2.6] with X' in place of X. □

REMARK 8.6. Theorem 8.5 will be a principal tool for constructing finitely generated modules M from their localizations. Note that the resulting M is a unique submodule of X, not just unique up to isomorphism.

We think of this theorem as constructing M in two approximations: First, by specifying the Ω-module L, we get the desired M almost everywhere. Second, by adjusting the (finitely many) incorrect localizations, we obtain the module M exactly. *Note that the use of this theorem requires that we start with an appropriate first approximation L.* There is usually a natural choice for the module L. For example, as we are implicitly requiring that the Ω_Q-module X_Q be finitely generated (since $L_Q = X_Q$), we can take L to be the Ω-submodule of X generated by the numerators of some finite Ω_Q-generating set of X_Q.

THEOREM 8.7. *Let Ω be a reduced noetherian ring of Krull dimension 1. Then $\Omega_{\mathfrak{m}}$ is an integral domain* $\big(\text{a}\forall\mathfrak{m} \in \text{maxspec}(\Omega)\big)$.

PROOF. First note that since Ω — and hence every $\Omega_{\mathfrak{m}}$ — is reduced, $\Omega_{\mathfrak{m}}$ is an integral domain if (and only if) $\mathfrak{m}$ contains only one minimal prime ideal. Next let P, Q be distinct minimal prime ideals. Then $\Omega/(P+Q)$ is a proper homomorphic image of the noetherian domain Ω/P of dimension 1. Therefore the ring $\Omega/(P+Q)$ is artinian, and hence has only finitely many maximal ideals. Hence only finitely many maximal ideals of Ω contain both P and Q. Since Ω has only finitely many minimal prime ideals, we are done. □

9. General Cancellation and Direct Summands in Dimension 1

Throughout this memoir we make frequent use of the following general direct-summand and cancellation facts.

LEMMA 9.1. *Let $(\Omega, \mathfrak{m})$ be a local (noetherian) ring, and $M, N \in$* fingen(Ω). *Then:*

(i) *N is isomorphic to a direct summand of M if and only if $\hat{N}_{\mathfrak{m}}$ is isomorphic to a direct summand of $\hat{M}_{\mathfrak{m}}$.*
(ii) *$M \cong N$ if and only if $\hat{M}_{\mathfrak{m}} \cong \hat{N}_{\mathfrak{m}}$.*

PROOF. See [**W2**] for a short proof of (i). Since M and N are noetherian, (ii) follows from (i). □

LEMMA 9.2. *Let Ω be a commutative noetherian ring, and suppose that $M \oplus X \cong N \oplus X$ for $M, N, X \in$* fingen(Ω). *Then:*

(i) ("Local cancellation") genus$(M) =$ genus(N). [**E**]
(ii) ("Cancellation holds in every genus") *If Ω is reduced, has Krull dimension 1, and* genus$(M) =$ genus(X), *then $M \cong N$.* [**GL2**, Corollary 5.10]

LEMMA 9.3. *Let $M, N \in$* fingen(Ω), *where Ω is noetherian of Krull dimension 1.*

(i) *Suppose* $\big(\forall\mathfrak{m} \in \text{maxspec}(\Omega)\big)$ *that $N_{\mathfrak{m}}$ is isomorphic to a direct summand of $M_{\mathfrak{m}}$. Then some $N' \in$* genus(N) *is a direct summand of M, and N is a direct summand of some $M' \in$* genus(M) [**GL2**, 4.1].
(ii) ("Jacobinski's" direct summand theorem) *Suppose, in addition to the hypothesis in (i), that every indecomposable direct summand of N_Q occurs strictly more often in a decomposition of M_Q. Then N itself is isomorphic to a direct summand of M* [**GL2**, 6.4].

(iii) *Suppose that* $M = \oplus_i M_i$ *and* $N \in \text{genus}(M)$. *Then there is a decomposition* $N = \oplus_i N_i$ *with each* $N_i \in \text{genus}(M_i)$ [**GL2**, 4.3].

Caution on notation in Lemma 9.3(ii). $(\ldots)_Q$ denotes the localization that inverts the complement of the union of the minimal prime ideals of Ω. If Ω is reduced — the situation in which we use this "Jacobinski" theorem — $(\ldots)_Q$ also equals the localization that inverts the regular elements of Ω.

CHAPTER 1

Dedekind-like Rings

10. Definition and Characterizations

DEFINITION 10.1. We call a commutative noetherian ring Λ *Dedekind-like* if Λ is reduced (no nonzero nilpotent elements) and its normalization Γ (in its total quotient ring) has the following properties.

(i) Γ is a direct sum of Dedekind domains.

(ii) $(\Gamma/\Lambda)_{\mathfrak{m}}$ is either a simple $\Lambda_{\mathfrak{m}}$-module or 0 $\big(\forall\mathfrak{m} \in \operatorname{maxspec}(\Lambda)\big)$.

(iii) $\mathfrak{m}_{\mathfrak{m}} = \operatorname{rad}(\Gamma_{\mathfrak{m}})$ in $\Gamma_{\mathfrak{m}}$ (the Jacobson radical) $\big(\forall\mathfrak{m} \in \operatorname{maxspec}(\Lambda)\big)$.

We do not consider fields to be Dedekind domains. As previously mentioned [§3.1], Λ *always denotes a Dedekind-like ring unless otherwise specified.* Essentially the only "otherwise specified" situation is that we regularly use it for rings that we are proving to be Dedekind-like.

In all of our tameness results — section 15 through the end of this memoir — we also make Additional Hypothesis 10.2 part of the definition of "Dedekind-like", because we do not know whether our tameness results hold without it. The "Additional Hypothesis" can be ignored in this section through Section 14.

ADDITIONAL HYPOTHESIS 10.2. *If, for some* $\mathfrak{m} \in \operatorname{maxspec}(\Lambda)$, *the ring* $\bar{\Gamma}(\mathfrak{m}) = \Gamma/\mathfrak{m}\Gamma$ *is a 2-dimensional field extension of* $\bar{\Lambda}(\mathfrak{m}) = \Lambda/\mathfrak{m}$, *then this extension is separable.*

EXAMPLES 10.3. Additional Hypothesis 10.2 is always satisfied unless $\bar{\Lambda}(\mathfrak{m})$ has charactistic 2.

Suppose that $\bar{\Lambda}(\mathfrak{m})$ has charactistic 2. Then the hypothesis still holds if $\bar{\Lambda}(\mathfrak{m})$ is finite, or $\bar{\Lambda}(\mathfrak{m})$ is algebraically closed.

The simplest Dedekind-like ring for which it fails is $\Lambda = k + xF[[x]]$, the formal power-series ring where k is any field (necessarily of characteristic 2) that has an inseparable extension field F of dimenion 2, all constant terms are in k and all other coefficients are in F.

In Corollary 10.7 we prove that Λ is Dedekind-like if and only if every localization $\Lambda_{\mathfrak{m}}$ is Dedekind-like. Then, in Theorem 11.9, we prove that Λ is Dedekind-like if and only if every completion $\hat{\Lambda}_{\mathfrak{m}}$ is Dedekind-like.

REMARK 10.4 (Terminology in [**L2**]). The rings called "Dedekind-like" in [**L2**] are — in the terminology of the Definitions 11.3 of the present paper — Dedekind like rings whose singular maximal ideals are split and finite in number (and hence ${}_{\Lambda}\Gamma$ is finitely generated, by Proposition 10.11). We mention some other slight changes in terminology when introducing the relevant ideas.

LEMMA 10.5. *Let* Ω *be a reduced noetherian ring with normalization* Γ. *Then condition (i) of Definition 10.1 is equivalent to:*

(i)$'$ *Every maximal ideal of* Ω *has height 1.*

PROOF. (i)$\Rightarrow$(i)$'$. All maximal ideals of Dedekind domains, and hence of Γ, have height 1. This property descends to Ω by the "lying-over" and "going up" theorem for integral overrings. (See e.g. [**M**, 9.3 and 9.4, respectively].)

(i)$'$ $\Rightarrow$(i). If $\mathfrak{p}_1' \subset \mathfrak{p}_2 \subset \dots$ is a strictly increasing chain of prime ideals of Γ, then the contractions $\mathfrak{p}_i = \mathfrak{p}_i' \cap \Omega$ form an increasing chain of prime ideals of Ω, and this chain is strictly increasing because of the "incomparability" part of the lying-over theorem. Therefore Γ has Krull dimension ≤ 1, and is therefore a direct sum of Dedekind domains and fields. Thus it now suffices to show that no fields occur in this direct sum. This is equivalent to showing that no minimal prime ideal $\mathfrak{p}'$ of Γ is a maximal ideal [Lemma 4.1(i), with $\Omega = \Gamma$].

Let $\mathfrak{p} = \mathfrak{p}' \cap \Omega$. By Lemma 4.1(iii), $\mathfrak{p}'$ is the unique minimal prime ideal of Γ lying over $\mathfrak{p}$. By hypothesis (i)$'$, we have $\mathfrak{p} \subset \mathfrak{q}$ for some maximal ideal of Ω, and by the going-up theorem we have $\mathfrak{p}' \subset \mathfrak{q}'$ for some prime ideal $\mathfrak{q}'$ of Γ such that $\mathfrak{q}' \cap \Omega = \mathfrak{q}$. This shows, as desired, that $\mathfrak{p}'$ is not a maximal ideal of Γ. □

The following slightly more abstract characterization of "Dedekind-like" is sometimes more convenient to use than the definition, and follows immediately from Lemma 10.5.

PROPOSITION 10.6. *A reduced noetherian ring* Λ *with normalization* Γ *is Dedekind-like if and only if the following conditions hold.*

(i)$'$ *Every maximal ideal of* Λ *has height 1.*

(ii) $(\Gamma/\Lambda)_{\mathfrak{m}}$ *is either a simple* $\Lambda_{\mathfrak{m}}$*-module or* 0 $\big(\forall \mathfrak{m} \in \operatorname{maxspec}(\Lambda)\big)$.

(iii) $\mathfrak{m}_{\mathfrak{m}} = \operatorname{rad}(\Gamma_{\mathfrak{m}})$ *in* $\Gamma_{\mathfrak{m}}$ $\big(\forall \mathfrak{m} \in \operatorname{maxspec}(\Lambda)\big)$.

COROLLARY 10.7. *A noetherian ring* Λ *is Dedekind-like if and only if every localization* $\Lambda_{\mathfrak{m}}$ $\big(\mathfrak{m} \in \operatorname{maxspec}(\Lambda)\big)$ *is Dedekind-like.*

PROOF. Λ is reduced if and only if each $\Lambda_{\mathfrak{m}}$ is reduced. Moreover, if Λ is reduced with normalization Γ, then the normalization of each $\Lambda_{\mathfrak{m}}$ is $\Gamma_{\mathfrak{m}}$ [Remarks 5.3(i)]. The rest of the proof follows from the fact that conditions (i)$'$–(iii) of Proposition 10.6 hold if and only if they hold locally. □

In view of Lemma 10.5, the definition of a local Dedekind-like ring given in [**KL1**, 2.5] is equivalent to the following.

COROLLARY 10.8. *A reduced local ring* $(\Lambda, \mathfrak{m}, k)$ *with normalization* Γ *is Dedekind-like if and only if* Λ *has Krull dimension 1,* $\mathfrak{m} = \operatorname{rad}(\Gamma)$, *and the minimum number of generators of* Γ *as a* Λ*-module satisfies* $\mu_\Lambda(\Gamma) \le 2$.

PROOF. We use the local case of Proposition 10.6 for both the "if" and "only if" assertions. If Λ is Dedekind-like, the only thing that we need to prove is that $\mu_\Lambda(\Gamma) \le 2$, and this is clearly satisfied since ${}_\Lambda(\Gamma/\Lambda)$ is either simple or zero.

Conversely, suppose that the conditions stated in the present corollary are satisfied. What we need to prove is that ${}_\Lambda(\Gamma/\Lambda)$ is simple or zero. Since $\mathfrak{m}$ is an ideal of Γ we have $\Gamma/\mathfrak{m} = \Gamma/\mathfrak{m}\Gamma$, which is a vector space over the field $\Lambda/\mathfrak{m}$. Since $\mu_\Lambda(\Gamma) \le 2$, the vector space $\Gamma/\mathfrak{m}$ has dimension ≤ 2. Therefore its homomorphic image Γ/Λ has dimension ≤ 1, as desired. □

We return to the case of non-local Dedekind-like rings.

PROPOSITION 10.9. *Every ideal of* Λ *can be generated by 2 elements.*

PROOF. For local Λ this is proved in [**KL1**, Lemma 2.7]. Thus, every ideal of $\Lambda_{\mathfrak{m}}$ can be generated by 2 elements $(\forall \mathfrak{m} \in \operatorname{maxspec}(\Lambda))$. Since Λ has Krull dimension 1, the lemma therefore follows from the Forster-Swan theorem [**EE**]. □

LEMMA 10.10. *Every Γ-module of finite length has finite length as a Λ-module.*

PROOF. This is an easy consequence of the Krull-Akizuki Theorem [**N**, Theorem 33.2], because Λ is a noetherian reduced ring of Krull dimension 1. (In fact, it is not difficult to prove that every simple Γ-module has length ≤ 2 as a Λ-module.) □

PROPOSITION 10.11. *The following conditions are equivalent for any (Dedekind-like ring) Λ.*

(i) ${}_{\Lambda}\Gamma$ *is finitely generated (i.e. Λ has "finite normalization").*
(ii) $d\Gamma \subseteq \Lambda$ *for some regular element $d \in \Lambda$.*
(iii) $\operatorname{singspec}(\Lambda)$ *is finite.*

PROOF. Note that only the proof of (iii)⇒(i) uses the fact that the noetherian reduced ring Λ of Krull dimension 1 is Dedekind-like.

(i)⇒(ii) This holds since $\Lambda_Q = \Gamma_Q$.

(ii)⇒(iii) This holds because $d/1$ is a unit in $\Lambda_{\mathfrak{m}}$ except for the finite number of maximal ideals that contain d [Lemma 8.1].

(iii)⇒(i) Since Λ is Dedekind-like, $\Gamma_{\mathfrak{m}}$ is a finitely generated $\Lambda_{\mathfrak{m}}$-module for every $\mathfrak{m} \in \operatorname{maxspec}(\Lambda)$. Choose a finite $\Lambda_{\mathfrak{m}}$-generating set of $\Gamma_{\mathfrak{m}}$ for each of the finitely many $\mathfrak{m}$ such that $\mathfrak{m} \in \operatorname{singspec}(\Lambda)$, that is, for each $\mathfrak{m}$ such that $\Lambda_{\mathfrak{m}} \neq \Gamma_{\mathfrak{m}}$. Then set of numerators of the elements in the union of these generating sets is a Λ-generating set of Γ. □

For examples showing that ${}_{\Lambda}\Gamma$ need not be finitely generated, see [**HL**].

11. Maximal Ideals: Residue Inclusions, Localizations, Completions

LEMMA 11.1. *Let ${}_{\Lambda}S \subseteq {}_{\Gamma}X = \Gamma S$ be an inclusion of modules $S \in \operatorname{fingen}(\Lambda)$ and $X \in \operatorname{fingen}(\Gamma)$, and let $\mathfrak{m} \in \operatorname{maxspec}(\Lambda)$.*

(i) *The natural map $\bar{S}(\mathfrak{m}) = S/(\mathfrak{m}S) \to X/(\mathfrak{m}X) = \bar{X}(\mathfrak{m})$ is an injection.*
(ii) *The natural localization maps $\bar{S}(\mathfrak{m}) \to \bar{S}(\mathfrak{m})_{\mathfrak{m}}$ and $\bar{X}(\mathfrak{m}) \to \bar{X}(\mathfrak{m})_{\mathfrak{m}}$ are bijections.*
(iii) *$\bar{X}(\mathfrak{m})$ is a Λ-module of finite length, equivalently, a finite dimensional $\bar{\Lambda}(\mathfrak{m}) = \Lambda/\mathfrak{m}$-vector space (even if X is not finitely generated as a Λ-module).*

PROOF. Since $\mathfrak{m}$ has height 1 [Proposition 10.6] it is not a minimal prime ideal. Therefore $\mathfrak{m}$ contains regular elements of Λ, and hence of Γ [Lemma 4.1(ii) and (iv)]. Therefore $\bar{\Gamma}(\mathfrak{m}) = \Gamma/(\mathfrak{m}\Gamma)$ is an artinian ring [Lemma 8.1].

Since ${}_{\Gamma}X$ is finitely generated, $\bar{X}(\mathfrak{m}) = \Gamma/(\mathfrak{m}\Gamma) \otimes_{\Gamma} X$ is a finitely generated module over the artinian ring $\bar{\Gamma}(\mathfrak{m})$, and hence has finite length as a Γ-module. Therefore $\bar{X}(\mathfrak{m})$ has finite length as a Λ-module [Lemma 10.10], and this proves (iii).

Statement (ii) holds since $\bar{S}(\mathfrak{m})$ is a Λ-module of finite length annihilated by the maximal ideal $\mathfrak{m}$ of Λ [Lemma 6.2(i)].

Statement (i). In view of statement (ii) we may replace $\bar{S}(\mathfrak{m})$ and $\bar{X}(\mathfrak{m})$ by their $\mathfrak{m}$-localizations in the proof of (i). In other words, we may assume that our

Dedekind-like ring is local, $\mathfrak{m}$ is its maximal ideal, and Γ remains its normalization [Remarks 5.3(i)]. But then $\mathfrak{m}$ becomes an ideal of Γ, and hence $\mathfrak{m}S = \mathfrak{m}\Gamma S = \mathfrak{m}X$. Therefore the natural map in (i) becomes the obvious injection $S/(\mathfrak{m}X) \to X/(\mathfrak{m}X)$. □

DEFINITION 11.2 (Residue inclusion). Let ${}_\Lambda S \subseteq {}_\Gamma X = \Gamma S$ be an inclusion of modules $S \in \text{fingen}(\Lambda)$ and $X \in \text{fingen}(\Gamma)$ respectively, and let $\mathfrak{m} \in \text{maxspec}(\Lambda)$. Since the natural map $\bar{S}(\mathfrak{m}) \to \bar{X}(\mathfrak{m})$ is an injection [Lemma 11.1(i)], we usually identify $\bar{S}(\mathfrak{m})$ with its image in $\bar{X}(\mathfrak{m})$. Thus this map becomes $\bar{S}(\mathfrak{m}) \subseteq \bar{X}(\mathfrak{m})$, the $\mathfrak{m}$-*residue inclusion of* $S \subseteq X$. A property that we use repeatedly is:

(11.2.1) The $\mathfrak{m}$-residue inclusion $\bar{S}(\mathfrak{m}) \subseteq \bar{X}(\mathfrak{m})$ remains unchanged under $\mathfrak{m}$-localization and $\mathfrak{m}$-adic completion.

Lack of change under $\mathfrak{m}$-localization is statement Lemma 11.1(ii), and $\mathfrak{m}$-adic completion of this localization causes no change, because modules of finite length over a local ring do not change under such completion.

The inclusion $\Lambda \subseteq \Gamma$ occurs so often in this memoir that we attach the following special notation to its residue inclusions. Let

$$k(\mathfrak{m}) = \bar{\Lambda}(\mathfrak{m}) \tag{11.2.2}$$

Then the $\mathfrak{m}$-residue inclusion of $\Lambda \subseteq \Gamma$ becomes $k(\mathfrak{m}) \subseteq \bar{\Gamma}(\mathfrak{m})$. Since this inclusion remains unchanged under $\mathfrak{m}$-localization, $k(\mathfrak{m})$ *is also the residue field of the local Dedekind-like ring* $\Lambda_\mathfrak{m}$. This agrees with the notation in [**KL1, KL2**], where the residue field of the complete local Dedekind-like ring Λ is called k. We extend this consistency with the local case to our notation for $\bar{\Gamma}(\mathfrak{m})$ in Notation 11.5.

Four types of local Dedekind-like rings are defined in [**KL1**, 2.5]. We name the "type" of each maximal ideal $\mathfrak{m}$ of an arbitrary Dedekind-like ring Λ according to the type of the local Dedekind-like ring $\Lambda_\mathfrak{m}$. The following theorem and definitions spell this out in detail.

THEOREM AND DEFINITIONS 11.3 (Types of maximal ideals and local Dedekind-like rings). *For each* $\mathfrak{m} \in \text{maxspec}(\Lambda)$, *exactly one of the following holds.*

(i) $\bar{\Gamma}(\mathfrak{m}) \cong k(\mathfrak{m}) \times k(\mathfrak{m})$ *as* $k(\mathfrak{m})$*-algebras, and* $\Gamma_\mathfrak{m}$ *is the direct sum of two DVRs. (Here we call* $\mathfrak{m}$ *and* $\Lambda_\mathfrak{m}$ "strictly split"*.)*

(ii) $\bar{\Gamma}(\mathfrak{m}) \cong k(\mathfrak{m}) \times k(\mathfrak{m})$ *as* $k(\mathfrak{m})$*-algebras, and* $\Gamma_\mathfrak{m}$ *is a principal ideal domain with exactly two maximal ideals. (Here we call* $\mathfrak{m}$ *and* $\Lambda_\mathfrak{m}$ "nonstrictly split"*.)*

(iii) $\bar{\Gamma}(\mathfrak{m})$ *is a 2-dimensional extension field of* $k(\mathfrak{m})$*, and* $\Gamma_\mathfrak{m}$ *is a DVR. (Here we call* $\mathfrak{m}$ *and* $\Lambda_\mathfrak{m}$ "unsplit"*.)*

(iv) $\bar{\Gamma}(\mathfrak{m}) \cong k(\mathfrak{m})$ *as* $k(\mathfrak{m})$*-algebras, and* $\Gamma_\mathfrak{m}$ *is a DVR. (Here* $\Gamma_\mathfrak{m} = \Lambda_\mathfrak{m}$ *in* $\Gamma_\mathfrak{m}$*, and we have already called such* $\mathfrak{m}$ *"nonsingular".)*

PROOF. Recall the identifications of $k(\mathfrak{m}) = k(\mathfrak{m})_\mathfrak{m}$ and $\bar{\Gamma}(\mathfrak{m}) = \bar{\Gamma}(\mathfrak{m})_\mathfrak{m}$ in Lemma 11.1(ii), and the fact that $\Gamma_\mathfrak{m}$ is the normalization of $\Lambda_\mathfrak{m}$ [Remarks 5.3(i)]. For the rest of this proof, we may therefore assume that Λ is local, and shorten $k(\mathfrak{m})$ and $\bar{\Gamma}(\mathfrak{m})$ to k and $\bar{\Gamma}$ respectively. What has been gained is that ${}_\Lambda\Gamma$ is now generated by 2 elements [Corollary 10.8]. Consequently, $\bar{\Gamma}$ is a k-vector space of dimension ≤ 2.

Consider Γ. Since Λ is local, the module-finite Λ-algebra Γ is semilocal. Normality and Krull dimension 1 therefore show Γ to be a direct sum of semilocal principal ideal domains. Moreover $\mathfrak{m} = \operatorname{rad}(\Gamma)$, an ideal of Γ. Therefore $\bar{\Gamma} = \Gamma/\mathfrak{m}$, and the number of maximal ideals of Γ therefore equals the number of maximal ideals of the reduced ring $\bar{\Gamma}$ of k-dimension ≤ 2. We break the remainder of the proof into several cases.

Suppose, first, that $\dim_k(\bar{\Gamma}) = 2$. Therefore the reduced k-algebra $\bar{\Gamma}$ can only be $k \times k$ or a 2-dimensional extension field of k, as stated in (i)—(iii).

If $\bar{\Gamma} = k \oplus k$, then Γ itself has exactly two maximal ideals, and is therefore either is single principal ideal domain or else the direct sum of two DVRs, as stated in (i) and (ii). If $\bar{\Gamma}$ is a field, then the number of maximal idealsof Γ is 1, and hence Γ is a DVR, as stated in (iii).

The remaining case is that $\dim_k(\bar{\Gamma}) = 1$, and hence $\Gamma = \Lambda$, with details as stated in (iv). □

PROPOSITION 11.4. *Only finitely many maximal ideals of* Λ *can be strictly split.*

PROOF. This is a special case of the fact that, for any reduced noetherian ring Ω of Krull dimension 1, $\Omega_{\mathfrak{m}}$ is an integral domain for almost all maximal ideals $\mathfrak{m}$ [Theorem 8.7]. □

When $\operatorname{singspec}(\Lambda)$ is infinite, the number of nonstrictly split and unsplit maximal ideals can both be infinite. The extreme case is that Λ has infinitely many maximal ideals and all of them are singular [**HL**].

A complete local Dedekind-like ring can never be nonstrictly split, since the completion of any semilocal ring is a direct sum of local rings. Therefore nonstrictly split local Dedekind-like rings play no role in [**KL1, KL2**], although they were defined in [**KL1**, 2.5].

NOTATION 11.5. To enhance our consistency with the notation used in the complete local case, we write:

$$\bar{\Gamma}(\mathfrak{m}) = \begin{cases} F(\mathfrak{m}) & \text{if } \mathfrak{m} \text{ is unsplit.} \\ k(\mathfrak{m}) \times k(\mathfrak{m}) = k(\mathfrak{m})_1 \times k(\mathfrak{m})_2 & \text{if } \mathfrak{m} \text{ is split.} \end{cases} \tag{11.5.1}$$

and we identify the residue field $k(\mathfrak{m}) = \bar{\Lambda}(\mathfrak{m})$ with its diagonal image in $k(\mathfrak{m})_1 \times k(\mathfrak{m})_2$

For $\gamma \in \Gamma$, we often write its natural image in $\bar{\Gamma}(\mathfrak{m})$ as $\bar{\gamma}(\mathfrak{m})$. If $\mathfrak{m}$ is split, $\bar{\gamma}(\mathfrak{m})_1$ and $\bar{\gamma}(\mathfrak{m})_2$ denote the two coordinates of $\bar{\gamma}(\mathfrak{m})$.

Recall that $\operatorname{maxsupp}_\Lambda(H)$ denotes the set of maximal ideals $\mathfrak{m}$ of Λ such that $H_{\mathfrak{m}} \neq 0$.

LEMMA 11.6. *Let* $\mathfrak{m} \in \operatorname{maxspec}(\Lambda)$.

(i) *If* $\mathfrak{m}$ *is strictly split, then* $\mathfrak{m}$ *is in the support of precisely two coordinate rings of* Γ. *Otherwise* $\mathfrak{m}$ *is in the support of a unique coordinate ring* Γ_h *of* Γ.

(ii) *Let* Γ_h *be a coordinate ring of* Γ *and* $\mathfrak{m} \in \operatorname{maxsupp}_\Lambda(\Gamma_h)$. *Then* $\Gamma_h\mathfrak{m}$ *is either a maximal ideal or the intersection of two distinct maximal ideals of* Γ_h. *In more detail:*

$$\Gamma_h\mathfrak{m} = \begin{cases} \ker\big(\Gamma_h \to F(\mathfrak{m})\big) & \textit{if } \mathfrak{m} \textit{ is unsplit;} \\ \ker\big(\Gamma_h \to k(\mathfrak{m})_1\big) \cap \ker\big(\Gamma_h \to k(\mathfrak{m})_2\big) & \textit{if } \mathfrak{m} \textit{ is nonstrictly split;} \\ \ker\big(\Gamma_h \to k(\mathfrak{m})_i\big) \quad (i = 1 \textit{ or } 2) & \textit{if } \mathfrak{m} \textit{ is strictly split.} \\ \ker\big(\Gamma_h \to k(\mathfrak{m})\big) & \textit{if } \mathfrak{m} \textit{ is nonsingular.} \end{cases} \tag{11.6.1}$$

(iii) *For $H \in \operatorname{fingen}(\Gamma)$, we have $\mathfrak{m} \in \operatorname{maxsupp}_\Lambda(H)$ if and only if $\bar{H}(\mathfrak{m}) \neq 0$.*

PROOF. (i) By Remarks 5.3(i), $\Gamma_\mathfrak{m}$ is the normalization of the local Dedekind-like ring $\Lambda_\mathfrak{m}$. Thus, if $\Lambda_\mathfrak{m}$ is strictly split, then $\Gamma_\mathfrak{m}$ has two coordinate rings; otherwise $\Gamma_\mathfrak{m}$ is an integral domain and hence has only one coordinate ring.

(ii) These assertions follow from the fact that $\Gamma_h/(\Gamma_h\mathfrak{m})$ is unchanged by $\mathfrak{m}$-localization, because it is annihilated by $\mathfrak{m}$. Thus we may assume that Λ is local, whence the statements are obvious from the conductor square for Λ if $\Lambda \neq \Gamma$ [**KL1**, 2.13], and trivial if $\Lambda = \Gamma$.

(iii) We prove the equivalent assertion that $H_\mathfrak{m} = 0 \iff \bar{H}(\mathfrak{m}) = 0$. Since Λ-modules annihilated by $\mathfrak{m}$ are unchanged by $\mathfrak{m}$-localization, the assertion we are proving is equivalent to $H_\mathfrak{m} = 0 \iff H_\mathfrak{m} = \mathfrak{m}_\mathfrak{m} H_\mathfrak{m}$. The nontrivial implication ($\Leftarrow$) follows from Nakayama's lemma, since $\Gamma_\mathfrak{m}$ is the normalization of the local Dedekind-like ring $\Lambda_\mathfrak{m}$ [Remarks 5.3(i)] and $\mathfrak{m}_\mathfrak{m}$ is the Jacobson radical of $\Gamma_\mathfrak{m}$ [Definition 10.1]. □

The preceding lemma used the fact that, for every local Dedekind-like ring Λ, its normalization Γ has at most two coordinate rings. On the other hand, when Λ is not local, Γ can have any finite number of coordinate rings, even if Λ is an indecomposable ring [Example 33.4].

LEMMA 11.7. *Let Γ_h be a coordinate ring of Γ and $\mathfrak{m} \in \operatorname{maxsupp}_\Lambda(\Gamma_h)$, and let $\mathfrak{n} = \Gamma_h\mathfrak{m}$. Then $(\Gamma_h)_\mathfrak{m} = (\Gamma_h)_\mathfrak{n}$ (natural isomorphism), where $(\Gamma_h)_\mathfrak{n}$ denotes the localization that inverts the elements of the complement (in Γ_h) of the union of the one or two maximal ideals of Γ_h whose intersection is $\mathfrak{n}$ [Lemma 11.6(i)].*

PROOF. It suffices to show that the same maximal ideals of Γ_h survive in both localizations.

First consider the case that $\mathfrak{n}$ is nonstrictly split, and therefore Γ_h is the unique coordinate ring of Γ with $\mathfrak{m}$ in its support [Lemma 11.6(i)]. The localization $(\Gamma_h)_\mathfrak{m}$ is formed by inverting the projections in Γ_h of the elements of $\Lambda - \mathfrak{m}$. All such elements are outside of the two maximal ideals of Γ_h whose intersection is $\mathfrak{n}$ [Lemma 11.6(ii)]. Therefore, the two maximal ideals of Γ_h that contain $\mathfrak{n}$ survive in $(\Gamma_h)_\mathfrak{m}$. On the other hand, since Γ_h is the unique coordinate ring of Γ supported by $\mathfrak{m}$, we have $(\Gamma_h)_\mathfrak{m} = \Gamma_\mathfrak{m}$, a ring with exactly two maximal ideals. Thus, the two maximal ideals of Γ_h that contain $\mathfrak{n}$ are precisely the maximal ideals of Γ_h that survive in $(\Gamma_h)_\mathfrak{m}$, as desired.

The remaining cases are similar, and we omit the details. □

LEMMA 11.8. *Let $(\Lambda, \mathfrak{m}, k)$ be any local ring, and $(\hat{\Lambda}, \hat{\mathfrak{m}}, k)$ its ($\mathfrak{m}$-adic) completion. Then:*

(i) *Λ is Dedekind-like if and only if $\hat{\Lambda}$ is Dedekind-like.*
(ii) *When the conditions in (i) hold, Λ and $\hat{\Lambda}$ are both unsplit, both split, or both DVRs. (But $\hat{\Lambda}$ is never nonstrictly split!)*
(iii) *If Λ is Dedekind-like with normalization Γ, then the normalization of $\hat{\Lambda}$ is $\hat{\Gamma}$, the completion in its $\mathfrak{m}$-adic topology. (Note that since $\mathfrak{m} = \operatorname{rad}(\Gamma)$, the $\mathfrak{m}$-adic completion of Γ as a Λ-module coincides with its $\mathfrak{m}$-adic completion as a Γ-module.)*

PROOF. The "only if" half of (i) was proved in [**KL1**, Lemma 2.21], where it was also shown that Λ and $\hat{\Lambda}$ have the same type and that $\hat{\Gamma}$ is the normalization of $\hat{\Lambda}$. Thus what remains is to prove that if $\hat{\Lambda}$ is Dedekind-like, so is Λ.

Since Λ is a subring of $\hat{\Lambda}$ [Notation 5.1(ii)], and local rings have the same Krull dimension as their completion [**N**, 17.12], we have:

(11.8.1) Λ is reduced and has Krull dimension 1.

Let Γ be the normalization of Λ, which makes sense because Λ is reduced. Let $\hat{\Gamma}$ be the $\mathfrak{m}$-adic completion of the Λ-algebra Γ, that is, the completion of Γ with respect to its filter of ideals $\{\mathfrak{m}^n\Gamma\}$. We claim that both of the following statements hold.

(11.8.2)
(i) Γ and $\hat{\Gamma}$ are module-finite over Λ and $\hat{\Lambda}$, respectively.
(ii) $\hat{\Gamma}$ is the normalization of $\hat{\Lambda}$.

By [**M**, bottom of page 263], since the completion $\hat{\Lambda}$ of the reduced local ring Λ is also reduced, the normalization Γ of Λ must be module-finite over Λ, proving the first assertion of (i). Clearly the second assertion of (i) follows from the first.

Statement (ii). Since ${}_\Lambda\Gamma$ is finitely generated, we can identify $\hat{\Gamma}$ with the rad(Γ)-adic completion of Γ [Lemma 5.2]. Since the unique maximal ideal of Λ has height 1, Γ is a direct sum of Dedekind domains [Lemma 10.5] — actually, semilocal principal ideal ideal domains in the present situation since ${}_\Lambda\Gamma$ is finitely generated. Therefore the completion of Γ with respect to its Jacobson radical is a direct sum of DVRs. In particular, $\hat{\Gamma}$ is a normal ring of Krull dimension 1. This, together with statement (i), yields statement (ii).

To complete the proof that Λ is Dedekind-like, it suffices to prove that $\mathfrak{m} = \text{rad}(\Gamma)$ and $\mu_\Lambda(\Gamma) \leq 2$ [Theorem 10.8 and (11.8.1)].

First, we claim that $\mathfrak{m}$ is an ideal of Γ. This is equivalent to showing that the finitely generated Λ-module $A = (\Gamma\mathfrak{m})/\mathfrak{m}$ equals zero. By flatness of $\hat{\Lambda}$ over Λ, we have $\hat{A} = (\hat{\Gamma}\hat{\mathfrak{m}})/\hat{\mathfrak{m}}$. But $\hat{\Gamma}$ is the normalization of $\hat{\Lambda}$, by (11.8.2)(ii), and $\hat{\Lambda}$ is Dedekind-like, by assumption. So the maximal ideal $\hat{\mathfrak{m}}$ of $\hat{\Lambda}$ is also an ideal of $\hat{\Gamma}$. Therefore, $\hat{A} = (\hat{\Gamma}\hat{\mathfrak{m}})/\hat{\mathfrak{m}} = 0$, and faithful flatness of $\hat{\Lambda}$ over Λ then yields $A = 0$, proving the claim.

Next, we claim that $\mathfrak{m} = \text{rad}(\Gamma)$. For every $x \in \mathfrak{m}$, the element $1 - x$ is invertible in Λ and hence also in Γ. This, together with the fact that $\mathfrak{m}$ is an ideal of Γ, shows that $\mathfrak{m} \subseteq \text{rad}(\Gamma)$. Conversely, $\Gamma/\mathfrak{m}$ is a finite dimensional vector space over $k = \Lambda/\mathfrak{m}$, by (11.8.2)(i), and is therefore a Λ-module of finite length. Thus, $\Gamma/\mathfrak{m}$ is unchanged by $\mathfrak{m}$-adic completion. Therefore $\Gamma/\mathfrak{m} \cong \hat{\Gamma}/\hat{\mathfrak{m}}$. This last ring has radical zero. For since $\hat{\Lambda}$ is local Dedekind-like with normalization $\hat{\Gamma}$, we have $\hat{\mathfrak{m}} = \text{rad}(\hat{\Gamma})$. Therefore, $\mathfrak{m} \supseteq \text{rad}(\Gamma)$, completing the proof that $\mathfrak{m} = \text{rad}(\Gamma)$.

Finally, we claim that $\mu_\Lambda(\Gamma) \leq 2$. By Nakayama's Lemma, this is equivalent to showing that the $k = \Lambda/\mathfrak{m}$-vector space $\Gamma/\mathfrak{m}$ has dimension at most 2. The latter holds because of the reasoning in the previous paragraph, since $\Gamma/\mathfrak{m} \cong \hat{\Gamma}/\hat{\mathfrak{m}}$. □

Putting Proposition 11.8 together with the fact that an arbitrary noetherian ring Λ is Dedekind-like if and only if every localization $\Lambda_\mathfrak{m}$ $(\mathfrak{m} \in \text{maxspec}(\Lambda))$ is Dedekind-like [Corollary 10.7] we get the following result, which allows us to apply the results of [**KL1, KL2**] about the complete local case to the non-local situation.

THEOREM 11.9. *A noetherian ring Λ is Dedekind-like if and only if $\hat{\Lambda}_\mathfrak{m}$ is Dedekind-like* $(\forall \mathfrak{m} \in \text{maxspec}(\Lambda))$.

PROPOSITION 11.10. *Let $\mathfrak{m}$ be a maximal ideal of the Dedekind-like ring Λ. If $\mathfrak{m}$ is (strictly or nonstrictly) split, then $\hat{\Gamma}_{\mathfrak{m}}$ has two coordinate rings. (Both are DVRs.) Otherwise $\hat{\Gamma}_{\mathfrak{m}}$ is a DVR.*

If $\mathfrak{m}$ is strictly split, in the support of Γ_h and Γ_k, then the two coordinate rings of $\hat{\Gamma}_{\mathfrak{m}}$ are $(\Gamma_h)\hat{}_{\mathfrak{m}}$ and $(\Gamma_k)\hat{}_{\mathfrak{m}}$.

PROOF. Since $\Gamma_{\mathfrak{m}}$ is the normalization of $\Lambda_{\mathfrak{m}}$ [Remarks 5.3(i)], we may assume that Λ is local with maximal ideal $\mathfrak{m}$ and normalization Γ. Then $\mathfrak{m} = \operatorname{rad}(\Gamma)$. Since $\mathfrak{m}$ is now an ideal of Γ, the $\mathfrak{m}$-adic completion $\hat{\Gamma}_{\mathfrak{m}}$ is the same whether we consider $\mathfrak{m}$ to be an ideal of Λ or an ideal of Γ. In this proof we consider $\mathfrak{m}$ to be an ideal of Γ, and we apply Lemma 11.6.

If $\mathfrak{m}$ is nonstrictly split, then Γ is a principal ideal domain with exactly two maximal ideals. Therefore $\hat{\Gamma}_{\mathfrak{m}}$ is the direct sum of two DVRs. In the remaining cases, Γ is a direct sum of one or two DVRs, and therefore $\hat{\Gamma}_{\mathfrak{m}}$ is the direct sum of the completions of these one or two DVRs. □

12. Surjective Pullback Squares

NOTATION 12.1 (Surjective pullback squares). Consider a commutative diagram of abelian groups — or of rings or modules over some ring — and homomorphisms of the following form.

(12.1.1)
$$\begin{array}{ccccc} & & \Omega & & \\ & \swarrow & & \searrow & \\ \Omega_1 & & & & \Omega_2 \\ & \searrow & & \swarrow & \\ & & \bar{\Omega} & & \end{array}$$

We call (12.1.1) a *surjective pullback square* and refer to Ω as the *pullback* of the square if all maps are surjective and

(12.1.2)
$$\Omega \cong \{(x, y) \in \Omega_1 \oplus \Omega_2 \mid (\Omega_1 \to \bar{\Omega})(x) = (\Omega_2 \to \bar{\Omega})(y)\}$$
$$\text{via} \quad \lambda \to \big((\Omega \to \Omega_1)(\lambda), (\Omega \to \Omega_2)(\lambda)\big)$$

(We use the slightly unusual notation $(\Omega_1 \to \bar{\Omega})(x)$ for the image of x under the map from Ω_1 to $\bar{\Omega}$, in order to reduce the amount of notation that one must remember for the maps in such diagrams.)

We normally identify Ω with its image in $\Omega_1 \oplus \Omega_2$, whence the isomorphism in (12.1.2) becomes equality. We often use the following formulas, which are valid in any surjective pullback square and follow immediately from the definition of "pullback."

(12.1.3)
$$\ker(\Omega \to \Omega_1) = \big(0, \ker(\Omega_2 \to \bar{\Omega})\big)$$
$$\ker(\Omega \to \Omega_2) = \big(\ker(\Omega_1 \to \bar{\Omega}), 0\big)$$

Because of commutativity of square (12.1.1), we can refer unambiguously to the map $\Omega \to \bar{\Omega}$.

When we refer to the ring or module Ω defined by the surjective pullback square (12.1.1), we assume tacitly that the homomorphisms in it are ring or module homomorphisms, respectively.

REMARK 12.2. The reason for our interest in surjective pullback squares is that strictly split Dedekind-like rings are precisely those rings Ω that are pullbacks of surjective pullback squares of the form (12.1.1), where Ω_1 and Ω_2 are DVRs with the same residue field $\bar{\Omega}$ [**KL1**, Lemma 2.15]. Moreover, all non-artinian homomorphic images of strictly split Dedekind-like rings are pullbacks of surjective pullback squares, as we show in Lemma 13.2.

LEMMA 12.3. *Given a commutative square of the form* (12.1.1)*, in which all maps are surjective, the square is a surjective pullback square if and only if:*

$$\begin{aligned} &\ker(\Omega \to \Omega_1) \cap \ker(\Omega \to \Omega_2) = 0 \quad \textit{and} \\ &\ker(\Omega \to \Omega_1) + \ker(\Omega \to \Omega_2) = \ker(\Omega \to \bar{\Omega}) \end{aligned} \tag{12.3.1}$$

PROOF. The natural map from Ω to $\Omega_1 \oplus \Omega_2$ is an embedding if and only if the first equation in (12.3.1) holds; the result then follows from [**L**, Lemma 3.1]. □

LEMMA 12.4. *Tensoring any surjective pullback square of modules over some ring R with a flat R-module yields another surjective pullback square. In particular this applies to $\mathfrak{m}$-localization (tensor with $R_{\mathfrak{m}}$) and, if R and the modules are noetherian, $\mathfrak{m}$-adic completion (tensor with $\hat{R}_{\mathfrak{m}}$).*

PROOF. This follows easily from the definition of flatness and Lemma 12.3, together with the fact that, for finitely generated modules over a noetherian ring R, the $\mathfrak{m}$-adic completion can be obtained by tensoring with the flat R-module $\hat{R}_{\mathfrak{m}}$. (See the Remarks in 5.1.) □

LEMMA 12.5. *Let the ring Ω be the pullback of surjective pullback square* (12.1.1)*, and $\mathfrak{m}$ a maximal ideal of Ω. Then:*

(i) *For at least one index i, the projection $p_i(\mathfrak{m})$ of $\mathfrak{m}$ in Ω_i is a maximal ideal of Ω_i. In this case we say that $\mathfrak{m}$* "is proper over" *Ω_i. Otherwise we say that $\mathfrak{m}$* "is improper over" *Ω_i.*
(ii) *If $\mathfrak{m}$ is proper over Ω_i ($i = 1$ or 2), then $\mathfrak{m} = \{(x_1, x_2) \in \Omega \mid x_i \in p_i(\mathfrak{m})\}$.*
(iii) *If $\mathfrak{m}$ is improper over Ω_i, then $p_i(\mathfrak{m}) = \Omega_i$.*
(iv) *For each index i, $\mathfrak{m}$ is proper over Ω_i if and only if $\mathfrak{m} \supseteq \ker(\Omega \to \Omega_i)$.*
(v) *$\mathfrak{m}$ is proper over both Ω_1 and Ω_2 if and only if $\mathfrak{m} \supseteq \ker(\Omega \to \bar{\Omega})$.*

PROOF. (i)–(iii) See [**L**, Lemma 3.5].

(iv) Since $\mathfrak{m}$ is maximal, we have either $\mathfrak{m} \supseteq \ker(\Omega \to \Omega_i)$ or $\mathfrak{m} + \ker(\Omega \to \Omega_i) = \Omega$. In the first situation $\mathfrak{m}$ is proper over Ω_i; in the second situation it is not.

(v) This follows from (iv) and (12.3.1). □

LEMMA 12.6. *Let the ring Ω be the pullback of surjective pullback square* (12.1.1)*, and suppose that the maximal ideal $\mathfrak{m}$ of Ω is improper over Ω_2. Then the $\mathfrak{m}$-localization $(\Omega_2)_{\mathfrak{m}} = 0$, while the $\mathfrak{m}$-localization $\Omega_{\mathfrak{m}} = (\Omega_1)_{\mathfrak{m}} = (\Omega_1)_{p_1(\mathfrak{m})}$, where p_1 is the projection map onto Ω_1, and $(\Omega_1)_{p_1(\mathfrak{m})}$ denotes the localization of the ring Ω_1 at its maximal ideal $p_1(\mathfrak{m})$.*

PROOF. For the first assertion, it suffices to show that there is an element of the form $(s, 0) \in \Omega - \mathfrak{m}$, for then $(s, 0)\Omega_2 = 0$, and hence invertibility of $(s, 0)$ in $\Omega_{\mathfrak{m}}$ shows that $(\Omega_2)_{\mathfrak{m}} = 0$. But the existence of such an element $(s, 0)$ follows immediately from Lemma 12.5(iv), since by assumption, $\mathfrak{m}$ is improper over Ω_2.

For the second assertion, recall that we obtain a surjective pullback square for $\Omega_{\mathfrak{m}}$ by taking the $\mathfrak{m}$-localization of the entire square (12.1.1) [Lemma 12.4]. In the present situation, this localization replaces Ω_2 by zero, and since we are working with *surjective* pullback squares, $\bar{\Omega}$ is also replaced by zero. This proves that $\Omega_{\mathfrak{m}} = (\Omega_1)_{\mathfrak{m}}$, which can be identified with $(\Omega_1)_{p_1(\mathfrak{m})}$ since the map $p_1 : \Omega \to \Omega_1$ is surjective. □

13. Homomorphic Images: Local versus Complete Local

In this section we show that local ring Ω is a homomorphic image of a Dedekind-like ring if and only if its completion is also a homomorphic image of a Dedekind-like ring [Corollary 13.6].

NOTATION 13.1. We denote by $N(\Omega)$ the nilradical of a ring Ω. By analogy with "DVR", which means "discrete valuation ring", an AVR (*artinian valuation ring*) is a nonzero artinian ring A whose ideals are totally ordered by inclusion. Therefore, every ideal of an AVR is a power of its Jacobson radical. Note that every proper homomorphic image of a DVR (including a field) is an AVR, according to this definition. We do not consider a field to be a DVR.

LEMMA 13.2. *Let Ω be a proper, non-artinian homomorphic image of a local Dedekind-like ring Λ with residue field k, and suppose that Ω is not a DVR. Then Λ is strictly split, and Ω is the pullback of a surjective pullback square*

(13.2.1)
$$\begin{array}{ccccc} & & \Omega & & \\ & \swarrow & & \searrow & \\ W & & & & A \\ & \searrow & & \swarrow & \\ & & k & & \end{array}$$

where:

(i) *W is a DVR;*
(ii) *A is an AVR; and*
(iii) $\ker(\Omega \to W) = \big(0, \ker(A \to k)\big) = N(\Omega) \neq 0$.

PROOF. First note that Λ must be strictly split, because the other types of local Dedekind-like rings are integral domains of Krull dimension 1, and therefore their proper homomorphic images are artinian rings.

Since Λ is strictly split, it is the pullback of the surjective pullback square in (13.2.2) [**KL1**, Lemma 2.15].

(13.2.2)
$$\begin{array}{ccccc} & & \Lambda & & \\ & \swarrow & & \searrow & \\ \Lambda_1 & & & & \Lambda_2 \\ & \searrow & & \swarrow & \\ & & k & & \end{array}$$

where Λ_1 and Λ_2 are DVRs and k is the residue field of Λ. In what follows we identify Λ with its image in $\Lambda_1 \oplus \Lambda_2$. Note that the regular elements of Λ are those elements $(y_1, y_2) \in \Lambda$ whose coordinates y_i are both nonzero, because both Λ_1 and Λ_2 are both integral domains and each $\ker(\Lambda_i \to k)$ is nonzero. Therefore, if an ideal I of Λ consists of zero divisors, then either the first coordinate of every element of I is zero, or the second coordinate of every element of I is zero.

Define the ring A in (13.2.1) as follows. Let $I = \ker(\Lambda \to \Omega)$. Since Λ has Krull dimension 1 and Ω is not artinian, I must consist of zero divisors. Therefore, as just noted, we may suppose that the first coordinate of every element of I is zero. Write $I = (0, I_2)$, where $I_2 \neq 0$ is an ideal of Λ_2. Let $A = \Lambda_2/I_2$, an AVR because Λ_2 is a DVR and $I_2 \neq 0$.

Build a surjective pullback square similar to (13.2.1) as follows. Let $W = \Lambda_1$, map W and A onto their common residue field k, and let Ω' be the pullback of these two maps. Then it is easily verified that we can map Λ onto Ω' by $(y_1, y_2) \to (y_1, y_2 + I_2)$. The kernel of this map is $I = (0, I_2)$, and therefore $\Omega' \cong \Omega$.

Thus we have proved all of the assertions of the Lemma except statement (iii). The first equality has already been noted in (12.1.3). To get the second equality, note that $\big(0, \ker(A \to k)\big)$ is a nilpotent ideal because the ring A is an AVR. Therefore $\ker(\Omega \to W) = \big(0, \ker(A \to k)\big) \subseteq N(\Omega)$. Equality holds because the integral domain $W = \operatorname{im}(\Omega \to W)$ has no nonzero nilpotent elements. Finally, $\ker(\Omega \to W)$ is nonzero because W is a DVR while, by assumption, Ω is not. $\square$

Next we show that the pullback structure described in the previous lemma descends from the completion $\hat{\Omega}$ to Ω itself.

LEMMA 13.3. *Let $(\Omega, \mathfrak{m}, k)$ be a local ring. Suppose that the completion $\hat{\Omega}$ is the pullback of the second pullback diagram displayed below*

$$\text{(13.3.1)} \qquad \begin{array}{ccc} & \Omega & \\ \swarrow & & \searrow \\ V & & A \\ \searrow & & \swarrow \\ & k & \end{array} \qquad\qquad \begin{array}{ccc} & \hat{\Omega} & \\ \swarrow & & \searrow \\ W & & A \\ \searrow & & \swarrow \\ & k & \end{array}$$

where W is a DVR, A is an AVR, and $\ker(\hat{\Omega} \to W) = \big(0, \ker(A \to k)\big) = N(\hat{\Omega}) \neq 0$.

Then Ω itself is the pullback of the first surjective pullback square in (13.3.1), *where V is a DVR, $\hat{V} \cong W$, and* $\ker(\Omega \to V) = \big(0, \ker(A \to k)\big) = N(\Omega) \neq 0$ *(and both maps $A \to k$ are the same).*

PROOF. Let the map $\Omega \to A$ be the composition $\Omega \to \hat{\Omega} \to A$, where the first map is the natural map $\Omega \to \hat{\Omega}$. Since the artinian ring A has finite length as an $\hat{\Omega}$-module, the map $\Omega \to A$ is surjective [Lemma 6.5(ii)].

Let the map $\Omega \to V \subseteq W$ be the composition $\Omega \to \hat{\Omega} \to W$, and let V be the image of Ω under this map. Since $\ker(\Omega \to V)$ is the contraction to Ω of the prime ideal $\ker(\hat{\Omega} \to W)$ of $\hat{\Omega}$, it follows that $\ker(\Omega \to V)$ is prime and V is a (local) domain. We do not yet know that V is a DVR, nor do we know that $\hat{V} \cong W$.

Now the intersection $\ker(\Omega \to A) \cap \ker(\Omega \to V)$ is the contraction to Ω of the intersection $\ker(\hat{\Omega} \to A) \cap \ker(\hat{\Omega} \to W) = 0$. Therefore, the maps $\Omega \to A$ and $\Omega \to V$ embed Ω into $V \oplus A$. Let $\bar{\Omega} = \ker(\Omega \to V) + \ker(\Omega \to A)$, and let the two remaining maps in the first square below be the unique maps (necessarily surjective) that make the square commute.

$$\text{(13.3.2)} \qquad \begin{array}{ccc} & \Omega & \\ \swarrow & & \searrow \\ V & & A \\ \searrow & & \swarrow \\ & \bar{\Omega} & \end{array} \qquad\qquad \begin{array}{ccc} & \hat{\Omega} & \\ \swarrow & & \searrow \\ \hat{V} & & A \\ \searrow & & \swarrow \\ & \bar{\Omega} & \end{array}$$

Then the first square in (13.3.2) is a surjective pullback square, and Ω is its pullback [Lemma 12.3]. Taking the $\mathfrak{m}$-adic completion of this square yields a surjective pullback square for $\hat{\Omega}$ [Lemma 12.4], shown in the second square of (13.3.2).

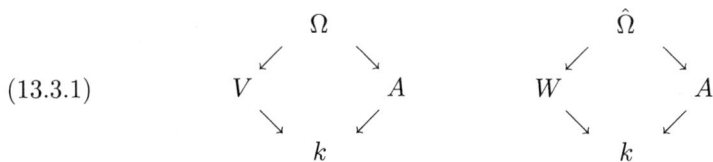

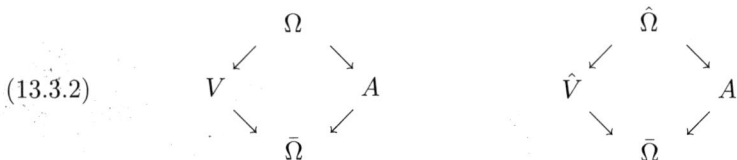

We claim that $\bar{\Omega} \cong k$ as Ω-algebras, and hence we may replace $\bar{\Omega}$ by k in (13.3.2). For this, it suffices to prove that $\ker(A \to \bar{\Omega})$ is a prime ideal (necessarily the maximal ideal of the artinian local ring A). This holds since $\big(0, \ker(A \to k)\big) = \ker(\Omega \to V)$, by (12.1.3), which is a prime ideal.

We claim that $\ker(\Omega \to V) = \big(0, \ker(A \to k)\big) = N(\Omega) \neq 0$. The first equality holds by (12.1.3), as just noted. As in the proof of Lemma 13.2(iii), we have that $\ker(\Omega \to V)$ is both nilpotent and a prime ideal, and therefore $\ker(\Omega \to V) = N(\Omega)$. To see that this is nonzero, it suffices to show that $\big(0, \ker(A \to k)\big) \neq 0$. But this holds by hypothesis.

We claim that $\hat{V} \cong W$. Since $\hat{\Omega} \to \hat{V}$ is surjective, it suffices to show that $\ker(\hat{\Omega} \to V) = \ker(\hat{\Omega} \to W)$. But this holds because both equal $\big(0, \ker(A \to k)\big)$, (recall that $\bar{\Omega} = k$). Finally, to see that V is a DVR it suffices to see that $\hat{V}$ is a DVR [Proposition 11.8]; and this holds because $\hat{V} \cong W$. □

THEOREM 13.4. *Let Ω be a local ring whose completion is a proper, non-artinian homomorphic image of a complete local Dedekind-like ring. Then Ω is a homomorphic image of a strictly split local Dedekind-like ring.*

PROOF. If Ω is a DVR, then it is a homomorphic image of a strictly split local Dedekind-like ring, as shown in [**KL1**, Lemma 2.19]. Thus, we can suppose that Ω is not a DVR. Then $\hat{\Omega}$ is not a DVR [Proposition 11.8]. Therefore $\hat{\Omega}$ has the type of pullback structure described in Lemma 13.2, and by Lemma 13.3, Ω itself also has this type of pullback structure. Thus, for the remainder of this proof, we may assume that Ω is a pullback as described in Lemma 13.2.

By a theorem of Hungerford [**H**], the AVR A in square (13.2.1) is a homomorphic image of some complete DVR X, by a homomorphism $X \to A$ which we call the "Hungerford map." Let Λ be the pullback of the surjective pullback square formed by replacing A by X in diagram (13.2.1), and by mapping X onto k by means of the composition $X \to A \to k$. Then Λ is strictly split local Dedekind-like [Remark 12.2] and maps onto Ω by sending each ordered pair $(w, x) \in \Lambda$ to the ordered pair $(w, \bar{x}) \in \Omega$, where $\bar{x}$ is the image of x in A under the Hungerford map. □

The following theorem summarizes our results on local rings whose completions are homomorphic images of Dedekind-like rings.

THEOREM 13.5. *Suppose that the completion $\hat{\Omega}$ of a local ring Ω is a homomorphic image of a complete local Dedekind-like ring.*

(i) *If the homomorphism is an isomorphism, then Ω is Dedekind-like.*
(ii) *If the homomorphism is proper and Ω has no nonzero ideals of finite length, then Ω is a DVR (and hence a homomorphic image of a strictly split local Dedekind-like ring).*
(iii) *If the homomorphism is proper and Ω is a non-artinian ring with a nonzero submodule of finite length, then Ω is a homomorphic image of a strictly split local Dedekind-like ring.*
(iv) *If the homomorphism is proper and Ω is an artinian ring, then $\Omega = \hat{\Omega}$ (and hence is a homomorphic image of a complete local Dedekind-like ring).*

PROOF. Statement (i) is Proposition 11.8(i).

(ii) and (iii) By Theorem 13.4 there is a surjective homomorphism $\Lambda \to \Omega$, with Λ local Dedekind-like. Then by Lemma 13.2, either Ω is a DVR or else Λ is strictly split. (Note that these situations do not overlap, as DVRs have no ideals of finite length, while $N(\Omega)$ has finite length when the pullback structure of Lemma 13.2 applies.)

(iv) is obvious. □

COROLLARY 13.6. *The local ring* Ω *is a homomorphic image of a local Dedekind-like ring if and only if its completion is.*

PROOF. The "only if" assertion holds because the completion of a local Dedekind-like ring is again Dedekind-like [Proposition 11.8(i)]. The "if" assertion is Theorem 13.5. □

CHAPTER 2

Wildness

14. Global Dichotomy, Global Wildness

In this section we give our formal definition of wildness, and prove our main theorems on ring-theoretic dichotomy and global wildness. We begin by showing that the property of being a homomorphic image of a Dedekind-like ring is determined locally.

PROPOSITION 14.1. *Suppose that Ω is a noetherian ring such that $\hat{\Omega}_{\mathfrak{m}}$ is a homomorphic image of a complete local Dedekind-like ring for all maximal ideals $\mathfrak{m}$ of Ω. Then Ω itself is a homomorphic image of a Dedekind-like ring.*

PROOF. We may assume that Ω is an indecomposable ring. Then, by Corollary 13.6, every localization $\Omega_{\mathfrak{m}}$ $\big(\mathfrak{m} \in \operatorname{maxspec}(\Omega)\big)$ is a homomorphic image of a local Dedekind-like ring. We split the rest of our proof into three cases.

Case 1. For some $\mathfrak{m}$, the ring $\Omega_{\mathfrak{m}}$ is artinian. Then $\Omega = \Omega_{\mathfrak{m}} = \hat{\Omega}_{\mathfrak{m}}$ by Lemma 6.6. Thus, in this case Ω is a homomorphic image of a complete local Dedekind-like ring.

Case 2. No $\Omega_{\mathfrak{m}}$ has a nonzero submodule of finite length. Then every $\Omega_{\mathfrak{m}}$ is a local Dedekind-like ring, by Theorem 13.5, and hence Ω is itself a Dedekind-like ring [Corollary 10.7].

Case 3. No $\Omega_{\mathfrak{m}}$ is artinian, but some $\Omega_{\mathfrak{m}}$ has a nonzero submodule of finite length. This is the most interesting case, and we give the structure of Ω in detail. Indeed, the proof consists of constructing two surjective pullback squares and a surjection between them, shown in diagram (14.1.1) below, and showing that the first pullback is Dedekind-like, and the second is isomorphic to Ω.

$$(14.1.1)\qquad \begin{array}{ccccccccc} & \Lambda & & \xrightarrow{\quad\phi\quad} & & \Omega & & \\ \swarrow & & \searrow & & \swarrow & & \searrow^{\mu} & \\ \Upsilon & & V = \oplus_i V_i & & \Upsilon = \Omega/N(\Omega) & & & A = \oplus_i A_i \\ \searrow & & \swarrow & & \searrow & & \swarrow & \\ & \oplus_i k(\mathfrak{m}_i) & & & & \oplus_i k(\mathfrak{m}_i) & & \end{array}$$

Structure of Ω $\big($second pullback square in diagram (14.1.1)$\big)$. By Lemma 6.3, there are only finitely many maximal ideals $\mathfrak{m}$ of Ω such that $\Omega_{\mathfrak{m}}$ has a nonzero ideal of finite length. Call these maximal ideals $\mathfrak{m}_1, \dots, \mathfrak{m}_n$ and, for each index i, let $k(\mathfrak{m}_i)$ denote the residue field $\Omega/\mathfrak{m}_i$, which is also the residue field of $\Omega_{\mathfrak{m}_i}$. Then for every maximal ideal $\mathfrak{m} \notin \{\mathfrak{m}_1, ..., \mathfrak{m}_n\}$, the localization $\Omega_{\mathfrak{m}}$ is a local Dedekind-like ring, again by Theorem 13.5.

For each index i, the ring $\Omega_{\mathfrak{m}_i}$ is a non-artinian homomorphic image of a local Dedekind-like ring and has a nonzero submodule of finite length. Therefore, $\Omega_{\mathfrak{m}_i}$ is the pullback of a surjective pullback square

$$\text{(14.1.2)} \qquad \begin{array}{ccccc} & & \Omega_{\mathfrak{m}_i} & & \\ & \swarrow & & \searrow & \\ W(\mathfrak{m}_i) & & & & A_i \\ & \searrow & & \swarrow & \\ & & k(\mathfrak{m}_i) & & \end{array}$$

as described in Lemma 13.2. In particular, the ring $W(\mathfrak{m}_i)$ is a DVR, and $\ker\big(\Omega_{\mathfrak{m}_i} \to W(\mathfrak{m}_i)\big) = N(\Omega_{\mathfrak{m}_i})$ is a nonzero nilpotent ideal of finite length.

We begin the construction of the second square in diagram (14.1.1) by setting $A = \oplus_i A_i$. For each index i, we also let $\mu_i\colon \Omega \to A_i$ be the composition $\Omega \to \Omega_{\mathfrak{m}_i} \to A_i$, where the first map is the natural localization map $\nu_{\mathfrak{m}_i}$, and the second map is the projection to A_i in (14.1.2). This defines the map $\mu\colon \Omega \to A$ in diagram (14.1.1) whose coordinate maps are the μ_i.

We claim that the map μ is surjective. First note that, for each index i, the map $\mu_i\colon \Omega \to \Omega_{\mathfrak{m}_i} \to A_i$ is surjective by Lemma 6.2(iv), because its second factor comes from surjective pullback square (14.1.2), and A_i has finite length [Lemma 6.2]. Surjectivity of μ now follows from the Chinese Remainder Theorem and the fact that, for each index i, there exists an integer $n(i)$ such that $\mathfrak{m}_i^{n(i)} A_i = 0$.

Let $\Upsilon = \Omega/N(\Omega)$, as in the second square in diagram (14.1.1), and define a map $\Omega \to \Upsilon \oplus A$ by

$$\text{(14.1.3)} \qquad x \to \big(x + N(\Omega), \mu(x)\big) = \big(x + N(\Omega), \mu_1(x), \ldots, \mu_n(x)\big)$$

We claim that this map is an injection. Let x be in its kernel. To show that $x = 0$, it suffices to show that $\nu_{\mathfrak{m}}(x) = 0$ for every maximal ideal $\mathfrak{m}$ of Ω. By (14.1.3) we have $x \in N(\Omega)$, and therefore every $\nu_{\mathfrak{m}}(x)$ is nilpotent. If $\mathfrak{m} \notin \{\mathfrak{m}_1, ..., \mathfrak{m}_n\}$, then $\Omega_{\mathfrak{m}}$ is a local Dedekind-like ring and therefore reduced, and so $\nu_{\mathfrak{m}}(x) = 0$. For given $\mathfrak{m}_i$, $\nu_{\mathfrak{m}_i}(x)$ is an ordered pair (u, v) by (14.1.2), where both u and v must be nilpotent. Therefore the element u of the integral domain $W(\mathfrak{m}_i)$ must zero, while $v = \mu_i(x) = 0$, by (14.1.3). Thus $x = 0$, and the claim is proved.

The projection map $\Omega \to \Upsilon$ in (14.1.3) is clearly a surjection, we have already proved that $\mu\colon \Omega \to A$ is a surjection, and $\ker(\Omega \to \Upsilon) \cap \ker(\Omega \to A) = 0$ by the claim following (14.1.3). It now follows from Lemma 12.3 that Ω is the pullback of the diagram formed by mapping the two coordinate rings Υ and A onto the ring Ω/K where $K = N(\Omega) + \ker(\Omega \to A)$. Thus, in order to show that the second square in diagram (14.1.1) is a surjective pullback square and Ω is its pullback, it suffices to verify that

$$\text{(14.1.4)} \qquad N(\Omega) + \ker(\Omega \to A) = \mathfrak{m}_1 \cap \cdots \cap \mathfrak{m}_n$$

We establish this equation by checking that, for every maximal ideal $\mathfrak{m}$, both sides localize to the same ideal of $\Omega_{\mathfrak{m}}$. If $\mathfrak{m} \notin \{\mathfrak{m}_1, ..., \mathfrak{m}_n\}$, then $A_{\mathfrak{m}} = 0$ (since $\mathfrak{m}_i^{n(i)} A_i = 0$), and the $\mathfrak{m}$-localization of (14.1.4) becomes $\Omega_{\mathfrak{m}} = \Omega_{\mathfrak{m}}$. For given $\mathfrak{m}_i$, (14.1.4) becomes

$$\text{(14.1.5)} \qquad N(\Omega_{\mathfrak{m}_i}) + \ker(\Omega_{\mathfrak{m}_i} \to A_i) = (\mathfrak{m}_i)_{\mathfrak{m}_i}$$

But as noted above, $N(\Omega_{\mathfrak{m}_i}) = \ker\big(\Omega_{\mathfrak{m}_i} \to W(\mathfrak{m}_i)\big)$, by Lemma 13.2. Then since (14.1.2) is a surjective pullback square, Lemma 12.3 implies that the left hand side of (14.1.5) equals $\ker\big(\Omega_{\mathfrak{m}_i} \to k(\mathfrak{m}_i)\big)$, which in turn equals $(\mathfrak{m}_i)_{\mathfrak{m}_i}$. This proves (14.1.5) for each $\mathfrak{m}_i$ and hence completes the proof of (14.1.4). Thus, we have now shown that the second square in (14.1.1) is a surjective pullback square and Ω is its pullback.

Finally, we claim that $\Upsilon = \Omega/N(\Omega)$ is a Dedekind-like ring. By Corollary 10.7, it suffices to prove that the localization $\Upsilon_{\mathfrak{n}}$ is a local Dedekind-like ring for every maximal ideal $\mathfrak{n}$ of Υ. Let $\mathfrak{m}$ be the maximal ideal of Ω that maps onto $\mathfrak{n}$; then $\Upsilon_{\mathfrak{n}} = \Upsilon_{\mathfrak{m}} \cong \Omega_{\mathfrak{m}}/N(\Omega_{\mathfrak{m}})$. If $\mathfrak{m} \notin \{\mathfrak{m}_1, ..., \mathfrak{m}_n\}$, then $\Omega_{\mathfrak{m}}$ is a local Dedekind-like ring, so that $N(\Omega_{\mathfrak{m}}) = 0$, and hence $\Upsilon_{\mathfrak{n}} \cong \Omega_{\mathfrak{m}}$ is a local Dedekind-like ring, as desired. On the other hand, if $\mathfrak{m} = \mathfrak{m}_i$ for some i, then $\Omega_{\mathfrak{m}_i}$ is the pullback of square (14.1.2), and by Lemma 13.2, $\Omega_{\mathfrak{m}_i}/N(\Omega_{\mathfrak{m}_i}) \cong W(\mathfrak{m}_i)$ is a DVR and hence a local Dedekind-like ring.

Definition of Λ (first pullback square in diagram (14.1.1)). Since each A_i is an AVR [see Notation 13.1], a theorem of Hungerford [**H**] shows that A_i is a homomorphic image of some complete DVR V_i. We then set $V = \oplus_i V_i$ in the first square in diagram (14.1.1). The direct sum of the surjections $V_i \to A_i$ yields a surjective homomorphism $V \to A$, which we call the *Hungerford map.* Composing the Hungerford map with the map $A \to \oplus_i k(\mathfrak{m}_i)$ from the second square defines the map $V \to \oplus_i k(\mathfrak{m}_i)$ displayed in the first square of diagram (14.1.1). Finally, we define the map $\Upsilon \to \oplus_i k(\mathfrak{m}_i)$ in the first square of diagram (14.1.1) to be the same as the corresponding map in the second square. We can then form the pullback of the first square in (14.1.1), calling it Λ.

The surjective map ϕ: $\Lambda \to \Omega$. We can map the lower triangle of the first square in (14.1.1) to the lower triangle in the second square by using the identity map on Υ and $\oplus_i k(\mathfrak{m}_i)$, and the Hungerford map $V \to A$. Since all three of these maps are surjective and the resulting diagram commutes, it induces a surjection ϕ of the pullback Λ of the first square onto the pullback Ω of the second square, as desired.

Structure of Λ. We claim that Λ is a Dedekind-like ring. By Corollary 10.7, it suffices to prove that the localization $\Lambda_{\mathfrak{n}}$ is a local Dedekind-like ring for every maximal ideal $\mathfrak{n}$ of Λ. According to the characterization of maximal ideals in pullback rings given in Lemma 12.5, the maximal ideal $\mathfrak{n}$ is proper over least one of the two coordinate rings of Λ. We break up the remainder of the proof into three cases.

Case 3a. The maximal ideal $\mathfrak{n}$ is proper over only the first coordinate ring Υ of Λ in diagram (14.1.1). Then $\Lambda_{\mathfrak{n}} = \Upsilon_{p_1(\mathfrak{n})}$, where p_1 is the projection to the coordinate ring Υ of Λ [Lemma 12.6]. But we already showed that Υ is a Dedekind-like ring, so $\Lambda_{\mathfrak{n}} = \Upsilon_{p_1(\mathfrak{n})}$ is a local Dedekind-like ring, as required.

Case 3b. The maximal ideal $\mathfrak{n}$ is proper over only the second coordinate ring V of Λ. Then, as in the previous case, $\Lambda_{\mathfrak{n}} = V_{p_2(\mathfrak{n})}$, where p_2 is the projection to the coordinate ring V of Λ. The proof of this case is completed by noting that V is a direct sum of DVRs, and hence its $p_2(\mathfrak{n})$-localization is one of these DVRs.

Case 3c. The maximal ideal $\mathfrak{n}$ is proper over both coordinate rings of Λ. In this case we prove that $\Lambda_{\mathfrak{n}}$ is a strictly split local Dedekind-like ring.

Our first claim is that $\phi(\mathfrak{n})$, the image of $\mathfrak{n}$ in the ring Ω, must be one of the maximal ideals $\mathfrak{m}_i$. By the definition of ϕ, it follows easily that $\phi(\mathfrak{n})$ is proper over both coordinate rings Υ and A of the ring Ω. Then Lemma 12.5(v) shows that $\phi(\mathfrak{n}) \supseteq \mathfrak{m}_1 \cap \cdots \cap \mathfrak{m}_n$, from which the claim follows. We fix $\mathfrak{m}_i = \phi(\mathfrak{n})$.

Recall that one can localize pullback rings by localizing all of the rings and maps in the pullback square [Lemma 12.4]. Then $\Omega_{\mathfrak{n}} = \Omega_{\mathfrak{m}_i}$, and taking the $\mathfrak{m}_i$-localization of the pullback square for Ω in (14.1.1) yields precisely the pullback square in (14.1.2), because $\Upsilon_{\mathfrak{m}_i} \cong \Omega_{\mathfrak{m}_i}/N(\Omega_{\mathfrak{m}_i})$ and $N(\Omega_{\mathfrak{m}_i}) = \ker(\Omega_{\mathfrak{m}_i} \to W(\mathfrak{m}_i))$ by Lemma 13.2, as noted above.

Similarly, taking the $\mathfrak{n}$-localization of the pullback square for Λ in (14.1.1) yields a pullback square for $\Lambda_{\mathfrak{n}}$, in which $\Upsilon_{\mathfrak{n}}$ replaces Υ, $V_{\mathfrak{n}}$ replaces V, and $(\oplus_i k(\mathfrak{m}_i))_{\mathfrak{n}}$ replaces $\oplus_i k(\mathfrak{m}_i)$. Now in diagram (14.1.1), ϕ restricted to Υ is the identity, by definition, so localizing at $\mathfrak{n}$ shows that $\Upsilon_{\mathfrak{n}} = \Upsilon_{\mathfrak{m}_i}$ is a DVR, by the previous paragraph. Likewise, $(\oplus_i k(\mathfrak{m}_i))_{\mathfrak{n}} = (\oplus_i k(\mathfrak{m}_i))_{\mathfrak{m}_i} = k(\mathfrak{m}_i)$ is a field. Finally, as in case 3b above, $V_{\mathfrak{n}} = V_{p_2(\mathfrak{n})}$, where p_2 is projection to the coordinate ring V of Λ; but V is a direct sum of DVRs, and hence its $p_2(\mathfrak{n})$-localization is one of these DVRs.

The preceding computations show that $\Lambda_{\mathfrak{n}}$ is the pullback formed by mapping two DVRs onto their common residue field. Therefore $\Lambda_{\mathfrak{n}}$ is a strictly split local Dedekind-like ring, as claimed [Remark 12.2]. □

We now have the tools we need to prove our two main theorems. But first we review the definitions of the three types of rings, other than Dedekind-like rings, that occur in these results. For a more detailed account of these types of rings, including examples, see [**KL1**, §2].

DEFINITIONS 14.2 (Artinian triad, Drozd ring, Klein ring). Suppose that Ω is a local ring with maximal ideal $\mathfrak{m}$. Denote by $\mu_\Omega(M)$ the minimum number of generators required by the Ω-module M. Now let Ω be artinian. We call Ω:

(i) An *artinian triad* if $\mu_\Omega(\mathfrak{m}) = 3$ and $\mathfrak{m}^2 = 0$.
(ii) A *Drozd ring* if $\mu_\Omega(\mathfrak{m}) = \mu_\Omega(\mathfrak{m}^2) = 2$, $\mathfrak{m}^3 = 0$, and $x^2 = 0$ for some $x \in \mathfrak{m} - \mathfrak{m}^2$.
(iii) A *Klein ring* if $\mu_\Omega(\mathfrak{m}) = 2$, $\mu_\Omega(\mathfrak{m}^2) = 1$, $\mathfrak{m}^3 = 0$, and $x^2 = 0$ for all $x \in \mathfrak{m}$.

THEOREM 14.3 (Ring-theoretic Dichotomy). *Let Ω be an indecomposable noetherian ring. Then exactly one of the following two possibilities occurs.*

(i) *Ω has an artinian triad or a Drozd ring as a homomorphic image.*
(ii) *Ω is a Klein ring or a homomorphic image of a Dedekind-like ring.*

PROOF. The complete local case of this theorem is proved in [**KL1**, Theorem 3.1].

First assume that possibility (i) does not hold for Ω. Since every artinian ring-homomorphic image of every completion $\hat{\Omega}_{\mathfrak{m}}$ (at every maximal ideal $\mathfrak{m}$ of Ω) is also a homomorphic image of Ω [Lemma 6.5(ii)], no $\hat{\Omega}_{\mathfrak{m}}$ can map onto a ring of the form (i). Therefore, by the complete local case of the theorem, every $\hat{\Omega}_{\mathfrak{m}}$ is either a Klein ring or a homomorphic image of a complete local Dedekind-like ring. If some $\hat{\Omega}_{\mathfrak{m}}$ is a Klein ring, then indecomposability of Ω implies that $\Omega = \hat{\Omega}_{\mathfrak{m}}$, by Lemma 6.6. Otherwise every $\hat{\Omega}_{\mathfrak{m}}$ is a homomorphic image of a Dedekind-like ring, and hence Ω is a homomorphic image of a Dedekind-like ring, by Proposition 14.1. Thus Ω is as in (ii).

It remains to show that Ω cannot be as in both (i) and (ii). Therefore suppose that Ω has an artinian triad or a Drozd ring A as a homomorphic image. Since artinian triads and Drozd rings are local, and hence complete, there is a maximal ideal $\mathfrak{m}$ of Ω such that $\hat{A}_{\mathfrak{m}} \cong A$ and this is a homomorphic image of $\hat{\Omega}_{\mathfrak{m}}$. By the complete local case, $\hat{\Omega}_{\mathfrak{m}}$ can be neither a Klein ring nor a homomorphic image of a complete Dedekind-like ring.

Consider Ω itself. If Ω were a Klein ring, its maximal ideal would be $\mathfrak{m}$; and since artinian local rings are complete, we would have $\Omega = \hat{\Omega}$ which, as we have just observed, cannot be a Klein ring. The remaining possibility is that Ω is a homomorphic image of a Dedekind-like ring. But then $\hat{\Omega}_{\mathfrak{m}}$ would also be a homomorphic image of a Dedekind-like ring [Proposition 11.8(i)], which is again a contradiction. □

DEFINITION 14.4 (Wildness). Let $k = \Omega/\mathfrak{m}$ be some residue field of a commutative noetherian ring Ω. We say that Ω is finlen-*wild* (wrt $\mathfrak{m}$) — or more completely, say that finlen(Ω) has *wild representation type (with respect to* $\mathfrak{m}$*)* — if the following condition holds for every finite dimensional (noncommutative!) k-algebra A. There exist $\mathcal{W} = \mathcal{W}_A$, a full subcategory of finlen(Ω) and there exists an additive functor: $\Phi = \Phi_A\colon \mathcal{W} \to$ finlen(A) (say, left A-modules) such that Φ is a *representation equivalence.* In other words:

- Φ is onto all isomorphism classes;
- For $M, N \in \mathcal{W}$: $\quad M \cong N \iff \Phi(M) \cong \Phi(N)$ in finlen(A); and
- Φ is a surjection on Hom groups.

It follows easily that ${}_{\Omega}M$ is indecomposable $\iff$ $\Phi(M)$ is indecomposable.

For an introduction to wildness from the standpoint of commutative noetherian rings, see [**KL1**, Remarks 2.3].

THEOREM 14.5 (Main Wildness Theorem). *Let Ω be an indecomposable noetherian ring. Then (at least) one of the following two possibilities occurs.*

(i) *Ω is* finlen-*wild (wrt some maximal ideal* $\mathfrak{m}$*).*
(ii) *Ω is a Klein ring or a homomorphic image of a Dedekind-like ring.*

PROOF. Any ring that maps onto a finlen-wild ring is obviously finite-length wild. Therefore, by our Ring-theoretic Dichotomy Theorem 14.3, it suffices to prove that artinian triads and Drozd rings are finite-length wild. This is accomplished in [**KL1**, Theorem 2.10]. □

The next remark summarizes the relation between the previous theorem and our tameness-versus-wildness problem.

FINAL REMARK 14.6 (Tame or wild). Let Ω be an indecomposable noetherian ring such that $\big(\forall \mathfrak{m} \in \text{maxspec}(\Omega)\big)$ Ω is not finlen-wild (wrt $\mathfrak{m}$). Then, according to the previous theorem, Ω either a Klein ring or a homomorphic image of a Dedekind-like ring.

(i) To establish fingen-tameness for homomorphic images of Dedekind-like rings, it suffices to describe fingen(Ω) when Ω is Dedekind-like.
  (a) Most of the rest of this memoir is spent describing the structure of fingen(Ω) when the Dedekind-like ring Ω satisfies Additional Hypothesis 10.2. (Recall, from Definition 10.1, that this Additional Hypothesis is an unstated, implicit part of all of our tameness results.)
  (b) If the Dedekind-like ring Ω does not satisfy Additional Hypothesis 10.2, then we can neither describe fingen(Ω) nor prove that finlen(Ω) is wild.
(ii) The structure of finitely generated modules over Klein rings (= finite length modules, since Klein rings are artinian) is proved in [**KL2**, §11]: Every finitely generated module, here, is the direct sum of a free module

and a module over a Dedekind-like ring that satisfies Additional Hypothesis 10.2.

CHAPTER 3

Structure of a Genus

For a complete local Dedekind-like ring Λ, [**KL2**] describes the structure of finitely generated Λ-modules in detail. The questions answered in the present chapter are the following. (i) The *package deal question:* For each $\mathfrak{m} \in \text{maxspec}(\Lambda)$ let a module $H(\mathfrak{m}) \in \text{fingen}(\hat{\Lambda}_\mathfrak{m})$ be given. When is the collection of modules $\{H(\mathfrak{m})\}$ the "package of completions" $\{\hat{M}_\mathfrak{m}\}$ of some $M \in \text{fingen}(\Lambda)$? This is answered in Section 15. (ii) When is the resulting Λ-module M indecomposable? This is answered in Section 16. It turns out that the answers to these questions require remarkably little detail from [**KL2**], which we review when it is needed.

15. Consistent (Torsionfree) Ranks

The answer to the package deal question mentioned above consists of two parts. The first is a general property of reduced noetherian rings of Krull dimension 1: The given $\hat{\Lambda}_\mathfrak{m}$-module $H(\mathfrak{m})$ must be $\hat{\Lambda}_\mathfrak{m}$-free (a$\forall\mathfrak{m}$). See Lemma 8.2 for necessity of the analogous condition for localizations. This obviously implies necessity for completions. The second is that the (torsionfree) ranks of the $H(\mathfrak{m})$ must satisfy a consistency consistency condition. See Package Deal Theorem 15.6 for the statement of this condition.

Recall our standard decomposition $\Gamma = \oplus_{h\in\mathcal{H}} \Gamma_h$ where each coordinate ring Γ_h of Γ is a Dedekind domain, as in (3.1.1), and recall that each localization $(\Gamma_h)_Q$ is the field of quotients of Γ_h [Lemma 4.4]. Almost everything in this section refers to the terms Γ_h of this decomposition. Recall also that, for every $\mathfrak{m} \in \text{maxsupp}_\Lambda(\Gamma_h)$, $(\Gamma_h)\hat{}_\mathfrak{m}$ has either one or two coordinate rings, necessarily DVRs. See Proposition 11.10 and Lemma 11.6 for details.

DEFINITION 15.1 (Rank). Let M be a Λ-module and $h \in \mathcal{H}$. We define the *(torsionfree)* Γ_h*-rank of* M to be the dimension of the $(\Gamma_h)_Q$-vector space $(\Gamma_h)_Q \otimes_\Lambda M$. Thus, M has a rank for each coordinate ring Γ_h. If Γ has only one coordinate ring (equivalently, if Λ is an integral domain), we often use the simpler terminology *rank of* M, instead of the more cumbersome "Γ-rank of M."

If $(\Lambda, \mathfrak{m})$ is local and strictly split, then Γ has two coordinate rings, say $\Gamma = \Upsilon \oplus \Upsilon'$, and we specify the corresponding ranks by saying that (a, b) *is the* (Υ, Υ')*-rank of* M. When the distinction between Υ and Υ' is not important, we simply say that (a, b) *is the rank of* M.

Other authors define torsionfree ranks (of modules over reduced rings) in terms of localizations at minimal prime ideals. This is equivalent to the present definition because, if $\mathfrak{p}_h$ is the minimal prime ideal $\mathfrak{p}_h = \ker(\Lambda \to \Gamma_h)$ of Λ associated with Γ_h [Lemma 4.1(i)], then

(15.1.1) $$(\Gamma_h)_Q \otimes_\Lambda M = M_{\mathfrak{p}_h} \qquad \text{(canonical isomorphism)}$$

which follows easily from Lemma 4.1, and which we shall use again.

LEMMA 15.2 (Ranks are complete-locally determined). *Let $M \in \operatorname{fingen}(\Lambda)$, $\mathfrak{m} \in \operatorname{maxsupp}_\Lambda(\Gamma_h)$, and let Υ be a coordinate ring of $(\Gamma_h)\hat{}_{\mathfrak{m}}$. Then the Γ_h-rank of M equals the Υ-rank of $\hat{M}_{\mathfrak{m}}$.*

PROOF. Suppose first that $\mathfrak{m}$ is not nonstrictly split. We claim that:

$$\text{(15.2.1)} \qquad \begin{aligned} \Upsilon_Q \otimes_{\hat{\Lambda}_{\mathfrak{m}}} \hat{M}_{\mathfrak{m}} &\cong \Gamma_h \otimes_\Lambda \Lambda_Q \otimes_\Lambda \hat{\Lambda}_{\mathfrak{m}} \otimes_\Lambda M \\ &\cong \Upsilon_Q \otimes_{(\Gamma_h)_Q} \left[(\Gamma_h)_Q \otimes_\Lambda M\right] \end{aligned}$$

Note that we may interpret the subscript Q in Υ_Q to mean either $Q(\Upsilon)$ or $Q(\Lambda)$ [Lemma and Definition 4.3], while the subscript Q in Λ_Q can only mean $Q(\Lambda)$. In proving (but not in applying!) (15.2.1) we take $Q = Q(\Lambda)$. Recall that $\Upsilon \cong (\Gamma_h)\hat{}_{\mathfrak{m}}$, because $\mathfrak{m}$ is not nonstrictly split. Then by Remarks 5.3(iii), we get that $(\Gamma_h)\hat{}_{\mathfrak{m}} \cong \Gamma_h \otimes_\Lambda \hat{\Lambda}_{\mathfrak{m}}$, so that $\Upsilon_Q \cong \Gamma_h \otimes_\Lambda \hat{\Lambda}_{\mathfrak{m}} \otimes_\Lambda \Lambda_Q$. Similarly, $\hat{M}_{\mathfrak{m}} \cong M \otimes_\Lambda \hat{\Lambda}_{\mathfrak{m}}$. Also, $\hat{\Lambda}_{\mathfrak{m}} \otimes_{\hat{\Lambda}_{\mathfrak{m}}} \hat{\Lambda}_{\mathfrak{m}} \cong \hat{\Lambda}_{\mathfrak{m}}$ so the first isomorphism in (15.2.1) follows. The proof of the second isomorphism is similar.

Now we apply (15.2.1). Interpreting the first Q as $Q(\Upsilon)$, we see that the Υ_Q-dimension of the left-hand side is the Υ-rank of $\hat{M}_{\mathfrak{m}}$. On the other hand, the Γ_h-rank of M is by definition the $(\Gamma_h)_Q$-dimension of $(\Gamma_h)_Q \otimes_\Lambda M$, which by a change of scalars is also the Υ_Q-dimension of the right-hand side of (15.2.1). Therefore, the isomorphisms in (15.2.1) show that the Υ-rank of $\hat{M}_{\mathfrak{m}}$ equals the Γ_h-rank of M, as desired.

Suppose next that $\mathfrak{m}$ is nonstrictly split. Then $\mathfrak{m}$ is in the support of a unique coordinate ring Γ_h of Γ, and $(\Gamma_h)\hat{}_{\mathfrak{m}} = \Upsilon \oplus \Upsilon'$ for some DVR Υ'. A minor modification of the proof of (15.2.1) establishes:

$$\text{(15.2.2)} \qquad \begin{aligned} (\hat{\Gamma}_{\mathfrak{m}})_Q \otimes_{\hat{\Lambda}_{\mathfrak{m}}} \hat{M}_{\mathfrak{m}} &\cong \Gamma_h \otimes_\Lambda \Lambda_Q \otimes_\Lambda \hat{\Lambda}_{\mathfrak{m}} \otimes_\Lambda M \\ &\cong (\hat{\Gamma}_{\mathfrak{m}})_Q \otimes_{(\Gamma_h)_Q} \left[(\Gamma_h)_Q \otimes_\Lambda M\right] \end{aligned}$$

As above, the Γ_h-rank of M is the $(\Gamma_h)_Q$-dimension of $(\Gamma_h)_Q \otimes_\Lambda M$, say d. This implies that the right-hand side (and therefore also the left-hand side) of (15.2.2) is a free $(\hat{\Gamma}_{\mathfrak{m}})_Q$-module of rank d.

Since $\mathfrak{m}$ is nonstrictly split, $\hat{\Gamma}_{\mathfrak{m}}$ has two coordinate rings Υ and Υ', and $\hat{\Gamma}_{\mathfrak{m}} = \Upsilon \oplus \Upsilon'$. Multiplying (15.2.2) by the idempotent $(1, 0)$ of $\Upsilon \oplus \Upsilon'$ then shows that the Υ-rank of $\hat{M}_{\mathfrak{m}}$ also equals d, the Γ_h-rank of M. □

LEMMA 15.3. *A finitely generated Λ-module has finite length if and only if all of its ranks are zero.*

PROOF. For every Λ-module M, the localization M_Q is the direct sum of the localizations $M_{\mathfrak{p}}$ of M at the finite number of minimal prime ideals of Λ [Corollary 6.4(i)]. Since the ring Λ is reduced, and every maximal ideal has height 1 [Lemma 10.5] the present lemma follows from Lemma 4.5. □

LEMMA 15.4. *Let $X \in \operatorname{fingen}(\hat{\Lambda}_{\mathfrak{m}})$ for some $\mathfrak{m} \in \operatorname{maxspec}(\Lambda)$.*

(i) *If $\mathfrak{m}$ is nonstrictly split, then X is the completion of a finitely generated $\Lambda_{\mathfrak{m}}$-module if and only if its ranks with respect to the two coordinate rings of $\hat{\Gamma}_{\mathfrak{m}}$ are equal, equivalently, if and only if X_Q is a free $(\hat{\Gamma}_{\mathfrak{m}})_Q$-module.*
(ii) *If $\mathfrak{m}$ is not nonstrictly split, then every such X is the completion of some finitely generated $\Lambda_{\mathfrak{m}}$-module.*

PROOF. By Lemma 5.4, X is the completion of some finitely generated $\Lambda_{\mathfrak{m}}$-module if and only if $X_Q \cong \hat{\Lambda}_{\mathfrak{m}} \otimes_{\Lambda} V$ for some $V \in \text{fingen}(\Lambda_Q)$. Moreover, this holds if and only if $X_Q \cong \hat{\Lambda}_{\mathfrak{m}} \otimes N_Q$ for some $N \in \text{fingen}(\Lambda_{\mathfrak{m}})$. (For the slightly less obvious "only if" assertion, take N' to be the Λ-module generated by the numerators of some finite set of Λ_Q-generators of V, and then let $N = N'_{\mathfrak{m}}$.)

Statement (i). If X_Q is free, we can take N to be $\Lambda_{\mathfrak{m}}$-free of the same rank. Conversely, suppose $X_Q \cong \hat{\Lambda}_{\mathfrak{m}} \otimes_{\Lambda_{\mathfrak{m}}} N_Q$ for some N. Since $\mathfrak{m}$ is nonstrictly split, $\Gamma_{\mathfrak{m}}$ is an integral domain, and hence $(\Gamma_{\mathfrak{m}})_Q = (\Lambda_{\mathfrak{m}})_Q$ is a field. Since every $(\Lambda_{\mathfrak{m}})_Q$-module is free, in this situation, we may take $N = \Lambda_{\mathfrak{m}}^{(d)}$ for some d. Then X_Q is $(\hat{\Lambda}_{\mathfrak{m}})_Q$-free, as desired.

Since $\mathfrak{m}$ is nonstrictly split, $\Gamma_{\mathfrak{m}}$ is a principal ideal domain with exactly two maximal ideals. Therefore $\hat{\Gamma}_{\mathfrak{m}}$ is the direct sum of two DVRs, and hence $(\hat{\Gamma}_{\mathfrak{m}})_Q$ is the direct sum of two fields. Therefore an arbitrary $(\hat{\Gamma}_{\mathfrak{m}})_Q$-module is the direct sum of a vector space over each of these fields, and is free if and only if the dimensions of these two vector spaces are equal. This proves the equivalence of the statement about equality of ranks with the statement about freeness.

To prove statement (ii) we consider two cases. See Lemma 4.3 for our seemingly ambiguous use of $(\dots)_Q$.

Case 1: $\mathfrak{m}$ is unsplit or nonsingular. Then $\hat{\Gamma}_{\mathfrak{m}}$ has only one coordinate ring [Lemma 11.10]. Therefore $\Gamma_{\mathfrak{m}}$ is an integral domain, and its field of quotients is $(\Gamma_{\mathfrak{m}})_Q$. Let d be the dimension of the $(\hat{\Gamma}_{\mathfrak{m}})_Q$-vector space X_Q. Then we can take $N = \Gamma_{\mathfrak{m}}^{(d)}$.

Case 2: $\mathfrak{m}$ is strictly split. Then each of $\Gamma_{\mathfrak{m}}$ and $\hat{\Gamma}_{\mathfrak{m}}$ has two coordinate rings, say $\Gamma_{\mathfrak{m}} = \Psi \oplus \Phi$, so that $\hat{\Gamma}_{\mathfrak{m}} = \hat{\Psi}_{\mathfrak{m}} \oplus \hat{\Phi}_{\mathfrak{m}}$ is the decomposition of $\hat{\Gamma}_{\mathfrak{m}}$ into the direct sum of its two coordinate rings [see Proposition 11.10]. In this situation, X_Q must be the direct sum of two vector spaces, say of dimensions d and e over $(\hat{\Psi}_{\mathfrak{m}})_Q$ and $(\hat{\Phi}_{\mathfrak{m}})_Q$, respectively. Thus, we can take $N = \Psi^{(d)} \oplus \Phi^{(e)}$. □

DEFINITION 15.5 (Package of localizations, package of completions). For each $\mathfrak{m} \in \text{maxspec}(\Lambda)$, let $H(\mathfrak{m})$ be a finitely generated $\Lambda_{\mathfrak{m}}$- (respectively $\hat{\Lambda}_{\mathfrak{m}}$-) module. We say that the set $\{H(\mathfrak{m})\}$ is the *package of localizations (respectively package of completions)* of some $M \in \text{fingen}(\Lambda)$ if, for every $\mathfrak{m} \in \text{maxspec}(\Lambda)$, we have $M_{\mathfrak{m}} \cong H(\mathfrak{m})$ as $\Lambda_{\mathfrak{m}}$-module (respectively $\hat{M}_{\mathfrak{m}} \cong H(\mathfrak{m})$ as $\hat{\Lambda}_{\mathfrak{m}}$-modules).

THEOREM 15.6 (Package Deal Theorem for Completions). $(\forall \mathfrak{m} \in \text{maxspec}(\Lambda))$, *let* $H(\mathfrak{m}) \in \text{fingen}(\hat{\Lambda}_{\mathfrak{m}})$. *Then the set* $\{H(\mathfrak{m})\}$ *is the package of completions of some finitely generated* Λ*-module if and only if both of the following conditions hold.*

(i) $H(\mathfrak{m})$ *is* $\hat{\Lambda}_{\mathfrak{m}}$*-free* (a$\forall\mathfrak{m}$).

(ii) (Consistent ranks) *The* Υ*-rank of* $H(\mathfrak{m})$ *equals the* Υ'*-rank of* $H(\mathfrak{n})$ *whenever* $\mathfrak{m}, \mathfrak{n} \in \text{maxsupp}_{\Lambda}(\Gamma_h)$ *for some* h*, and* Υ *and* Υ' *are coordinate rings of* $(\Gamma_h)\hat{}_{\mathfrak{m}}$ *and* $(\Gamma_h)\hat{}_{\mathfrak{n}}$ *respectively.*

Before beginning the proof we note that condition (ii), above, allows the possibility that $\mathfrak{m} = \mathfrak{n}$, in which case it states that $(\Gamma_h)\hat{}_{\mathfrak{m}} = \Upsilon \oplus \Upsilon'$ implies that the Υ-rank of $H(\mathfrak{m})$ equals the Υ'-rank of $H(\mathfrak{m})$.

PROOF. In this proof, $\mathfrak{m}$ and $\mathfrak{n}$ always denote maximal ideals of Λ.

Suppose first that $\{H(\mathfrak{m})\}$ is the package of completions of some $M \in \text{fingen}(\Lambda)$. Condition (i) holds because $M_{\mathfrak{m}}$ is $\Lambda_{\mathfrak{m}}$-free (a$\forall\mathfrak{m}$) [Lemma 8.2]. Condition (ii) holds because, by Lemma 15.2, both the Υ-rank of $H(\mathfrak{m})$ and the Υ'-rank of $H(\mathfrak{n})$ equal the Γ_h-rank of M.

Conversely, suppose that conditions (i) and (ii) hold. We need to find $M \in \text{fingen}(\Lambda)$ such that $\{H(\mathfrak{m})\}$ is the package of completions of M.

First we claim that, for each maximal ideal $\mathfrak{m}$ of Λ, there exists $M(\mathfrak{m}) \in \text{fingen}(\Lambda_{\mathfrak{m}})$ such that $M(\mathfrak{m})\hat{}_{\mathfrak{m}} \cong H(\mathfrak{m})$ as $\hat{\Lambda}_{\mathfrak{m}}$-modules. If $\mathfrak{m}$ is not nonstrictly split, then the claim holds by Lemma 15.4(ii). If $\mathfrak{m}$ is nonstrictly split, then $\hat{\Gamma}_{\mathfrak{m}} = \Upsilon \oplus \Upsilon'$ and therefore by condition (ii) (with $\mathfrak{n} = \mathfrak{m}$), $H(\mathfrak{m})$ has the same rank with respect to each coordinate ring of $\hat{\Gamma}_{\mathfrak{m}}$, so the claim follows from Lemma 15.4(i).

Therefore, to complete the proof of the theorem, it now suffices to show that the set $\{M(\mathfrak{m})\}$ is the package of localizations of some $M \in \text{fingen}(\Lambda)$. To show this, it suffices to find $L \in \text{fingen}(\Lambda)$ such that:

(a) $L_{\mathfrak{m}} \cong M(\mathfrak{m})$ (a$\forall\mathfrak{m}$); and
(b) $(L_{\mathfrak{m}})_Q \cong M(\mathfrak{m})_Q$ ($\forall\mathfrak{m}$).

because Lemma 8.4 allows us to correct the finite number of incorrect localizations in (a). Before constructing the module L, we establish a few needed facts.

Claim 1: $M(\mathfrak{m})$ is $\Lambda_{\mathfrak{m}}$-free (a$\forall\mathfrak{m}$). We note that $M(\mathfrak{m})\hat{}_{\mathfrak{m}} \cong H(\mathfrak{m})$ for each maximal ideal $\mathfrak{m}$, by definition, and $H(\mathfrak{m})$ is almost always $\hat{\Lambda}_{\mathfrak{m}}$-free, by hypothesis. That is, $M(\mathfrak{m})\hat{}_{\mathfrak{m}}$ is $\hat{\Lambda}_{\mathfrak{m}}$-free (a$\forall\mathfrak{m}$). Therefore, $M(\mathfrak{m})$ is $\Lambda_{\mathfrak{m}}$-free (a$\forall\mathfrak{m}$) [Lemma 9.1].

Claim 2. For each coordinate ring Γ_h, there exists a nonnegative integer $d(h)$ such that:

(15.6.1)
$$M(\mathfrak{m})_Q \cong \begin{cases} (\Gamma_h)_Q^{(d(h))} & \text{if } \mathfrak{m} \in \text{maxsupp}_\Lambda(\Gamma_h), \\ & \text{and } (\Gamma_k)_{\mathfrak{m}} = 0 \text{ when } k \neq h. \\ (\Gamma_h)_Q^{(d(h))} \oplus (\Gamma_k)_Q^{(d(k))} & \text{if } \mathfrak{m} \in \text{maxsupp}_\Lambda(\Gamma_h) \\ & \text{and } \mathfrak{m} \in \text{maxsupp}_\Lambda(\Gamma_k) \end{cases}$$

To prove this claim, let $\mathfrak{m} \in \text{maxsupp}_\Lambda(\Gamma_h)$, and let Υ be a coordinate ring of $(\Gamma_h)\hat{}_{\mathfrak{m}}$. By condition (ii), the Υ-rank of $H(\mathfrak{m})$ depends only on Γ_h (not on $\mathfrak{m}$ or Υ), so let $d(h)$ be this Υ-rank of $H(\mathfrak{m})$. Then (15.6.1) follows immediately from (15.2.1) and (15.2.2), with M replaced by $M(\mathfrak{m})$.

We are now ready to construct the desired $L \in \text{fingen}(\Lambda)$. For each coordinate ring Γ_h of Γ, let Λ_h be the projection of Λ in Γ_h. Then Λ_h is a cyclic Λ-module, and $(\Lambda_h)_Q = (\Gamma_h)_Q$, because $\Lambda_Q = \Gamma_Q$. Let

(15.6.2)
$$L = \oplus_h \Lambda_h^{(d(h))}$$

To complete the proof of the theorem, we show that L has the required properties (a) and (b) above.

Property (a). Only finitely many maximal ideals of Λ are strictly split [Lemma 11.4]. Each remaining maximal ideal is in the support of a unique coordinate ring of Γ [Lemma 11.6]. Consider some maximal ideal $\mathfrak{m}$ that is in the support of a unique coordinate ring, say Γ_h. On the one hand, since Λ_h is the projection of Λ in Γ_h, we see that $(\Lambda_h)_{\mathfrak{m}} \cong \Lambda_{\mathfrak{m}}$ while $(\Lambda_k)_{\mathfrak{m}} = 0$ for $k \neq h$, and hence by (15.6.2), $L_{\mathfrak{m}} \cong \Lambda_{\mathfrak{m}}^{(d(h))}$ is free of rank $d(h)$. On the other hand, for almost all such $\mathfrak{m}$, $M(\mathfrak{m})$ is free (Claim 1) of rank $d(h)$ (Claim 2). Therefore, $L_{\mathfrak{m}} \cong M(\mathfrak{m})$ for almost all maximal ideals $\mathfrak{m}$ of Λ.

Property (b). If $\mathfrak{m} \in \operatorname{maxsupp}_\Lambda(\Gamma_h)$, then $((\Lambda_h)_\mathfrak{m})_Q = (\Lambda_h)_Q = (\Gamma_h)_Q$ is the field of quotients of Γ_h. Thus, localizing (15.6.2) at $\mathfrak{m}$ and then at Q yields exactly (15.6.1), so that $(L_\mathfrak{m})_Q \cong M(\mathfrak{m})_Q$, as required. □

16. Indecomposable Λ-modules; Connections Graph

A general property rings of Krull dimension 1 is that if some finitely generated module M is indecomposable, then every module in genus(M) is indecomposable [a special case of Lemma 9.3(iii)]. In this section we determine when the modules in arbitrary Λ-genera are indecomposable. This indecomposability question is answered purely in terms of the ranks of indecomposable summands of the completions $\hat{M}_\mathfrak{m}$ except if M has finite length, in which case M is an indecomposable module over some completion $\hat{\Lambda}_\mathfrak{m}$ and is therefore described in [**KL2**]. (See Proposition 16.1.) Thus, there is no loss of generality in assuming that the indecomposable M is in $\operatorname{fingen}_\infty(\Lambda)$. Most of the work in this section involves working with ranks.

We introduce a graph, called the "connections graph" of $M \in \operatorname{fingen}_\infty(\Lambda)$ [Definition 16.5] and relate indecomposability of M to connectedness of this graph [Theorem 16.8].

One main consequence is that all ranks of indecomposable Λ-modules are at most 2, and when this maximum occurs, M cannot be torsionfree [Corollary 16.9]. Another is that if M is indecomposable, then every completion $\hat{M}_\mathfrak{m}$ is the direct sum of at most four indecomposable summands [Theorem 16.10]. The section closes with a proof that an arbitrary $M \in \operatorname{fingen}(\Lambda)$ has at most finitely many genera of direct summands (indecomposable or not) [Theorem 16.11]. This is actually a general property of reduced noetherian rings of Krull dimension 1, and is obvious in the classical situation that singspec is finite.

In this section we repeatedly use the fact that if $\mathfrak{m}$ is strictly split, supporting Γ_h and Γ_k, then the two coordinate rings of $\hat{\Gamma}_\mathfrak{m}$ are $(\Gamma_h)\hat{}_\mathfrak{m}$ and $\Gamma_k)\hat{}_\mathfrak{m}$. [Proposition 11.10]

PROPOSITION 16.1. *If $M \in \operatorname{finlen}(\Lambda)$ is an indecomposable Λ-module of finite length, then for some $\mathfrak{m} \in \operatorname{maxspec}(\Lambda)$, we have natural identifications $M = M_\mathfrak{m} = \hat{M}_\mathfrak{m}$. In particular, every indecomposable Λ-module of finite length is an indecomposable module of finite length over $\hat{\Lambda}_\mathfrak{m}$ for some $\mathfrak{m} \in \operatorname{maxspec}(\Lambda)$.*

PROOF. See Corollary 6.4 for the identification $M = M_\mathfrak{m}$, for appropriate $\mathfrak{m}$. It is obvious that $\Lambda_\mathfrak{m}$-modules of finite length are unchanged by completion. □

The next lemma contains what we need to know about indecomposable modules in $\operatorname{fingen}_\infty(\Lambda)$ in the complete local case, from [**KL2**], translated into the language of the present paper. For the corresponding facts in the incomplete local case, see Corollary 16.4

LEMMA 16.2. *Let Λ be a complete local Dedekind-like ring, and let $M \in \operatorname{fingen}_\infty(\Lambda)$ be indecomposable.*

(i) *If Λ is unsplit, then M has rank 1 or 2. In the latter case, the separated cover of M has nonzero kernel and M has a nonzero submodule of finite length.*
(ii) *If Λ is (necessarily strictly) split, then M has rank (1,0), (0,1), or (1,1), with respect to the two coordinate rings of Γ.*
(iii) *If Λ is a DVR, then M has rank 1.*

Moreover, for every Λ of the specified type, all of these possibilities actually occur.

PROOF. We omit the details of the well-known situation (iii). Moreover in case (ii) recall that complete split Dedekind-like rings must be strictly split [Lemma 11.8].

Each indecomposable Λ-module M is described in [**KL2**] as a certain combination of indecomposable over the (one or two) DVR coordinate rings of Γ. The Γ-modules needed for M, as well as the actual combinations used in the construction, are specified by a "standard" diagram denoted by $\mathcal{D}$ [**KL2**, Theorems 2.7 and 3.5]. In this description the indecomposable $M \in \text{fingen}(\Lambda)$ is obtained in the form $M \cong S/K$ where $S = S(\mathcal{D})$ and $K = K(\mathcal{D})$ are determined by $\mathcal{D}$. Moreover, the natural homomorphism ϕ: $S \twoheadrightarrow M$ is a separated cover [**KL2**, subsections 9.6 and 10.2].

Let Υ be one of the one or two coordinate rings of Γ. It follows easily from the definitions of S and K that:

(16.2.1) The Υ-rank of M equals the number of copies of Υ in $\mathcal{D}$.

The parts of statements (i) and (ii) about ranks follow immediately from (16.2.1). For by [**KL2**, Theorem 2.7], if Λ is unsplit, the rank of the Γ-free part of $\mathcal{D}$ must be 0, 1 or 2, and all three possibilities can occur. (But rank 0 yields a Λ-module of finite length, which does not interest us here.) Similarly, by [**KL2**, Theorem 3.5], if Λ is (strictly) split, so that $\Gamma = \Upsilon \oplus \Upsilon'$, the rank of the Γ-projective part of $\mathcal{D}$ must be (0,0), (1,0), (0,1), or (1,1), and all four possibilities can occur. (Again, rank (0,0) yields a Λ-module of finite length, which does not interest us here.)

To obtain the part of statement (i) about the separated cover, consider the case that Λ is unsplit and M has rank 2. The the fact that M has rank 2 requires that exactly two of the Γ-modules in the diagram for M be $\cong \Gamma$, which has infinite length. The only standard diagram that allows this is the non-reduced diagram $\mathcal{D}_{\text{Nrd}}$ shown in [**KL2**, (2.4.1)] and described in [**KL2**, (2.4.2))]; and in this situation we always have $K(\mathcal{D}) \neq 0$, as desired.

Finally, we obtain the statement in (i) about a submodule of finite length when M has rank 2. We have $M \cong S(\mathcal{D})/K(\mathcal{D})$, and we are still talking about the non-reduced diagram $\mathcal{D}_{\text{Nrd}}$. It is obvious from the definition of this diagram that $K(\mathcal{D})$ contains part, but never all, of the maximum finite-length Λ-submodule of $S(\mathcal{D})$; and hence M contains a nonzero submodule of finite length. □

As a corollary of the Package Deal Theorem 15.6, we show how to assemble a direct summand of a Λ-module from direct summands of its completions.

COROLLARY 16.3 (Existence of direct summands). *Given $M \in \text{fingen}(\Lambda)$, for each coordinate ring Γ_h of Γ let the Γ_h-rank of M be d_h, and let e_h be an integer such that $0 \leq e_h \leq d_h$. Then there exists a direct summand N of M such that the Γ_h-rank of N equals e_h for every h if and only if both of the following conditions hold.*

(i) *For every h and every unsplit $\mathfrak{m} \in \text{maxsupp}_\Lambda(\Gamma_h)$, $\hat{M}_\mathfrak{m}$ has a direct summand of rank e_h.*

(ii) *For every pair $h \neq k$ and every strictly split $\mathfrak{m} \in \text{maxspec}(\Lambda)$ supporting both Γ_h and Γ_k, $\hat{M}_\mathfrak{m}$ has a direct summand of $\big((\Gamma_h)\hat{_\mathfrak{m}}, (\Gamma_k)\hat{_\mathfrak{m}}\big)$-rank (e_h, e_k).*

Moreover, if these conditions hold and if, for each $\mathfrak{m} \in \text{maxspec}(\Lambda)$, $X(\mathfrak{m})$ *is any direct summand of* $\hat{M}_\mathfrak{m}$ *whose rank equals that of* $\hat{N}_\mathfrak{m}$, *then* $\{X(\mathfrak{m})\}$ *is the package of completions of some direct summand of* M.

PROOF. For the nontrivial implication of the Corollary, let the nonnegative integers $e_h \leq d_h$ be given, such that conditions (i) and (ii) hold. We claim that, for each $\mathfrak{m} \in \text{maxspec}(\Lambda)$, there is a direct summand $X(\mathfrak{m})$ of $\hat{M}_\mathfrak{m}$ such that the following conditions hold.

(i)′ If $\mathfrak{m}$ is unsplit or nonsingular, then the rank of $X(\mathfrak{m})$ is e_h.
(ii)′ If $\mathfrak{m}$ is strictly split and in the support of Γ_h and Γ_k (where $h \neq k$), then $X(\mathfrak{m})$ has the rank stated in condition (ii).
(iii)′ If $\mathfrak{m}$ is nonstrictly split and in the support of Γ_h, then the rank of $X(\mathfrak{m})$ is (e_h, e_h).

To prove this claim, we need to show that such summands $X(\mathfrak{m})$ exist when their existence is not guaranteed by hypotheses (i) and (ii). Thus, we may assume that $\mathfrak{m}$ is either nonsingular or nonstrictly split. In both situations, $\mathfrak{m}$ is in the support of a unique coordinate ring Γ_h of Γ. If $\mathfrak{m}$ is nonsingular, then $\hat{M}_\mathfrak{m}$ is the direct sum of a free module of rank d_h and a torsion module, and so we can take $X(\mathfrak{m})$ to be free of rank e_h. If $\mathfrak{m}$ is nonstrictly split, then $\hat{\Lambda}_\mathfrak{m}$ is strictly split [Lemma 11.8], and therefore $(\Gamma_h)\hat{}_\mathfrak{m}$ is the direct sum of two coordinate rings. In this situation $\hat{M}_\mathfrak{m}$ is the direct sum of indecomposable modules, each of rank (1,0), (0,1), and (1,1), and a module of finite length [Lemma 16.2]. Since $(\Gamma_h)\hat{}_\mathfrak{m}$ is the direct sum of two coordinate rings, the case $\mathfrak{m} = \mathfrak{n}$ of Lemma 15.2 shows that the rank of $\hat{M}_\mathfrak{m}$ is (d_h, d_h). It follows that $\hat{M}_\mathfrak{m}$ has a direct summand of rank (e_h, e_h), as needed for (iii)′. This completes the proof of the claim.

Next, we claim that there exists $N' \in \text{fingen}(\Lambda)$ such that $\hat{N}'_\mathfrak{m} \cong X(\mathfrak{m})$ $(\forall \mathfrak{m} \in \text{maxspec}(\Lambda))$. For this, it suffices to check the two conditions of Package Deal Theorem 15.6. By that theorem, applied to M and the set $\{\hat{M}_\mathfrak{m}\}$, we see that $\hat{M}_\mathfrak{m}$ is $\hat{\Lambda}_\mathfrak{m}$-free (a$\forall\mathfrak{m}$). Since projective modules over local rings are free, the direct summand $X(\mathfrak{m})$ of $\hat{M}_\mathfrak{m}$ is free (a$\forall\mathfrak{m}$). Thus, the freeness condition of that theorem holds. The "consistent ranks" condition holds by our present conditions (i)′–(iii)′, so this claim is proved.

Since $\hat{N}'_\mathfrak{m}$ is isomorphic to a direct summand of $\hat{M}_\mathfrak{m}$, we have that the localization $N'_\mathfrak{m}$ is isomorphic to a direct summand of $M_\mathfrak{m}$ [Lemma 9.1], and therefore there exists $N \in \text{genus}(N')$ such that N is isomorphic to a direct summand of M [Lemma 9.3(i)]. □

COROLLARY 16.4. *Suppose that* Λ *is local, but not complete, and let* $M \in \text{fingen}(\Lambda)$ *be indecomposable and not of finite length.*

(i) *If* Λ *is nonstrictly split, then the rank of* M *is 1, and the rank of* $\hat{M}_\mathfrak{m}$ *is* $(1, 1)$.
(ii) *If* Λ *is not nonstrictly split, then the one or two ranks of* M *are the same as the ranks of* $\hat{M}_\mathfrak{m}$, *and the possibilities are those listed in Lemma 16.2.*

PROOF. (i) Since Λ is an integral domain, the "consistent ranks" criterion of Package Deal Theorem 15.6 shows that, for every $M \in \text{fingen}(\Lambda)$, the rank of $\hat{M}_\mathfrak{m}$ is (d, d) for some d. Thus, is suffices to show that, when $d \neq 0$, M has a direct summand of rank (1,1). This follows immediately from Corollary 16.3, since conditions (i) and (ii) of that corollary are vacuously satisfied.

(ii) If Λ is not nonstrictly split, then every finitely generated $\hat{\Lambda}_{\mathfrak{m}}$-module is the completion of some Λ-module [Lemma 15.4(ii)], so by Lemma 9.1, $\hat{M}_{\mathfrak{m}}$ is indecomposable. Since completion does not alter ranks in this situation [Lemma 15.2], statement (ii) follows. □

For each $M \in \text{fingen}_{\infty}(\Lambda)$, we define a graph which displays how M ties together the coordinate rings of Γ. We state our conditions for indecomposability in terms of this graph, and we use the graph again in Section 29 when studying possible decompositions of an arbitrary M.

Definition 16.5 (Connections graph). Let $M \in \text{fingen}_{\infty}(\Lambda)$. We define the *connections graph* $\mathcal{K} = \mathcal{K}(M)$ of M to consist of vertices, edges, and labels, as defined below. [See (16.5.1) for an example.]

Vertices. There is one vertex for each coordinate ring Γ_h for which the Γ_h-rank of M is nonzero. We label each vertex with both the corresponding Γ_h and the Γ_h-rank of M. *(When the rank is 1, we omit the rank label.)*

Edges. Recall that each strictly split maximal ideal $\mathfrak{m}$ of Λ is in the support of exactly two coordinate rings of Γ, say Γ_h and Γ_k [Lemma 11.6]; $(\Gamma_h)\hat{}_{\mathfrak{m}}$ and $(\Gamma_k)\hat{}_{\mathfrak{m}}$ are the two coordinate rings of $\hat{\Gamma}_{\mathfrak{m}}$ [Proposition 11.10]; and the completion $\hat{M}_{\mathfrak{m}}$ is a direct sum of indecomposable modules each of which has rank (1,0), (0,1), or (1,1) with respect to the pair of coordinate rings $(\Gamma_h)\hat{}_{\mathfrak{m}}$ and $(\Gamma_k)\hat{}_{\mathfrak{m}}$ of $\hat{\Gamma}_{\mathfrak{m}}$ [Lemma 16.2]. If at least one direct summand of rank (1,1) occurs we say that $\mathfrak{m}$ *connects* Γ_h to Γ_k (in M). (Other maximal ideals have a unique coordinate ring in their support, and thus do no such connecting.)

We connect vertices $\Gamma_h \neq \Gamma_k$ in $\mathcal{K}(M)$ by an edge if and only if some strictly split maximal ideal connects these vertices. We label this edge with the maximum number d such that, for some strictly split maximal ideal $\mathfrak{m}$ connecting Γ_h to Γ_k (in M), d indecomposable summands of rank (1,1) occur in a decomposition of $\hat{M}_{\mathfrak{m}}$. *(Again, we omit this label in case $d = 1$.)* Recall that Λ has only finitely many strictly split maximal ideals [Proposition 11.4], so we need only consider finitely many maximal ideals in determining the edges and edge labels of the graph.

(16.5.1)

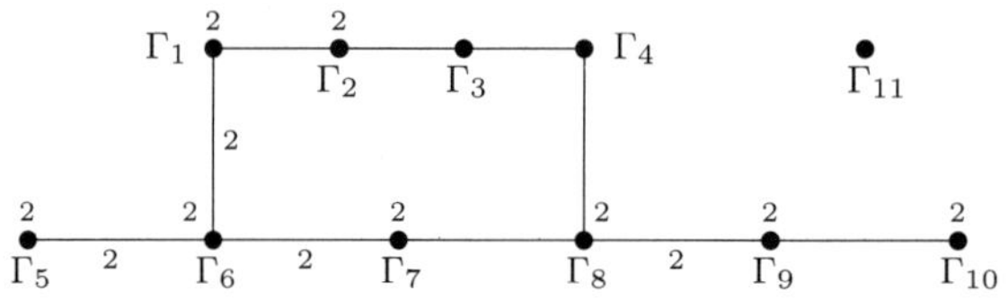

Note that vertices of rank 2 can be connected to other vertices of rank 2 by either unlabeled or 2-labeled edges [e.g., Γ_1 in (16.5.1)]. But connections between unlabeled vertices must be unlabeled edges. As we shall see below [Corollary 16.9], if the module M is indecomposable, then no vertex in $\mathcal{K}(M)$ can have rank label greater than 2. For this reason, if no vertex of the graph $\mathcal{K}$ has rank label greater than 2, then we define the *2-subgraph of* $\mathcal{K}$ to be the subgraph consisting of all vertices of rank 2 and all 2-labeled edges. For example, the 2-subgraph of (16.5.1) has four connected components, two of which (Γ_2 and Γ_{10}) are isolated points of that subgraph.

We note the significance of isolated points of $\mathcal{K}$. In (16.5.1), for example, for every strictly split maximal ideal $\mathfrak{m}$ supporting Γ_{11} and Γ_{10}, it must be that $\hat{M}_{\mathfrak{m}}$ is the direct sum of one indecomposable module of $(\Gamma_{11}, \Gamma_{10})$-rank (1,0) and two indecomposables of rank (0,1).

For some examples of connections graphs of easily described Λ-modules, see (31.3.1).

Let $\mathcal{K}$, $\mathcal{K}'$, and $\mathcal{K}''$ be connections graphs. We say that $\mathcal{K}$ is the *disjoint union* of $\mathcal{K}'$ and $\mathcal{K}''$ if every connected component of $\mathcal{K}$ is a connected component of exactly one of $\mathcal{K}'$ and $\mathcal{K}''$, and every connected component of each of $\mathcal{K}'$ and $\mathcal{K}''$ is a connected component of $\mathcal{K}$.

LEMMA AND DEFINITION 16.6 (Connections decomposition). *For all* $M \in \text{fingen}_\infty(\Lambda)$ *there is a decomposition* $M = \oplus_j B_j$ *such that:*

(i) *For each* j *the connections graph* $\mathcal{K}(B_j)$ *is a connected component of* $\mathcal{K}(M)$*; and*

(ii) $\mathcal{K}(M)$ *is the disjoint union of the graphs* $\mathcal{K}(B_j)$.

Moreover, the genus of each B_j *is determined by the isomorphism class of* M *and the graph* $\mathcal{K}(B_j)$.

We call any such direct sum decomposition a "connections decomposition" of M.

PROOF. *Existence of the decomposition.* First we prove a slightly more general statement that is needed later.

(16.6.1) Let $\mathcal{B}$ be a proper subset of the set of coordinate rings of Γ with the following two properties. (i) The Γ_h-rank of M is nonzero for all $\Gamma_h \in \mathcal{B}$. (ii) If $\Gamma_h \in \mathcal{B}$ and $\Gamma_k \notin \mathcal{B}$, then no strictly split maximal ideal connects Γ_h to Γ_k (in M). Then there is a decomposition $M = B \oplus D$ such that B has nonzero rank Γ_h-rank if and only if $\Gamma_h \in \mathcal{B}$, and D has Γ_h-rank zero when $\Gamma_h \in \mathcal{B}$.

For every coordinate ring Γ_h of Γ, let d_h be the Γ_h-rank of M. Thus, $d_h = 0$ if and only if Γ_h does not label any vertex in $\mathcal{K}(M)$. Let $e_h = d_h$ if $\Gamma_h \in \mathcal{B}$, and $e_h = 0$ otherwise. We claim that M has a direct summand B such that, for each h, the Γ_h-rank of B is e_h.

To prove the claim, we verify the two conditions of Corollary 16.3. Condition (i) is trivially satisfied. For condition (ii), suppose that $\mathfrak{m}$ is a strictly split maximal ideal supporting the two coordinate rings Γ_h and Γ_k. We consider four cases.

Case 1: $\Gamma_h \in \mathcal{B}$, $\Gamma_k \notin \mathcal{B}$. Here no edge connects Γ_h to Γ_k, and hence $\hat{M}_{\mathfrak{m}}$ is the direct sum of d_h indecomposable summands of rank (1,0) and d_k indecomposable summands of rank (0,1). Therefore, $\hat{M}_{\mathfrak{m}}$ has a direct summand of rank $(d_h, 0)$, as required by condition (ii) of Corollary 16.3.

Case 2: $\Gamma_k \in \mathcal{B}$, $\Gamma_h \notin \mathcal{B}$. The proof is the same as that of case 1.

Case 3: $\Gamma_h, \Gamma_k \in \mathcal{B}$. Here take the direct summand to be $\hat{M}_{\mathfrak{m}}$ itself.

Case 4: $\Gamma_h, \Gamma_k \notin \mathcal{B}$. Take the summand to be 0.

Corollary 16.3 now yields a direct summand whose Γ_h-rank is nonzero if and only if $\Gamma_h \in \mathcal{B}$, and the summand D complementary to B must have rank zero at these vertices. This completes the proof of (16.6.1).

Now we prove the existence of a connections decomposition of M, using (16.6.1). Let $\mathcal{B}$ be a connected component of $\mathcal{K}(M)$. Then by (16.6.1) there is a decomposition $M = B \oplus D$ such that $\mathcal{K}(B) = \mathcal{B}$ and $\mathcal{K}(D)$ is the union of the remaining connected components of $\mathcal{K}(M)$. Now give the same treatment to some connected component of $\mathcal{K}(D)$, and repeat the procedure until we have a connections decomposition of M.

Uniqueness of the decomposition. Consider another such decomposition $M = \oplus_j A_j$. We need to prove that $(A_j)\hat{}_{\mathfrak{m}} \cong (B_j)\hat{}_{\mathfrak{m}}$ for all $\mathfrak{m} \in \text{maxspec}(\Lambda)$ and all indices j. Thus, fix $\mathfrak{m} \in \text{maxspec}(\Lambda)$, and fix some index j. Now we have $\oplus_i (A_i)\hat{}_{\mathfrak{m}} \cong \oplus_i (B_i)\hat{}_{\mathfrak{m}}$. For each index i, denote by $\mathcal{B}_i$ the component of $\mathcal{K}(M)$ such that $\mathcal{K}(A_i) = \mathcal{B}_i = \mathcal{K}(B_i)$. Recall that $\mathfrak{m}$ is in the support of at most two coordinate rings of Γ.

Suppose first that $\mathfrak{m}$ is in the support of a unique coordinate ring, say Γ_h. Then we have $(A_i)\hat{}_{\mathfrak{m}} = 0 = (B_i)\hat{}_{\mathfrak{m}}$ unless $\Gamma_h \in \mathcal{B}_i$. Therefore, if $\Gamma_h \in \mathcal{B}_j$, then the isomorphism $\oplus_i (A_i)\hat{}_{\mathfrak{m}} \cong \oplus_i (B_i)\hat{}_{\mathfrak{m}}$ reduces to $(A_j)\hat{}_{\mathfrak{m}} \cong (B_j)\hat{}_{\mathfrak{m}}$, while if $\Gamma_h \notin \mathcal{B}_j$, then $(A_j)\hat{}_{\mathfrak{m}} = 0 = (B_j)\hat{}_{\mathfrak{m}}$, as desired.

Suppose next that $\mathfrak{m}$ is in the support of two coordinate rings, say Γ_h and Γ_k. If both coordinate rings belong to a single $\mathcal{B}_i$, then the same argument as in the previous paragraph shows that $(A_j)\hat{}_{\mathfrak{m}} \cong (B_j)\hat{}_{\mathfrak{m}}$.

The remaining case is that $\mathfrak{m}$ is in the support of two coordinate rings Γ_h and Γ_k, where say $\Gamma_h \in \mathcal{B}_j$ and $\Gamma_k \in \mathcal{B}_g$ for some $g \neq j$. Here the isomorphism $\oplus_i (A_i)\hat{}_{\mathfrak{m}} \cong \oplus_i (B_i)\hat{}_{\mathfrak{m}}$ reduces to $(A_j)\hat{}_{\mathfrak{m}} \oplus (A_g)\hat{}_{\mathfrak{m}} \cong (B_j)\hat{}_{\mathfrak{m}} \oplus (B_g)\hat{}_{\mathfrak{m}}$. Since the two coordinate rings supported by $\mathfrak{m}$ belong to distinct connected components of $\mathcal{K}(M)$, $\hat{M}_{\mathfrak{m}}$ is a direct sum of indecomposable modules whose ranks are (1,0) and (0,1), by Case 1 above. Moreover, all of the rank-(1,0) summands must occur in (say) $(A_j)\hat{}_{\mathfrak{m}}$ and $(B_j)\hat{}_{\mathfrak{m}}$, while all of the rank-(0,1) summands occur in $(A_g)\hat{}_{\mathfrak{m}}$ and $(B_g)\hat{}_{\mathfrak{m}}$. The Krull-Schmidt theorem for complete local rings therefore yields $(A_j)\hat{}_{\mathfrak{m}} \cong (B_j)\hat{}_{\mathfrak{m}}$, as desired. □

REMARKS 16.7. (i) *The terms B_j of a connections decomposition of any $M \in \text{fingen}_\infty(\Lambda)$ are unique up to isomorphism.*

Although we will not use this strengthening of Lemma 16.6, we observe that it is an easy consequence of results proved later in the paper: Choose another such decomposition $M = \oplus_j B'_j$. Since $\text{genus}(B_j) = \text{genus}(B'_j)$ for every j [Lemma 16.6] and the terms B_j have disjoint sets of torsionfree support sets [Definitions 22.1], the proof is completed by Lemma 28.1.

(ii) The hypothesis about no summands of finite length cannot be deleted, because such summands contribute neither vertices nor edges to connections graphs.

THEOREM 16.8 (Structure of indecomposables). *Let $M \in \text{fingen}_\infty(\Lambda)$ (equivalently, every completion $\hat{M}_{\mathfrak{m}} \in \text{fingen}_\infty(\hat{\Lambda}_{\mathfrak{m}})$ [Proposition 7.3]). Then M is indecomposable if and only if all of the following conditions hold.*

(i) *The connections graph $\mathcal{K} = \mathcal{K}(M)$ is connected.*
(ii) *The rank label of every vertex in $\mathcal{K}$ is less than or equal to 2 (and hence the label of every edge is less than or equal to 2).*
(iii) *In every connected component of the 2-subgraph of $\mathcal{K}$, there exists a vertex Γ_h and an unsplit $\mathfrak{m} \in \text{maxspec}(\Lambda)$ in the support of Γ_h, such that $\hat{M}_{\mathfrak{m}}$ is indecomposable.*

PROOF. In this proof, $\mathfrak{m}$ and $\mathfrak{n}$ always denote maximal ideals of Λ. Also, for each index h, let d_h be the Γ_h-rank of M.

Suppose first that at least one of the conditions fails; we show that M has a proper direct summand. We take each condition in turn.

Suppose that statement (i) fails. The existence of the connections decomposition [Lemma and Definition 16.6] yields the desired summand.

Suppose that statement (ii) fails, so that some vertex has rank label at least 3. We apply Corollary 16.3 with each $e_h = \min\{2, d_h\}$, showing that M has a nonzero direct summand with all ranks less than or equal to 2. Condition (i) of that corollary is satisfied because every indecomposable module over any complete local *unsplit* Dedekind-like ring has rank at most 2 [Lemma 16.2]. For condition (ii) of that corollary, let $\mathfrak{m}$ be a strictly split maximal ideal in the support of coordinate rings and Γ_h and Γ_k. We consider several cases, remembering that every indecomposable summand of $\hat{M}_\mathfrak{m}$ has rank (1,0), (0,1), or (1,1) [Lemma 16.2].

Case 1: The Γ_h-rank and Γ_k-rank of M are both greater than or equal to 2. Then it is easy to assemble a direct summand of rank $(2,2) = (e_h, e_k)$.

Case 2: The Γ_h-rank and Γ_k-rank of M are d_h and 1, respectively, with $d_h \geq 2$. Then $\hat{M}_\mathfrak{m}$ has at most one indecomposable summand of rank (1,1), and the rest all have rank (1,0). Again, it is easy to assemble a direct summand of rank $(2,1) = (e_h, e_k)$.

Remaining cases: These are similar to Case 2, because at least one of the ranks involved is 1 or 0.

By Corollary 16.3, we obtain a proper direct summand of M (with ranks $e_h \leq 2$, and at least one rank nonzero).

Suppose that statement (iii) fails. We may assume that all ranks of M are ≤ 2. Then there is a connected component $\mathcal{B}$ of the 2-subgraph of $\mathcal{K}$ such that, for every unsplit $\mathfrak{m}$ in the support of a coordinate ring in $\mathcal{B}$, $\hat{M}_\mathfrak{m}$ is the direct sum of two indecomposable summands of rank 1. For each index h, set $e_h = 1$ if $\Gamma_h \in \mathcal{B}$, and set $e_h = d_h$ otherwise. We show that M has a direct summand whose Γ_h-rank is e_h for each index h. As before, we verify the two conditions of Corollary 16.3.

Condition 16.3(i). If $\Gamma_h \in \mathcal{B}$ and $\mathfrak{m}$ is unsplit, then $\hat{M}_\mathfrak{m}$ must be the direct sum of two modules of rank 1, in which case $\hat{M}_\mathfrak{m}$ has a direct summand $X(\mathfrak{m})$ of rank $e_h = 1$. For any other unsplit $\mathfrak{m}$, we take $X(\mathfrak{m}) = \hat{M}_\mathfrak{m}$, which has rank $e_h = d_h$.

Condition 16.3(ii). Let $\mathfrak{m}$ be a strictly split maximal ideal in the support of coordinate rings Γ_h and Γ_k; we choose a direct summand $X(\mathfrak{m})$ of rank (e_h, e_k), by considering several cases.

Case 1: Both Γ_h and Γ_k are in $\mathcal{B}$ (and hence $d_h = d_k = 2$ and $e_h = e_k = 1$). If $\hat{M}_\mathfrak{m}$ has an indecomposable summand A of rank (1,1), let $X(\mathfrak{m}) = A$. Otherwise $M_\mathfrak{m}$ is the direct sum of two indecomposable summands of rank (1,0) and two of rank (0,1); let $X(\mathfrak{m})$ be the direct sum of one of each type.

Case 2: Exactly one of Γ_h or Γ_k is in $\mathcal{B}$, say $\Gamma_h \in \mathcal{B}$, $\Gamma_k \notin \mathcal{B}$. Then the two vertices are either connected by a 1-labeled edge or no edge at all. Thus, $\hat{M}_\mathfrak{m}$ has at most one indecomposable summand of rank (1,1), and the remaining summands have rank (1,0) or (0,1). If $\hat{M}_\mathfrak{m}$ has an indecomposable summand of rank (1,1), then let $X(\mathfrak{m})$ be the direct sum of this indecomposable rank (1,1)-summand with all of the rank (0,1)-summands; otherwise, we let $X(\mathfrak{m})$ be the direct sum of one rank (1,0)-summand with all of the (0,1)-summands.

Case 3: Neither Γ_h nor Γ_k is in $\mathcal{B}$. In this case, simply set $X(\mathfrak{m}) = M_\mathfrak{m}$.

This completes the proof that, if any of conditions (i)–(iii) fails, then M is decomposable.

Conversely, assume that all of conditions (i)–(iii) hold. By condition (ii), each edge in $\mathcal{K}$ is either unlabeled (i.e., has implicit label 1) or has label 2. To prove that

M is indecomposable, we suppose that $M = X \oplus Y$, and show that either X or Y must equal zero. To do this, we prove that every torsionfree rank of one of these summands, say Y, equals zero, and therefore Y has finite length [Lemma 15.3]. Since we are assuming that M has no nonzero direct summands of finite length, it follows that Y must equal zero. In the course of the proof, we repeatedly use the fact that

$$\hat{M}_{\mathfrak{m}} = \hat{X}_{\mathfrak{m}} \oplus \hat{Y}_{\mathfrak{m}} \tag{16.8.1}$$

for every maximal ideal $\mathfrak{m}$ of Λ.

Claim 1. Let $\mathcal{B}$ be a connected component of the 2-subgraph of $\mathcal{K}$. We claim that one of X or Y has rank 2 at every vertex in $\mathcal{B}$, and hence the other has rank zero at every vertex in $\mathcal{B}$. By condition (iii), there exists a vertex $\Gamma_h \in \mathcal{B}$, and an unsplit maximal ideal $\mathfrak{m} \in \operatorname{maxsupp}_\Lambda(\Gamma_h)$, such that $\hat{M}_{\mathfrak{m}}$ is indecomposable. Therefore by (16.8.1), one of $\hat{X}_{\mathfrak{m}}$ or $\hat{Y}_{\mathfrak{m}}$ equals $\hat{M}_{\mathfrak{m}}$ and the other equals 0; say $\hat{X}_{\mathfrak{m}} = \hat{M}_{\mathfrak{m}}$ and $\hat{Y}_{\mathfrak{m}} = 0$. That is, the Γ_h-ranks of X and Y equal 2 and 0, respectively [Lemma 15.2]. Because $\mathcal{B}$ is a connected component of the 2-subgraph of $\mathcal{K}$, to prove the claim it suffices to show that, if Γ_h and Γ_k are any two vertices in $\mathcal{B}$ that are connected by an edge, and if the Γ_h-ranks of X and Y equal 2 and 0, respectively, then the same is true of their Γ_k-ranks.

Since the edge connecting Γ_h and Γ_k must be 2-labeled, there is a strictly split maximal ideal $\mathfrak{m}$, in the support of Γ_h and Γ_k, such that $\hat{M}_{\mathfrak{m}}$ is the direct sum of two indecomposable modules, each of rank (1,1). Using (16.8.1) together with the fact that the Krull-Schmidt Theorem holds for complete local rings, it follows that the only possible decomposition of $\hat{M}_{\mathfrak{m}}$ in which the Γ_h-ranks of $\hat{X}_{\mathfrak{m}}$ and $\hat{Y}_{\mathfrak{m}}$ are 2 and 0, respectively, is the decomposition in which $\hat{X}_{\mathfrak{m}} = \hat{M}_{\mathfrak{m}}$ and $\hat{Y}_{\mathfrak{m}} = 0$. Thus, the Γ_k-ranks of X and Y equal their Γ_h-ranks, which completes the proof of claim 1.

Claim 2. Let Γ_h and Γ_k be vertices of rank 1 that are connected by an edge of $\mathcal{K}$. We claim that, if X has nonzero Γ_h-rank, then the Γ_h- and Γ_k-ranks of X are both 1, while the Γ_h- and Γ_k-ranks of Y are both 0.

Since an edge connects Γ_h and Γ_k, there must be a strictly split $\mathfrak{m}$ is in the support of these two coordinate rings such that $\hat{M}_{\mathfrak{m}}$ has an indecomposable direct summand of rank (1,1). Since the Γ_h- and Γ_k-ranks of M both equal 1, it follows that $\hat{M}_{\mathfrak{m}}$ is indecomposable. Therefore, (16.8.1) shows that $\hat{M}_{\mathfrak{m}} = \hat{X}_{\mathfrak{m}}$ and $\hat{Y}_{\mathfrak{m}} = 0$, from which the claim follows.

Claim 3. Let Γ_h and Γ_k be vertices of rank 1 and 2, respectively, connected by an edge. We claim that, if X has nonzero rank at one of these vertices, then X has ranks 1 and 2 at Γ_h and Γ_k, respectively, and Y has rank 0 at both vertices.

As in the proof of claim 2, there must be a strictly split $\mathfrak{m} \in \operatorname{maxsupp}_\Lambda(\Gamma_h)$ and Γ_k such that $\hat{M}_{\mathfrak{m}}$ has an indecomposable direct summand of rank (1,1). Since the ranks of these vertices are 1 and 2, respectively, we conclude that $\hat{M}_{\mathfrak{m}} = A(1,1) \oplus A(0,1)$, a direct sum of two indecomposable modules of ranks (1,1) and (0,1), respectively.

On the one hand, if X has nonzero rank at Γ_h, then this decomposition of $\hat{M}_{\mathfrak{m}}$ shows that X and Y have Γ_h-ranks 1 and 0, respectively. Therefore, the Krull-Schmidt Theorem shows that either $\hat{Y}_{\mathfrak{m}} \cong A(0,1)$ or $\hat{Y}_{\mathfrak{m}} = 0$. But Γ_k is in the 2-subgraph of $\mathcal{K}$, so claim 1 shows that Y must have Γ_k-rank 0 or 2. Thus, $\hat{Y}_{\mathfrak{m}} = 0$, and hence $\hat{X}_{\mathfrak{m}} = \hat{M}_{\mathfrak{m}}$, proving claim 3 in this case.

On the other hand, if X has nonzero rank at Γ_k, then again by claim 1, it follows that X and Y must have Γ_k-ranks 2 and 0, respectively. Then comparing the decomposition $\hat{M}_{\mathfrak{m}} = A(1,1) \oplus A(0,1)$ with (16.8.1) shows that $\hat{X}_{\mathfrak{m}} = \hat{M}_{\mathfrak{m}}$ and $\hat{Y}_{\mathfrak{m}} = 0$, completing the proof of claim 3.

Claim 4. Let Γ_h and Γ_k be vertices of rank 2, connected by an unlabeled edge. We claim that if X has nonzero rank at one of these vertices, then X and Y have ranks 2 and 0, respectively, at both vertices.

Since the edge connecting these rank-2 vertices is unlabeled, hence implicitly 1-labeled, we have a strictly split $\mathfrak{m}$ and a decomposition $\hat{M}_{\mathfrak{m}} = A(1,1) \oplus A(1,0) \oplus A(0,1)$, a direct sum of three indecomposable modules of ranks (1,1), (1,0), and (0,1), respectively. If say the Γ_h-rank of X is nonzero, then by claim 1, the Γ_h-ranks of X and Y must be 2 and 0, respectively. Moreover, by the Krull-Schmidt Theorem, $\hat{X}_{\mathfrak{m}}$ must then have a summand of rank (1,1), so that also the Γ_k-rank of X is nonzero. Therefore, again by claim 1, the Γ_k-ranks of X and Y must also be 2 and 0, respectively, proving claim 4.

We can now complete the proof of this half of the theorem. If $X \neq 0$, then the rank of X is nonzero at some vertex Γ_h. By claims 2 through 4, the rank of Y must be zero not only at Γ_h, but also at each vertex connected by an edge to Γ_h. Since $\mathcal{K}$ is connected, it follows that all ranks of Y are zero, as desired. □

COROLLARY 16.9 (Ranks of indecomposables). *Let $M \in \text{fingen}_\infty(\Lambda)$ be indecomposable. Then:*

(i) *All ranks of M are less than or equal to 2.*
(ii) *If there is no unsplit $\mathfrak{m}$ such that $\hat{M}_{\mathfrak{m}}$ is indecomposable of rank 2 (for example, if Λ has no unsplit maximal ideals), then all ranks of M are less than or equal to 1.*

When rank 2 occurs, M cannot be torsionfree (i.e. M necessarily has nonzero submodules of finite length).

PROOF. (i) This is statement (ii) of Theorem 16.8.

(ii) By statement (iii) of Theorem 16.8 and the hypothesis that there is no $\mathfrak{m}$ such that $\hat{M}_{\mathfrak{m}}$ is indecomposable of rank 2, the 2-subgraph must be empty; that is, no vertex of $\mathcal{K}$ can have rank-label 2. We also know, by statement (ii) of Theorem 16.8, that no vertex can have rank-label greater than 2. Therefore each vertex must have (undisplayed) rank-label 1, which is what we want to prove.

Supplementary statement. Let the Γ_h-rank of M be 2. By indecomposability of M, there exists $\mathfrak{m} \in \text{maxsupp}_\Lambda(\Gamma_h)$ such that $\hat{M}_{\mathfrak{m}}$ is indecomposable [Theorem 16.8]. And since M has Γ_h-rank 2 and $\mathfrak{m}$ is unsplit, $\hat{M}_{\mathfrak{m}}$ has $\hat{\Gamma}_{\mathfrak{m}}$-rank 2 [Lemma 15.2]. By Lemma 16.2(i), $\hat{M}_{\mathfrak{m}}$ has a nonzero submodule of finite length. Therefore the same is true of M itself [Proposition 7.4]. □

THEOREM 16.10. *Let $M \in \text{fingen}_\infty(\Lambda)$ be indecomposable. Then for each $\mathfrak{m} \in \text{maxspec}(\Lambda)$, $\hat{M}_{\mathfrak{m}}$ is the direct sum of 0, 1, 2, 3, or 4 indecomposable $\hat{\Lambda}_{\mathfrak{m}}$-modules. Moreover, there is an indecomposable Dedekind-like ring Λ for which all of these possibilities can occur for a single $M \in \text{fingen}(\Lambda)$.*

PROOF. We first construct the example. Let Γ have three coordinate rings Γ_1, Γ_2, Γ_3, and let Λ have at least seven maximal ideals, with the following properties.

$$(16.10.1)\qquad \begin{array}{ll} \mathfrak{m}_1, \mathfrak{m}_2, \mathfrak{m}_3 & \text{unsplit, supporting } \Gamma_1, \Gamma_2, \Gamma_3, \text{ respectively} \\ \mathfrak{n}, \mathfrak{p}, \mathfrak{q} & \text{strictly split, each supporting } \Gamma_1 \text{ and } \Gamma_2 \\ \mathfrak{m}_{23} & \text{strictly split, supporting } \Gamma_2 \text{ and } \Gamma_3 \end{array}$$

Assume also that Λ has no other strictly split maximal ideals. (Such a Dedekind-like ring Λ exists, by Construction 33.1.) Then Λ is an indecomposable ring (equivalently, an indecomposable Λ-module) because its connections graph is as shown in (16.10.2) below [Theorem 16.8].

For each $\mathfrak{m} \in \operatorname{maxspec}(\Lambda)$ choose $H(\mathfrak{m}) \in \operatorname{fingen}(\hat{\Lambda}_\mathfrak{m})$ as follows, where (a, b) denotes an indecomposable module of rank (a, b). (Make the choices so that no indecomposable summand of any of the modules $H(\mathfrak{m})$ has finite length.)

$$\begin{array}{lcl} H(\mathfrak{m}_1) & = & \text{indecomposable, of rank 2} \\ H(\mathfrak{m}_2) & = & \text{arbitrary, of rank 2} \\ H(\mathfrak{m}_3) & = & 0 \\ H(\mathfrak{n}) & = & (1,1) \oplus (1,1) \\ H(\mathfrak{p}) & = & (1,0) \oplus (1,0) \oplus (0,1) \oplus (0,1) \\ H(\mathfrak{q}) & = & (1,1) \oplus (1,0) \oplus (0,1) \\ H(\mathfrak{m}_{23}) & = & (1,0) \oplus (1,0) \end{array}$$

Note that, for $i = 1, 2$, or 3, the completion $\hat{\Lambda}_{\mathfrak{m}_i}$ is an unsplit Dedekind-like ring inside the ring $(\Gamma_i)\hat{}_{\mathfrak{m}_i}$, while for $\mathfrak{m} = \mathfrak{n}$, $\mathfrak{p}$, or $\mathfrak{q}$, the completion $\hat{\Lambda}_\mathfrak{m}$ is a strictly split Dedekind-like ring inside the ring $(\Gamma_1)\hat{}_\mathfrak{m} \oplus (\Gamma_2)\hat{}_\mathfrak{m}$, and the completion $\hat{\Lambda}_{\mathfrak{m}_{23}}$ is a strictly split Dedekind-like ring inside the ring $(\Gamma_2)\hat{}_{\mathfrak{m}_{23}} \oplus (\Gamma_3)\hat{}_{\mathfrak{m}_{23}}$. If $\mathfrak{m}$ is any other maximal ideal of Λ, then by assumption, $\mathfrak{m}$ is not strictly split, so $\mathfrak{m}$ is in the support of a unique coordinate ring Γ_h; if $h = 1$ or 2, then let $H(\mathfrak{m}) = \hat{\Lambda}_\mathfrak{m} \oplus \hat{\Lambda}_\mathfrak{m}$ (free module of rank 2), but if $h = 3$, then let $H(\mathfrak{m}) = 0$.

By Theorem 15.6, there exists $M \in \operatorname{fingen}(\Lambda)$ such that $\hat{M}_\mathfrak{m} \cong H(\mathfrak{m})$ $\big(\forall \mathfrak{m} \in \operatorname{maxspec}(\Lambda)\big)$; and $M \in \operatorname{fingen}_\infty(\Lambda)$ because every $\hat{M}_\mathfrak{m}) \in \operatorname{fingen}_\infty(\hat{\Lambda}_\mathfrak{m})$ [Proposition 7.3]. The Λ-module M is indecomposable by Theorem 16.8, because its connections graph is as shown in (16.10.2) below, and $H(\mathfrak{m}_1)$ is indecomposable.

(16.10.2) Λ : 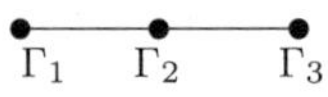

M : 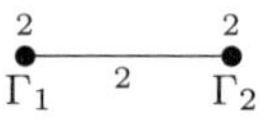

Finally, the number of indecomposable summands of $\hat{M}_\mathfrak{m}$, where $\mathfrak{m}$ runs through the ideals listed in (16.10.1), is 0, 1, 2, 3, 4 for $\mathfrak{m} = \mathfrak{m}_3, \mathfrak{m}_1, \mathfrak{n}, \mathfrak{q}, \mathfrak{p}$, respectively.

Conversely, if M is indecomposable, we show that the number of indecomposable summands of any completion $\hat{M}_\mathfrak{m}$ is at most 4. By Corollary 16.9, all ranks of M must be less than or equal to 2. Thus, if $\mathfrak{m}$ is unsplit or nonsingular, then the number of indecomposable summands of $\hat{M}_\mathfrak{m}$ must be less than or equal to 2. If $\mathfrak{m}$ is split, then the sum of the ranks at the two coordinate rings of $\hat{\Gamma}_\mathfrak{m}$ can be at most 4, and since each indecomposable summand of $\hat{M}_\mathfrak{m}$ must have nonzero rank for at least one of the coordinate rings, it follows that the number of indecomposable summands can be at most 4. □

If $\operatorname{maxspec}(\Lambda)$ is a finite set, the next theorem is well-known, and sufficiently obvious that we have never seen it stated anywhere.

THEOREM 16.11. *Let Ω be a reduced noetherian ring of Krull dimension 1, (e.g., any Dedekind-like ring), and let $M \in \text{fingen}(\Omega)$. Then there are only finitely many genera of direct summands of M.*

PROOF. Let the (finitely many) minimal prime ideals of Ω be $\mathfrak{p}_1, \ldots, \mathfrak{p}_n$. Since the ring Ω is reduced, each localization $\Omega_{\mathfrak{p}_h}$ is a field. For each index h, let r_h be the dimension of the $\Omega_{\mathfrak{p}_h}$-vector space $M_{\mathfrak{p}_h}$ (that is, the $\mathfrak{p}_h$-rank of M).

Let X be any direct summand of M. Then for each index h, the $\Omega_{\mathfrak{p}_h}$-vector space $X_{\mathfrak{p}_h}$ is a direct summand of $M_{\mathfrak{p}_h}$, so if we let ρ_h be the dimension of $X_{\mathfrak{p}_h}$, then $\rho_h \leq r_h$ for each index h. Thus, there are only finitely many choices for the n-tuple $(\rho_1, \ldots, \rho_n)$, and so it suffices to show that there are only finitely many genera of summands X of M with any fixed n-tuple of dimensions.

For almost all maximal ideals $\mathfrak{m}$, the localization $M_{\mathfrak{m}}$ must be $\Omega_{\mathfrak{m}}$-free. (When Ω is Dedekind-like, this is given in Lemma 8.2; for general reduced noetherian ring Ω of Krull dimension 1, see [**GL2**, 2.2].) Thus, as X is a summand of M, the localization $X_{\mathfrak{m}}$ must be projective and hence free for almost all maximal ideals $\mathfrak{m}$. Moreover, the rank of this free module must be the same as the $\Omega_{\mathfrak{p}_h}$-vector space dimension ρ_h of $X_{\mathfrak{p}_h}$ for any minimal prime $\mathfrak{p}_h$ contained in $\mathfrak{m}$. Therefore, the completion $\hat{X}_{\mathfrak{m}}$ is $\hat{\Omega}_{\mathfrak{m}}$-free of this same rank ρ_h. Thus, there is no choice at all for the isomorphism class of almost every $\hat{X}_{\mathfrak{m}}$ (once the isomorphism class of M and the ranks ρ_h are fixed). We can therefore restrict our consideration to one of the finitely many remaining nonfree $\hat{X}_{\mathfrak{m}}$.

Fix a decomposition of $\hat{M}_{\mathfrak{m}}$ as a direct sum of indecomposable modules. Since the Krull-Schmidt Theorem holds in the complete local case, any decomposition of $\hat{X}_{\mathfrak{m}}$ into a direct sum of indecomposable modules must be (up to isomorphisms and reordering of terms) a subsum of the fixed decomposition of $\hat{M}_{\mathfrak{m}}$. Therefore, there are only finitely many possibilities for the isomorphism class of $\hat{X}_{\mathfrak{m}}$, and hence there are only finitely many possibilities for the isomorphism class of $X_{\mathfrak{m}}$ [Lemma 9.1]. □

CHAPTER 4

Substitute for Conductor Squares

This chapter introduces our two tools, which we call "residue inclusions" and "separated covers", for viewing fingen(Γ) as an approximation to fingen(Λ). Let $S \in$ fingen(Λ), and suppose that S is contained in some finitely generated Γ-module (a condition always satisfied by the torsionfree modules studied in integral representation theory). Then the product $X = \Gamma S$ makes sense and is an element of fingen(Γ).

Let $\mathfrak{m} \in$ singspec(Λ). Then $\mathfrak{m}_\mathfrak{m}$ is an ideal of $\Gamma_\mathfrak{m}$ as well as $\Lambda_\mathfrak{m}$ and hence, as we have seen, the $\mathfrak{m}$-residue inclusion $\bar{S}(\mathfrak{m}) \subseteq \bar{X}(\mathfrak{m})$ is an inclusion of modules of finite length over artinian rings [Definition 11.2]. In the classical situation that ${}_\Lambda\Gamma$ is finitely generated, Λ and Γ themselves have a common "conductor ideal" C such that Λ/C and Γ/C are artinian rings, and hence $S/CS \subseteq X/CX$ is an inclusion of modules of finite length over the pair of artinian rings Λ/C and Γ/C; and the commutative square whose top and bottom rows are $S \subseteq C$ and $S/CS \subseteq X/CX$ is traditionally called a "conductor square" for S. Moreover, in this situation, singspec(Λ) is finite and the inclusion $(S/C)_\mathfrak{m} \subseteq (X/C)_\mathfrak{m}$ can be proper only for this finite number of $\mathfrak{m}$.

In place of this single bottom row, we use the (possibly infinite) family of residue inclusions $\bar{S}(\mathfrak{m}) \subseteq \bar{X}(\mathfrak{m})$ $\big(\mathfrak{m} \in \text{singspec}(\Lambda)\big)$. What makes this tractable is that, for fixed S, almost all residue inclusions $\bar{S}(\mathfrak{m}) \subseteq \bar{X}(\mathfrak{m})$ turn out to be trivial — identity maps, in our setting. But, unlike the classical setting, the finite set of $\mathfrak{m}$ that can yield nontrivial inclusions depends upon S.

The details of the above discussion are the subject of Section 17.

Our second tool arises from the fact that not every $M \in$ fingen(Λ) is contained in a Γ-module, because we do not restrict ourselves to studying torsionfree Λ-modules. Our tool for dealing with this is a "best approximation" to M by means of a Λ-submodule of some Γ-module. This approximation, called a "separated cover" of M, is the subject of Section 18.

17. Residue Inclusions of Separated Modules

Definitions 17.1 (Separated module, ($\Delta \subseteq \Upsilon$)-inclusion). Let $\Delta \subseteq \Upsilon$ be an inclusion of rings. We say that a Δ-module S is Υ-*separated* if S is a Δ-submodule of some Υ-module X'. This refines (by explicitly mentioning Υ) the definition of "separated" given in [**KL2**, §4]. In this situation we have $S \subseteq X = \Upsilon S$, the Υ-submodule of X' generated by S.

We define a ($\Delta \subseteq \Upsilon$)-*inclusion* to be an inclusion ($S \subseteq X = \Upsilon S$) of finitely generated modules over the rings Δ and Υ respectively. We call such an inclusion *standard* if the canonical map $\Upsilon \otimes_\Delta S \to \Upsilon S = X$ is an isomorphism of Υ-modules, and we write this as $X = \Upsilon S = \Upsilon \otimes_\Delta S$.

Let $(S \subseteq X)$ and $(S' \subseteq X')$ be $(\Delta \subseteq \Upsilon)$-inclusions. We say that $(S \subseteq X) \cong (S' \subseteq X')$ if some Υ-isomorphism $X \cong X'$ carries S onto S'.

The next lemma shows that every Υ-separated Δ-module is part of a standard $(\Delta \subseteq \Upsilon)$-inclusion. (For an example of a nonstandard inclusion involving Dedekind-like rings, see Example 17.11.)

LEMMA 17.2. *Let $\Delta \subseteq \Upsilon$ be an inclusion of rings, and $S \in \text{fingen}(\Delta)$ be Υ-separated. Then:*

(i) *The canonical map $S \to \Upsilon \otimes_\Delta S$ is injective.*
(ii) *If we identify S with its image in $X = \Upsilon \otimes_\Delta S$, then $(S \subseteq X)$ is a standard $(\Delta \subseteq \Upsilon)$-inclusion.*
(iii) *Let $(S \subseteq X)$ and $(S' \subseteq X')$ be standard $(\Delta \subseteq \Upsilon)$-inclusions. Then ${}_\Delta S \cong {}_\Delta S' \iff (S \subseteq X) \cong (S' \subseteq X')$.*

PROOF. (i) This map is injective because following it by the canonical map $\Upsilon \otimes_\Delta S \to \Upsilon S$ yields the identity map on S.

(ii) This follows immediately from (i).

(iii) To obtain the less trivial implication $(\Rightarrow)$, let f be any Δ-isomorphism $S \cong S'$, and note that $1 \otimes f$ is then the required Υ-isomorphism $X \cong X'$ that takes S onto S'. □

Now we return to our Dedekind-like ring Λ.

REMARK 17.3. *For a Γ-separated Λ-module S, the Γ-isomorphism class of $X = \Gamma S$ is not always determined by that of S.* For example, let Λ be local and let $(S \subseteq X')$ be the nonstandard $(\Lambda \subseteq \Gamma)$-inclusion in Example 17.11. In this example, S is an ideal of Λ and $X' = \Gamma S$, the product in Γ. Let $(S \subseteq X)$ be the standard inclusion given by Lemma 17.2(ii). We claim that $X' \not\cong X$. Since the canonical surjection $\tau \colon \Gamma \otimes_\Lambda S \twoheadrightarrow \Gamma S$ is not an isomorphism, we have $\ker(\tau) \neq 0$; and this kernel has finite length by (17.8.2). Thus X has a nonzero Γ-submodule of finite length, while the ideal X' of Γ does not, and hence $X' \not\cong X$.

Let $(S \subseteq X)$ be a $(\Lambda \subseteq \Gamma)$-inclusion. For $\mathfrak{m} \in \text{singspec}(\Lambda)$, recall that the "$\mathfrak{m}$-residue inclusion" of $(S \subseteq X)$ is the map $\bar{S}(\mathfrak{m}) = S/(\mathfrak{m}S) \to X/(\mathfrak{m}X) = \bar{X}(\mathfrak{m})$. It is an injection since Λ is Dedekind-like [Lemma 11.1], and we usually write it as actual inclusion: $\bar{S}(\mathfrak{m}) \subseteq \bar{X}(\mathfrak{m})$ [Definition 11.2]. When referring to residue inclusions, we always assume that $\mathfrak{m} \in \text{singspec}(\Lambda)$, because for nonsingular $\mathfrak{m}$, we have $\Lambda_\mathfrak{m} = \Gamma_\mathfrak{m}$, and hence $\bar{S}(\mathfrak{m}) = \bar{\Gamma}(\mathfrak{m})\bar{S}(\mathfrak{m}) = \bar{X}(\mathfrak{m})$.

THEOREM 17.4. *Let $(S \subseteq X)$ be a $(\Lambda \subseteq \Gamma)$-inclusion. Then $(S \subseteq X)$ is the pullback of its residue inclusions; that is,*

$$S = \{x \in X \mid \bar{x}(\mathfrak{m}) \in \bar{S}(\mathfrak{m}) \ \big(\forall \mathfrak{m} \in \text{singspec}(\Lambda)\big)\} \tag{17.4.1}$$

where $\bar{x}(\mathfrak{m})$ denotes the natural image of x in $\bar{X}(\mathfrak{m})$.

PROOF. The inclusion $(\subseteq)$ is clear. For the opposite inclusion, choose any x in the pullback (i.e. on the right-hand side). It suffices to show that $x \in S_\mathfrak{m}$ for all $\mathfrak{m} \in \text{maxspec}(\Lambda)$, where $S_\mathfrak{m}$ is viewed as a $\Lambda_\mathfrak{m}$-submodule of $X_\mathfrak{m}$. Thus, after localizing at $\mathfrak{m}$ and remembering that the $\mathfrak{m}$-residue inclusion is not changed by $\mathfrak{m}$-localization [see (11.2.1)], we may assume that $(\Lambda, \mathfrak{m}, k)$ is a local Dedekind-like ring with normalization Γ [Remarks (5.3)(i)].

Now (17.4.1) consists of only one residue inclusion, and it holds since (when Λ is local) the ideal $\mathfrak{m}$ of Λ is also an ideal of Γ. □

LEMMA 17.5. *The following conditions are equivalent for any* $(\Lambda \subseteq \Gamma)$*-inclusion* $(S \subseteq X)$.

(i) *The* $(\Lambda \subseteq \Gamma)$*-inclusion* $(S \subseteq X)$ *is standard.*

(ii) *For each* $\mathfrak{m} \in \operatorname{singspec}(\Lambda)$ *there is an integer* $d(\mathfrak{m})$ *such that* $\bar{S}(\mathfrak{m}) \cong k(\mathfrak{m})^{d(\mathfrak{m})}$ *and* $\bar{X}(\mathfrak{m}) \cong \bar{\Gamma}^{d(\mathfrak{m})}$ *as modules over* $k(\mathfrak{m})$ *and* $\bar{\Gamma}(\mathfrak{m})$*, respectively.*

PROOF. If Λ is a DVR, the lemma is obvious, while if Λ is local but not a DVR, the lemma is [**KL2**, 5.2]. Moreover, the non-local case of the lemma follows immediately from the local case because the canonical map $\Gamma \otimes_\Lambda S \to \Gamma S$ is an isomorphism whenever all of its localizations at maximal ideals of Λ are isomorphisms. □

Reversing our point of view, we now want to construct an arbitrary Γ-separated $S \in \operatorname{fingen}(\Lambda)$, by specifying an $X \in \operatorname{fingen}(\Gamma)$ together with a collection of residue inclusions. The next theorem gives the basic finiteness restriction which must be satisfied by the collection of residue inclusions. The succeeding theorem shows that this finiteness restriction, plus one more obvious condition, is the full set of constraints needed to define S in this way. Recall that (a$\forall\mathfrak{m}$) means "for almost all $\mathfrak{m}$."

THEOREM 17.6. *Let* $(S \subseteq X)$ *and* $(T \subseteq X)$ *be* $(\Lambda \subseteq \Gamma)$*-inclusions. Then* $\bar{S}(\mathfrak{m}) = \bar{T}(\mathfrak{m})$ *as submodules of* $\bar{X}(\mathfrak{m})$ $\big(\text{a}\forall\mathfrak{m} \in \operatorname{singspec}(\Lambda)\big)$.

PROOF. Since $\Gamma S = X = \Gamma T$ and $\Gamma_Q = \Lambda_Q$, we have $S_Q = X_Q = T_Q$. Therefore finite generation of S and T implies that $S_\mathfrak{m} = T_\mathfrak{m}$ in $X_\mathfrak{m}$ (a$\forall\mathfrak{m}$), by Lemma 8.2(iii), and hence $\bar{S}(\mathfrak{m}) = \bar{T}(\mathfrak{m})$ (a$\forall\mathfrak{m}$), by (11.2.1). □

THEOREM 17.7. *Let* $(S \subseteq X)$ *be a* $(\Lambda \subseteq \Gamma)$*-inclusion. For each* $\mathfrak{m} \in \operatorname{singspec}(\Lambda)$ *let* $V(\mathfrak{m})$ *be a* $k(\mathfrak{m})$*-subspace of* $\bar{X}(\mathfrak{m})$ *such that*

(i) $V(\mathfrak{m}) = \bar{S}(\mathfrak{m})$ (a$\forall\mathfrak{m}$), *and*

(ii) $\bar{\Gamma}(\mathfrak{m})V(\mathfrak{m}) = \bar{X}(\mathfrak{m})$ ($\forall\mathfrak{m}$).

Then there is a unique $(\Lambda \subseteq \Gamma)$*-inclusion* $(T \subseteq X)$ *such that* $\bar{T}(\mathfrak{m}) = V(\mathfrak{m})$ $\big(\forall\mathfrak{m} \in \operatorname{singspec}(\Lambda)\big)$.

Suppose also that $(S \subseteq X)$ *is standard. Then* $(T \subseteq X)$ *is standard if and only if*

$$V(\mathfrak{m}) \cong \bar{S}(\mathfrak{m}) \quad \text{as } k(\mathfrak{m})\text{-vector spaces} \quad \big(\forall\mathfrak{m} \in \operatorname{singspec}(\Lambda)\big) \tag{17.7.1}$$

PROOF. Uniqueness holds because T is the pullback of its residue inclusions [Theorem 17.4].

Existence. Recall that $\rho^\mathfrak{m}\colon Y \twoheadrightarrow \bar{Y}(\mathfrak{m})$ denotes the natural homomorphism for all Λ-modules Y, and we usually make the identification $Y_\mathfrak{m}/\mathfrak{m}_\mathfrak{m} Y_\mathfrak{m} = \bar{Y}(\mathfrak{m})$. For each $\mathfrak{m} \in \operatorname{singspec}(\Lambda)$, let $T(\mathfrak{m})$ be the pullback of the diagram

$$\begin{array}{ccc} T(\mathfrak{m}) & \subseteq & X_\mathfrak{m} \\ \Big\downarrow{\scriptstyle \rho^\mathfrak{m}} & & \Big\downarrow{\scriptstyle \rho^\mathfrak{m}} \\ V(\mathfrak{m}) & \subseteq & \bar{X}(\mathfrak{m}) \end{array} \tag{17.7.2}$$

That is, $T(\mathfrak{m}) = \{x \in X_\mathfrak{m} \mid \rho^\mathfrak{m}(x) \in V(\mathfrak{m})\}$. Then $T(\mathfrak{m})$ is a finitely generated $\Lambda_\mathfrak{m}$-submodule of $X_\mathfrak{m}$, and $\bar{T}(\mathfrak{m}) = V(\mathfrak{m})$.

For each maximal ideal $\mathfrak{m} \notin \operatorname{singspec}(\Lambda)$, let $T(\mathfrak{m}) = X(\mathfrak{m}) = S(\mathfrak{m})$. Then by (11.2.1) it suffices to find a finitely generated Λ-submodule $T \subseteq X$ such that

$T_{\mathfrak{m}} = T(\mathfrak{m})$ in $X_{\mathfrak{m}}$ $\big(\forall \mathfrak{m} \in \operatorname{maxspec}(\Lambda)\big)$. The existence of such a Λ-module T follows from Theorem 8.5, once we show: 8.5(i) $T(\mathfrak{m}) = S_{\mathfrak{m}}$ (a$\forall\mathfrak{m}$), and 8.5(ii) $T(\mathfrak{m})_Q = (S_{\mathfrak{m}})_Q$ ($\forall\mathfrak{m}$). By our definition of $T(\mathfrak{m})$, both conditions hold when $\mathfrak{m} \notin \operatorname{singspec}(\Lambda)$, so we may assume that $\mathfrak{m} \in \operatorname{singspec}(\Lambda)$.

Since (17.7.2) is a pullback diagram, it suffices to verify 8.5(i) modulo $\mathfrak{m}$, but this holds by hypothesis (i) of the present theorem.

To verify 8.5(ii), tensor pullback diagram (17.7.2) by the flat Λ-module Λ_Q, noting that the diagram remains a pullback diagram, and both entries in the bottom row become zero. Therefore, $T(\mathfrak{m})_Q = (X_{\mathfrak{m}})_Q$. On the other hand, $(S \subseteq X)$ is a $(\Lambda \subseteq \Gamma)$-inclusion, and hence $\Gamma S = X$. Since $(\Lambda_{\mathfrak{m}})_Q = (\Gamma_{\mathfrak{m}})_Q$, 8.5(ii) follows.

Supplementary statement. Let τ be the canonical map $\Gamma \otimes_\Lambda T \to \Gamma T$. We want to show that τ is an isomorphism if and only if (17.7.1) holds. We do this by showing that (17.7.1) holds if and only if the isomorphisms in Lemma 17.5(ii) hold. Note that these isomorphisms can be verified one maximal ideal at a time. Therefore, fix a maximal ideal $\mathfrak{m} \in \operatorname{singspec}(\Lambda)$, and let σ be the canonical map $\Gamma \otimes_\Lambda S \to \Gamma S$. By hypothesis, σ is an isomorphism, and therefore Lemma 17.5(ii) shows that $\bar{S}(\mathfrak{m}) \cong k(\mathfrak{m})^{(n)}$ and $\bar{X}(\mathfrak{m}) \cong \bar{\Gamma}(\mathfrak{m})^{(n)}$ for some n (depending on $\mathfrak{m}$). Because $\bar{T}(\mathfrak{m}) \cong V(\mathfrak{m})$, we see that $V(\mathfrak{m}) \cong \bar{S}(\mathfrak{m})$ if and only if $\bar{T}(\mathfrak{m}) \cong k(\mathfrak{m})^{(n)}$ as desired. □

THEOREM 17.8 (Almost always standard). *Let $(S \subseteq X)$ be a $(\Lambda \subseteq \Gamma)$-inclusion. Then the $(\bar{\Lambda}(\mathfrak{m}) \subseteq \bar{\Gamma}(\mathfrak{m}))$-inclusion $(\bar{S}(\mathfrak{m}) \subseteq \bar{X}(\mathfrak{m}))$ is standard* $\big(\text{a}\forall\mathfrak{m} \in \operatorname{maxspec}(\Lambda)\big)$.

PROOF. It suffices to prove

(17.8.1) The $(\Lambda_{\mathfrak{m}} \subseteq \Gamma_{\mathfrak{m}})$-inclusion $(S_{\mathfrak{m}} \subseteq X_{\mathfrak{m}})$ is standard $\big(\text{a}\forall\mathfrak{m} \in \operatorname{maxspec}(\Lambda)\big)$

since reducing modulo $\mathfrak{m}_{\mathfrak{m}}$ then gives the inclusion $(\bar{S}(\mathfrak{m}) \subseteq \bar{X}(\mathfrak{m}))$ and shows that it is standard.

First we prove a lemma. Let $\tau\colon \Gamma \otimes_\Lambda S \twoheadrightarrow \Gamma S = X$ be the natural surjection. We need to prove that $\ker(\tau_{\mathfrak{m}}) = 0$ (a$\forall\mathfrak{m}$). We claim:

(17.8.2) If Λ is local then $\ker(\tau)$ is a Λ-module of finite length.

Since the ring $\Gamma/\mathfrak{m}$ is artinian and the Γ-module X is finitely generated, it suffices to prove that $\mathfrak{m}\ker(\tau) = 0$. Choose elements $m \in \mathfrak{m}$ and $z = \sum_i \gamma_i \otimes s_i \in \ker(\tau)$, so that $\sum_i \gamma_i s_i = 0$. Since the maximal ideal of a local Dedekind-like ring is also an ideal of its normalization, we have $m\gamma_i \in \Lambda$ for each index i, and therefore

$$mz = \sum_i (m\gamma_i) \otimes s_i = 1 \otimes \sum_i m\gamma_i s_i = 1 \otimes m \sum_i \gamma_i s_i = 0$$

proving the claim.

Now return to the general, non-local Λ. We prove (17.8.1). Recall that $\ker(\tau_{\mathfrak{m}})$ is a free $\Lambda_{\mathfrak{m}}$-module (a$\forall\mathfrak{m}$) [Lemma 8.2(ii)], and at these maximal ideals has no nonzero submodules of finite length, since $\Lambda_{\mathfrak{m}}$ is Dedekind-like. This, together with the claim, completes the proof since Γ/Λ — like all Λ-modules — is contained in the full direct product of its localizations at maximal ideals. □

We have seen that any two inclusions of the form $(S \subseteq X)$ and $(T \subseteq X)$ have the same $\mathfrak{m}$-residue inclusions for almost all $\mathfrak{m}$ [Theorem 17.6] and almost all of these repeated residue inclusions are standard [Theorem 17.8]. The next result shows exactly what standard residue inclusions look like. A consequence of this simple form is that standardness of any $(\Lambda \subseteq \Gamma)$-inclusion is determined by that of its residue inclusions.

PROPOSITION 17.9. (i) *A* $\big(k(\mathfrak{m}) \subseteq \bar{\Gamma}(\mathfrak{m})\big)$*-inclusion* $\big(\bar{S}(\mathfrak{m}) \subseteq \bar{X}(\mathfrak{m})\big)$ *is standard if and only if it is isomorphic to a direct sum of copies of* $\big(k(\mathfrak{m}) \subseteq \bar{\Gamma}(\mathfrak{m})\big)$.

(ii) *A* $(\Lambda \subseteq \Gamma)$*-inclusion* $(S \subseteq X)$ *is standard if (and only if) the induced* $\big(k(\mathfrak{m}) \subseteq \bar{\Gamma}(\mathfrak{m})\big)$*-inclusion* $\big(\bar{S}(\mathfrak{m}) \subseteq \bar{X}(\mathfrak{m})\big)$ *is standard* $\big(\forall \mathfrak{m} \in \operatorname{singspec}(\Lambda)\big)$.

PROOF. (i) Since only a single $\mathfrak{m}$ is involved here, we simplify the notation by omitting it, writing the original inclusion of rings as $(k \subseteq \bar{\Gamma})$ and the module-inclusion as $(\bar{S} \subseteq \bar{X})$. To prove the nontrivial implication, assume that $(\bar{S} \subseteq \bar{X})$ is standard. The fact that $k = \bar{\Lambda}$ is a field greatly simplifies everything here.

Choose a decomposition $\bar{S} = \oplus_{i=1}^{d} V_i$ where each V_i is a 1-dimensional subspace of the k-vector space $\bar{S}$. Then standardness of the $(k \subseteq \bar{\Gamma})$-inclusion $(\bar{S} \subseteq \bar{X})$ yields the decomposition $\bar{X} = \oplus_{i=1}^{d} W_i$, where each $W_i = \bar{\Gamma}{\cdot}V_i = \bar{\Gamma} \otimes_k V_i$. It now suffices to prove the following: For each i there is a $\bar{\Gamma}$-isomorphism $g_i\colon W_i \cong \bar{\Gamma}$ that takes each V_i onto k.

Since each V_i is 1-dimensional over k there is a k-isomorphism $f_i\colon V_i \cong k$. Then $g_i = 1 \otimes f_i$, the $\bar{\Gamma}$-isomorphism $W_i \cong \bar{\Gamma} \otimes_k k = \bar{\Gamma}$ completes the proof.

(ii) To prove the nontrivial implication, suppose that every $\big(\bar{S}(\mathfrak{m}) \subseteq \bar{X}(\mathfrak{m})\big)$ is standard; and hence by statement (i), isomorphic to $\big(k(\mathfrak{m}) \subseteq \bar{\Gamma}(\mathfrak{m})\big)^{d(\mathfrak{m})}$ for some $d(\mathfrak{m})$. Standardness of $(S \subseteq X)$ now follows immediately from Lemma 17.5. □

THEOREM 17.10. *Let* $X \in \operatorname{fingen}(\Gamma)$.

(i) *If there exists a standard* $(\Lambda \subseteq \Gamma)$*-inclusion,* $(S \subseteq X)$ *then* $\bar{X}(\mathfrak{m})$ *is a free* $\bar{\Gamma}(\mathfrak{m})$*-module for every* $\mathfrak{m} \in \operatorname{singspec}(\Lambda)$.

(ii) *If* $\bar{X}(\mathfrak{m})$ *is a free* $\bar{\Gamma}(\mathfrak{m})$*-module for every split* $\mathfrak{m} \in \operatorname{singspec}(\Lambda)$*, then there exists a a standard* $(\Lambda \subseteq \Gamma)$*-inclusion of the form* $(S \subseteq X)$.

PROOF. *Complete local case.* If Λ is a DVR, the theorem is trivially true, while if Λ is complete local but not a DVR (hence unsplit or strictly split), the lemma is proved in [**KL2**, 5.4, and the remark after it].

General case. Statement (i) follows immediately from the complete local case.

Statement (ii).We first build a $(\Lambda \subseteq \Gamma)$-inclusion $(S \subseteq X)$ that "almost" works, and then refine this first approximation to obtain an inclusion which we show is standard. To verify that the inclusion we build is standard, it is necessary and sufficient to check that, for every $\mathfrak{m} \in \operatorname{singspec}(\Lambda)$, there exists $d(\mathfrak{m})$ such that the following condition holds [Lemma 17.5].

$$\bar{S}(\mathfrak{m}) \cong k(\mathfrak{m})^{d(\mathfrak{m})} \quad \text{and} \quad \bar{X}(\mathfrak{m}) \cong \bar{\Gamma}^{d(\mathfrak{m})} \tag{17.10.1}$$

For our first approximation to S, choose any finite set of Γ-generators of X, and let S be the Λ-submodule of X which they generate. Then $\Gamma S = X$, so that $(S \subseteq X)$ is a $(\Lambda \subseteq \Gamma)$-inclusion. Like all $(\Lambda \subseteq \Gamma)$-inclusions, almost all residue inclusions of $(S \subseteq X)$ are standard [Theorem 17.8]. Therefore (17.10.1) holds for *almost all* $\mathfrak{m} \in \operatorname{singspec}(\Lambda)$. Let $\mathcal{F}$ be the finite set of maximal ideals in $\operatorname{singspec}(\Lambda)$ at which (17.10.1) fails. To complete the proof, we need to change finitely many residue inclusions in such a way that the resulting inclusion $(S \subseteq X)$ satisfies (17.10.1) for all $\mathfrak{m} \in \mathcal{F}$.

Fix some $\mathfrak{m} \in \mathcal{F}$. Note that, if $\mathfrak{m}$ is split, then $\bar{X}(\mathfrak{m})$ is $\bar{\Gamma}(\mathfrak{m})$-free by hypothesis, while if $\mathfrak{m}$ is unsplit, then all modules over the field $\bar{\Gamma}(\mathfrak{m}) = F(\mathfrak{m})$ are free. Thus, in both cases, we have $\bar{X}(\mathfrak{m}) \cong \bar{\Gamma}(\mathfrak{m})^{d(\mathfrak{m})}$ for some integer $d(\mathfrak{m})$. Choose any $d(\mathfrak{m})$-element basis of the free $\bar{\Gamma}(\mathfrak{m})$-module $\bar{X}(\mathfrak{m})$, and let $V(\mathfrak{m})$ be the $k(\mathfrak{m})$-vector subspace of $\bar{X}(\mathfrak{m})$ which they generate. Then $V(\mathfrak{m}) \cong k(\mathfrak{m})^{d(\mathfrak{m})}$, and $\bar{\Gamma}(\mathfrak{m})V(\mathfrak{m}) = \bar{X}(\mathfrak{m})$. Therefore, by Theorem 17.7, we can find a new S whose $\mathfrak{m}$-residue inclusion is $\big(V(\mathfrak{m}) \subseteq \bar{X}(\mathfrak{m})\big)$, and whose residue inclusions for all other maximal ideals in singspec(Λ) remain unchanged. Making this adjustment for each $\mathfrak{m} \in \mathcal{F}$ completes the proof. □

EXAMPLE 17.11 ($\Gamma S \neq \Gamma \otimes_\Lambda S$, nonstandard $(\Lambda \subseteq \Gamma)$-inclusion). Let Λ be any local Dedekind-like ring such that $\Lambda \neq \Gamma$. Then Λ is not a principal ideal ring. Let S be any nonprincipal ideal of Λ, and let $X = \Gamma S$ (product in Γ). We claim that $\Gamma S \neq \Gamma \otimes_\Lambda S$, *and hence the* $(\Lambda \subseteq \Gamma)$*-inclusion* $(S \subseteq X)$ *is not standard.*

Since the ideal S is nonprincipal and Λ is local, we have $\bar{S}(\mathfrak{m}) \cong k(\mathfrak{m})^{(d)}$ for some $d \geq 2$, by Nakayama's lemma (actually $d = 2$, by Proposition 10.9). Therefore, $\bar{\Gamma}(\mathfrak{m}) \otimes_{k(\mathfrak{m})} \bar{S}(\mathfrak{m}) \cong \bar{\Gamma}(\mathfrak{m})^{(d)}$ as Γ-modules. Hence it suffices to show that $\bar{X}(\mathfrak{m}) \cong \bar{\Gamma}(\mathfrak{m})$ [Lemma 17.5]. But this holds because Γ is a principal ideal ring, since Λ is local. □

18. Separated Covers

We begin this section by showing that all module-finite overrings Ω of Λ in Λ_Q are obtained by "desingularizing" finitely many $\mathfrak{m} \in$ singspec(Λ). We need this for making the results on separated covers in [**KL2**, §4] available for use in studying Γ-separated covers of Λ-modules when ${}_\Lambda\Gamma$ is not finitely generated.

NOTATION 18.1 ($\Omega(\mathcal{F})$). Let $\mathcal{F}$ be a finite subset of singspec(Λ), and define a finitely generated Λ-submodule $\Omega = \Omega(\mathcal{F})$ of Γ_Q by

$$\Omega_{\mathfrak{m}} = \begin{cases} \Gamma_{\mathfrak{m}} & (\text{in } \Gamma_{\mathfrak{m}}) \quad \text{if } \mathfrak{m} \in \mathcal{F} \\ \Lambda_{\mathfrak{m}} & (\text{in } \Gamma_{\mathfrak{m}}) \quad \text{if } \mathfrak{m} \in \text{maxspec}(\Lambda) - \mathcal{F} \end{cases} \tag{18.1.1}$$

This Ω is obviously unique if it exists. It exists because we are increasing the size of only finitely many localizations of Λ at maximal ideals and not changing Λ_Q. [See a more precise statement of this reason in Theorem 8.5]. It is easily checked, locally, that Ω is a ring and $\Lambda \subseteq \Omega \subseteq \Gamma$.

Alternatively, $\Omega = \{x \in \Gamma \mid x\mathfrak{m} \subseteq \mathfrak{m} \ (\forall \mathfrak{m} \in \mathcal{F})\}$, as is easily checked locally.

PROPOSITION 18.2. *For every finite subset* $\mathcal{F} \subseteq$ singspec(Λ), *the ring* $\Omega(\mathcal{F})$ *is Dedekind-like. Conversely, if* Ω *is a ring module-finite over* Λ *such that* $\Lambda \subseteq \Omega \subseteq \Gamma$, *then* $\Omega = \Omega(\mathcal{F})$ *for some unique finite subset* $\mathcal{F} \subseteq$ singspec(Λ), *and* Ω *is Dedekind-like.*

PROOF. If $\Omega = \Omega(\mathcal{F})$ for some finite subset $\mathcal{F} \subseteq$ singspec(Λ), then ${}_\Lambda\Omega$ is finitely generated, and hence the ring Ω is noetherian. Moreover, Ω is locally Dedekind-like, by (18.1.1), the fact that there exist no rings strictly between $\Lambda_{\mathfrak{m}}$ and $\Gamma_{\mathfrak{m}}$, and the lying-over theorem for prime ideals. Therefore Ω itself is Dedekind-like [Corollary 10.7].

Before proceeding to the converse, recall that, if Λ is local and $\Lambda \neq \Gamma$, then Γ/Λ is a simple Λ-module [Definition 10.1]. Therefore, if Λ is local, there are no rings strictly between Λ and Γ.

Now let Λ be an arbitrary Dedekind-like ring, and let Ω be a ring module-finite over Λ such that $\Lambda \subseteq \Omega \subseteq \Gamma$. Then for each maximal ideal $\mathfrak{m}$, we have $\Lambda_{\mathfrak{m}} \subseteq \Omega_{\mathfrak{m}} \subseteq \Gamma_{\mathfrak{m}}$, so that by the local case, $\Omega_{\mathfrak{m}}$ must equal $\Lambda_{\mathfrak{m}}$ or $\Gamma_{\mathfrak{m}}$. Since ${}_{\Lambda}\Lambda$ and ${}_{\Lambda}\Omega$ are finitely generated, there can be only finitely many maximal ideals $\mathfrak{m}$ such that $\Lambda_{\mathfrak{m}} \neq \Omega_{\mathfrak{m}}$ in $\Gamma_{\mathfrak{m}}$ [Proposition 10.11]. For this finite set $\mathcal{F}$ of maximal ideals, we have $\Omega = \Omega(\mathcal{F})$ by Theorem 8.5. □

Finally, we determine the conductor-square structure of the rings $\Omega(\mathcal{F})$. Since the definition of "conductor ideal" is not completely standard in the literature, we state the definition we use. Consider a ring Ω such that $\Lambda \subseteq \Omega \subseteq \Gamma$. We define *a conductor ideal* for Λ and Ω to be any common ideal C of Λ and Ω such that C contains a unit of Λ_Q. We say that C is *the* conductor ideal for Λ and Ω if it is the largest conductor ideal for this pair of rings.

PROPOSITION 18.3. *Let $\Omega = \Omega(\mathcal{F})$ for some finite set $\mathcal{F} \subseteq \operatorname{singspec}(\Lambda)$. Then Λ has an ideal C such that, for each maximal ideal $\mathfrak{m}$,*

$$C_{\mathfrak{m}} = \begin{cases} \mathfrak{m}_{\mathfrak{m}} & (\text{in } \Lambda_{\mathfrak{m}}) \quad \text{if } \mathfrak{m} \in \mathcal{F} \\ \Lambda_{\mathfrak{m}} & (\text{in } \Lambda_{\mathfrak{m}}) \quad \text{otherwise.} \end{cases} \tag{18.3.1}$$

Moreover:

(i) *C is "the" conductor ideal for Λ and Ω; that is, C is the largest common ideal of Λ and Ω, and it contains a unit of Λ_Q.*
(ii) *Λ is the pullback of the conductor square*

$$\begin{array}{ccc} \Lambda & \subseteq & \Omega \\ \downarrow & & \downarrow \ (\ker = C) \\ \Lambda/C & \subseteq & \Omega/C \end{array} \tag{18.3.2}$$

(iii) *The inclusion $\Lambda/C \subseteq \Omega/C$ can be identified with the direct sum of inclusions of rings $\oplus_{\mathfrak{m}\in\mathcal{F}} \big(k(\mathfrak{m}) \subset \bar{\Gamma}(\mathfrak{m})\big)$.*

PROOF. As a first approximation to C, we use Λ itself, and then we use Theorem 8.5 to change $\Lambda_{\mathfrak{m}}$ to $\mathfrak{m}_{\mathfrak{m}}$ for the finitely many $\mathfrak{m} \in \mathcal{F}$. To verify that Theorem 8.5 applies, we need to check that each $(\mathfrak{m}_{\mathfrak{m}})_Q = (\Lambda_{\mathfrak{m}})_Q$ in $(\Lambda_{\mathfrak{m}})_Q$. Since $(\Lambda_{\mathfrak{m}})_Q$ is the total quotient ring of $\Lambda_{\mathfrak{m}}$ [Lemma and Definition 4.3], it suffices to check that $\mathfrak{m}_{\mathfrak{m}}$ contains a regular element of $\Lambda_{\mathfrak{m}}$ when $\mathfrak{m} \in \mathcal{F}$. This, in turn, holds because $\Lambda_{\mathfrak{m}}$ is a local Dedekind-like ring, and its maximal ideal $\mathfrak{m}_{\mathfrak{m}}$ therefore equals the Jacobson radical of its semi-local normalization $\Gamma_{\mathfrak{m}}$, which has no maximal ideals of height 0.

The fact that C is an ideal of Ω, and indeed the conductor from Ω into Λ, is easily checked locally at all maximal ideals, so that statement (i) holds. Statement (ii) follows because C is an ideal of both rings Λ and Ω.

(iii) First we claim that the Λ-module Λ/C has finite length. Since every Λ-module is contained in the full direct product of its localizations at all maximal ideals, it suffices to note that $(\Lambda/C)_{\mathfrak{m}}$ equals zero except for the finitely many $\mathfrak{m} \in \mathcal{F}$, at which $(\Lambda/C)_{\mathfrak{m}}$ is isomorphic to the simple Λ-module $\Lambda_{\mathfrak{m}}/\mathfrak{m}_{\mathfrak{m}} \cong \Lambda/\mathfrak{m} = k(\mathfrak{m})$.

Now since Λ/C has finite length, it is canonically isomorphic to the direct sum of its localizations at all maximal ideals of Λ. Thus, by the previous paragraph, $\Lambda/C = \oplus_{\mathfrak{m}\in\mathcal{F}}\, k(\mathfrak{m})$.

Finally, Ω/C has finite length as a Λ-module, because it is a finitely generated module over the artinian ring Λ/C. Therefore, as above, Ω/C is canonically isomorphic to the direct sum of its localizations at maximal ideals of Λ [Corollary 6.4]. By the definition of $\Omega(\mathcal{F})$, we conclude that $\Omega/C = \oplus_{\mathfrak{m}\in\mathcal{F}} \bar{\Gamma}(\mathfrak{m})$, as desired. □

We now introduce separated covers, our "best approximation" to Λ-modules by Λ-submodules of Γ-modules. More precisely, we extend the basic results and terminology developed in [**KL2**, §4], so that they cover the case in which Λ and Γ have no common conductor ideal — which happens when Λ has infinitely many singular maximal ideals. The key to handling this extension consists of direct-limit arguments: Since Γ is integral over Λ, every finite set of elements of Γ is contained in a ring Ω that is module-finite over Λ. Therefore, a conductor ideal exists for Λ and Ω [Proposition 18.3], a requirement for applying the results in [**KL2**, §4].

DEFINITION 18.4 (Separated cover). Let Υ be a ring such that $\Lambda \subseteq \Upsilon$, and recall [Definitions 17.1] that an Υ-separated Λ-module is a Λ-module that is contained in some Υ-module. An *Υ-separated cover of a Λ-module M* is a surjective Λ-module homomorphism $\phi\colon S \twoheadrightarrow M$, in which S is an Υ-separated Λ-module, and S is "as close as possible to M" in the following sense: If $S \twoheadrightarrow S' \twoheadrightarrow M$ is any factorization of ϕ, in which S' is also Υ-separated, then $S \twoheadrightarrow S'$ is one-to-one (and therefore the image of S in S' is no closer to M than is S).

This is the same terminology used in [**KL2**, §4], except that we now explicitly mention the secondary ring Υ (called Γ in that paper). The two situations in which we need these notions are when ${}_{\Lambda}\Upsilon$ is finitely generated (in which case we write Ω for Υ), and when Υ is the normalization Γ of Λ. The following fact is so fundamental to the rest of this paper that we repeat its simple proof from [**KL2**, 4.7].

(18.4.1) For every ring Υ such that $\Lambda \subseteq \Upsilon$, every $M \in \text{fingen}(\Lambda)$ has an Υ-separated cover $S \twoheadrightarrow M$ in which ${}_{\Lambda}S$ is finitely generated.

Proof. Let $f\colon F \twoheadrightarrow M$ be a map of a free (therefore Υ-separated) Λ-module of finite rank onto M. Since Λ is noetherian, there is a submodule A of $\ker(f)$ that is maximal with respect to the property that F/A is Υ-separated. Then the induced map $F/A \twoheadrightarrow M$ is obviously an Υ-separated cover. □

The present series of papers has relatively little to say about separated covers of infinitely generated modules. See [**KL2**, Remarks 4.8(ii)] and the rest of the present section. In particular, we have nothing to say about when such covers exist. See [**LT**] (in preparation) for more about separated covers of infinitely generated modules, and for the relationship between separated covers and the general theory of covering classes.

Let $S \twoheadrightarrow M$ be a Γ-separated cover of a Λ-module. We call $X = \Gamma \otimes_{\Lambda} S$ the *associated* Γ*-module* of this separated cover. Recall that we can consider S to be a Λ-submodule of X via the imbedding $s \to 1 \otimes s$, in which case we have $X = \Gamma S$ [Lemma 17.2, whose proof does not use the stated finite generation of S].

LEMMA 18.5. *Let Ω be a ring finitely generated as a Λ-module, such that $\Lambda \subset \Omega \subseteq \Gamma$, and let C be the conductor ideal for Λ and Ω [Proposition 18.3]. Then Λ/C and Ω/C are direct products of fields. Moreover, all of the results stated in* [**KL2**, §4], *for the pair of rings called Λ and Γ in that paper, hold for the pair of rings called Λ and Ω in the present lemma.*

In particular, uniqueness of Ω-separated covers of Λ-modules holds: If $\phi\colon S \twoheadrightarrow M$ and $\phi'\colon S' \twoheadrightarrow M$ are Ω-separated covers of any Λ-module M, then there is an isomorphism $\theta\colon S \cong S'$ such that $\phi = \phi'\theta$.

PROOF. The fact that Λ/C and Ω/C are direct products of fields is proved in Proposition 18.3. This is a special case of the standing hypotheses of [**KL2**, §4], stated in [**KL2**, (4.1.1)]. For uniqueness of separated covers, see [**KL2**, Corollary 4.13]. □

LEMMA 18.6. *Let Ω be a ring finitely generated as a Λ-module, such that $\Lambda \subset \Omega \subseteq \Gamma$, let C be the conductor ideal for Λ and Ω, and $\phi\colon S \twoheadrightarrow M$ a Λ-module surjection in which S is Ω-separated. Then the following two conditions are equivalent.*

(i) *ϕ is an Ω-separated cover of M.*
(ii) (a) $\ker(\phi)$ *has no nonzero Ω-submodules (so that $C\ker(\phi) = 0$, and hence* $\ker(\phi)$ *is canonically a Λ/C-module); and*
  (b) $\ker(\phi) \subseteq CS$.

PROOF. See Lemma 18.5 and [**KL2**, 4.9]. □

LEMMA 18.7 (Ω-separated covers localize and globalize). *Let Ω be a ring finitely generated as a Λ-module, with $\Lambda \subseteq \Omega \subseteq \Gamma$, and let $\phi\colon S \twoheadrightarrow M$ be a Λ-module surjection. Then:*

(i) *S is Ω-separated (respectively, Γ-separated) if and only if $S_{\mathfrak{m}}$ is $\Omega_{\mathfrak{m}}$-separated (respectively $\Gamma_{\mathfrak{m}}$-separated) $(\forall \mathfrak{m} \in \operatorname{maxspec}(\Lambda))$.*
(ii) *ϕ is an Ω-separated cover if and only if $\phi_{\mathfrak{m}}\colon S_{\mathfrak{m}} \twoheadrightarrow M_{\mathfrak{m}}$ is an $\Omega_{\mathfrak{m}}$-separated cover $(\forall \mathfrak{m} \in \operatorname{maxspec}(\Lambda))$.*

PROOF. (i) If S is Ω-separated, there is an Ω-module X such that $S \subseteq X$. But then $S_{\mathfrak{m}} \subseteq X_{\mathfrak{m}}$, and $X_{\mathfrak{m}}$ is an $\Omega_{\mathfrak{m}}$-module.

Conversely, suppose that $S_{\mathfrak{m}}$ is $\Omega_{\mathfrak{m}}$-separated $(\forall \mathfrak{m} \in \operatorname{maxspec}(\Lambda))$. Recall that S is Ω-separated if and only if the canonical map $\tau\colon S \to \Omega \otimes_{\Lambda} S$ is an injection [Lemma 17.2, whose simple proof does not require the stated finite generation of S]. For this it suffices to show that every localization $\tau_{\mathfrak{m}}$ is an injection, which holds by hypothesis.

The proof of the situation where Γ replaces Ω is exactly the same.

(ii)($\Rightarrow$) Suppose that $\phi\colon S \twoheadrightarrow M$ is an Ω-separated cover, and choose a maximal ideal $\mathfrak{m}$ of Λ. We know that $S_{\mathfrak{m}}$ is $\Omega_{\mathfrak{m}}$-separated by part (i). To prove that $\phi_{\mathfrak{m}}$ is an $\Omega_{\mathfrak{m}}$-separated cover, we verify the $\mathfrak{m}$-localized versions $(\mathrm{a})_{\mathfrak{m}}$ and $(\mathrm{b})_{\mathfrak{m}}$ of conditions (a) and (b) respectively, of Lemma 18.6. Since S is Ω-separated, we may choose an Ω-module X containing S; and after doing this, it makes sense to refer to the Ω-module generated by any subset of S (viewed as a submodule of X).

Condition $(\mathrm{a})_{\mathfrak{m}}$. Since ϕ is an Ω-separated cover, condition (a) shows that $\ker(\phi)$ is a Λ/C-modules, where C is the conductor ideal for Λ and Ω. Since Λ/C is a direct product of fields [Lemma 18.5], $\ker(\phi)$ is a direct sum of simple Λ-modules. Each simple Λ-module is isomorphic to $\Lambda/\mathfrak{n}$ for some $\mathfrak{n} \in \operatorname{maxspec}(\Lambda)$. It therefore coincides with its $\mathfrak{n}$-localization, and its localization at any other maximal ideal is zero. We conclude: $\ker(\phi)$ is the direct sum of its localizations $\ker(\phi)_{\mathfrak{m}} = \ker(\phi_{\mathfrak{m}})$ at the maximal ideals of Λ.

Now let W be any $\Omega_{\mathfrak{m}}$-submodule of $\ker(\phi_{\mathfrak{m}})$. By the previous paragraph, W is an $\Omega_{\mathfrak{m}}$-submodule of $\ker(\phi)$, and hence an Ω-submodule of $\ker(\phi)$. Therefore, by condition (a), we have $W = 0$, completing the proof of condition $(\mathrm{a})_{\mathfrak{m}}$.

Condition $(\mathrm{b})_{\mathfrak{m}}$. By condition (b), $\ker(\phi) \subseteq CS$. Therefore $\ker(\phi_{\mathfrak{m}}) \subseteq C_{\mathfrak{m}} S_{\mathfrak{m}}$ as desired.

(ii)($\Leftarrow$) Suppose that every $\phi_{\mathfrak{m}}$ is an $\Omega_{\mathfrak{m}}$-separated cover. Then S is Ω-separated by part (i). Consider a factorization $S \twoheadrightarrow S' \twoheadrightarrow M$ of ϕ in which S' is Ω-separated. Localizing at each maximal ideal yields a factorization of the $\Omega_{\mathfrak{m}}$-separated cover $\phi_{\mathfrak{m}}$ in which $S'_{\mathfrak{m}}$ is $\Omega_{\mathfrak{m}}$-separated. It follows that each map $S_{\mathfrak{m}} \to S'_{\mathfrak{m}}$ is one-to-one, and hence the map $S \to S'$ is one-to-one. Therefore ϕ is an Ω-separated cover. □

The next lemma gives the direct-limit argument which enables us to extend results on Ω-separated covers of Λ-modules developed in [**KL2**] — which require the existence of a conductor ideal — to Γ-separated covers of Λ-modules. Unfortunately, it requires us to assume that M is finitely generated.

LEMMA 18.8. *Let $M \in \mathrm{fingen}(\Lambda)$ and $\mathcal{F}$ a finite set of maximal ideals of Λ containing the (necessarily finite) set of maximal ideals $\mathfrak{m}$ at which $M_{\mathfrak{m}}$ is not $\Lambda_{\mathfrak{m}}$-free. Then:*

(i) *M has an $\Omega(\mathcal{F})$-separated cover ϕ: $S \twoheadrightarrow M$, with ${}_{\Lambda}S$ finitely generated.*
(ii) *Every $\Omega(\mathcal{F})$-separated cover ϕ: $S \twoheadrightarrow M$ is a Γ-separated cover .*
(iii) *Every Γ-separated cover $S' \twoheadrightarrow M$ is an $\Omega(\mathcal{F})$-separated cover.*

PROOF. (i) Since Λ is noetherian, this is a special case of (18.4.1).

(ii) First recall that, since M is finitely generated, the set of $\mathfrak{m}$ such that $M_{\mathfrak{m}}$ is not free, is finite [Lemma 8.2] and M has a finitely generated Ω-separated cover ϕ, as displayed in (i) [see (18.4.1)]. Now let ϕ be any $\Omega(F)$-separated cover of M.

Since $\Omega = \Omega(\mathcal{F})$ is a finitely generated Λ-module [see (18.1.1)], we know that Ω-separated covers localize and globalize [Lemma 18.7]. Unfortunately, we do not yet know that Γ-separated covers localize and globalize [Theorem 18.13]. However we do know [Lemma 18.7] that the properties of being Ω-separated or Γ-separated Λ-modules localize and globalize.

Finally, recall that since Ω is a finitely generated Λ-module, the properties of separated covers proved in [**KL2**, §4] apply to Ω-separated covers of Λ-modules, as explained in Lemma 18.5. In particular, uniquess of Ω-separated covers of Λ-modules holds [Lemma 18.5].

We claim that the Ω-separated Λ-module S is Γ-separated. Choose a maximal ideal $\mathfrak{m}$ of Λ. Since the property of being an Ω-separated cover localizes, $S_{\mathfrak{m}}$ is a $\Omega_{\mathfrak{m}}$-separated module. We consider two cases.

If $\mathfrak{m} \in \mathcal{F}$, then $\Omega_{\mathfrak{m}} = \Gamma_{\mathfrak{m}}$ [see (18.1.1)], and hence the $\Omega_{\mathfrak{m}}$-separated $\Lambda_{\mathfrak{m}}$-module $S_{\mathfrak{m}}$ is $\Gamma_{\mathfrak{m}}$-separated. On the other hand, if $\mathfrak{m} \notin \mathcal{F}$, then $M_{\mathfrak{m}}$ is $\Lambda_{\mathfrak{m}}$-free, and hence the identity map on $M_{\mathfrak{m}}$ is obviously an $\Omega_{\mathfrak{m}}$-separated cover of $M_{\mathfrak{m}}$. It follows from uniqueness of Ω-separated covers [Lemma 18.5] that the localized cover $\phi_{\mathfrak{m}}$ is just a copy of the identity map on $M_{\mathfrak{m}}$. In particular, $S_{\mathfrak{m}}$ is $\Lambda_{\mathfrak{m}}$-free, and hence $\Gamma_{\mathfrak{m}}$-separated.

Since the property of being a Γ-separated module globalizes, we conclude that S is Γ-separated, as claimed.

In order to complete the proof that ϕ is a Γ-separated cover, consider a factorization $S \twoheadrightarrow S' \twoheadrightarrow M$ of ϕ, in which S' is Γ-separated. We need to prove that the map $S \twoheadrightarrow S'$ is an injection. Since S' is Γ-separated, it is obviously Ω-separated. Since ϕ is an Ω-separated cover, the map $S \twoheadrightarrow S'$ is an injection, as desired.

(iii) Let $\phi'\colon S' \twoheadrightarrow M$ be a Γ-separated cover. By part (i), M has an $\Omega = \Omega(\mathcal{F})$-separated cover $\phi\colon S \twoheadrightarrow M$ with S finitely generated. Since S' is a Γ-separated Λ-module, it is obviously Ω-separated.

By Lemma 18.5 we can apply [**KL2**, Corollary 4.14]. It states that if $\phi\colon S \twoheadrightarrow M$ is an Ω-separated cover, then any surjective Λ-homomorphism $\phi'\colon S' \twoheadrightarrow M$, with S' Ω-separated, has a factorization $\phi'\colon S' \overset{\theta}{\twoheadrightarrow} S \overset{\phi}{\twoheadrightarrow} M$. On the other hand, S is Γ-separated by part (ii). Since ϕ' is a Γ-separated cover, the definition of Γ-separated cover implies that the surjective map θ is an injection, and hence an isomorphism. Thus ϕ' is an Ω-separated cover (a copy of ϕ). □

COROLLARY 18.9. *Let $\phi\colon S \twoheadrightarrow M$ be a Γ-separated cover in* fingen(Λ) *with associated Γ-module $X = \Gamma S = \Gamma \otimes_\Lambda S$. Then* $\ker(\phi)$ *and* $\Gamma \otimes_\Lambda \ker(\phi)$ *are semisimple modules of finite length over Λ and Γ respectively, and $\Gamma\cdot \ker\phi$ is a semisimple Γ-submodule of X.*

PROOF. First we prove that $\ker(\phi)$ is semisimple. By Lemma 18.8, ϕ is an Ω-separated cover for some Ω with $\Lambda \subseteq \Omega \subseteq \Gamma$ and ${}_\Lambda\Omega$ finitely generated. Therefore, in the notation of Lemma 18.6, we have that $C\cdot \ker\phi = 0$, and hence $\ker\phi$ is a module over Λ/C. By Lemma 18.5, Λ/C is a direct product of fields, and hence all of its finitely generated modules are semisimple of finite length.

Next we claim that $\Gamma \otimes_\Lambda \ker(\phi)$ is a semisimple Γ-module. In view of the previous paragraph, it suffices to prove that, for every simple Λ-module V, the Γ-module $\Gamma \otimes_\Lambda V$ is semisimple of finite length. We have $V \cong \Lambda/\mathfrak{m}$ for some $\mathfrak{m} \in \operatorname{maxspec}(\Lambda)$, and hence $\Gamma \otimes_\Lambda V \cong \Gamma/(\Gamma\mathfrak{m})$. Then by Lemma 11.6, we get that $\Gamma\mathfrak{m}$ is the intersection of the one or two maximal ideals of Γ containing $\mathfrak{m}$, and this completes the proof of the claim.

Finally, $\Gamma\cdot \ker(\phi)$ is a semisimple Γ-module of finite length since it is a homomorphic image of $\Gamma \otimes_\Lambda \ker(\phi)$. □

THEOREM 18.10 (Almost functorial property). *Let $f\colon N \to M$ be homomorphisms in* fingen(Λ), *and let ϕ', ϕ be Γ-separated covers. Then f can be lifted to a Λ-homomorphism θ such that the following diagram commutes.*

(18.10.1)
$$\begin{array}{ccc} S' & \overset{\theta}{\dashrightarrow} & S \\ \Big\downarrow{\scriptstyle \phi'} & & \Big\downarrow{\scriptstyle \phi} \\ N & \overset{f}{\longrightarrow} & M \end{array}$$

If f is one-to-one or onto, then any such θ has the same property.

If ${}_\Lambda\Gamma$ is finitely generated, then the same conclusions remain true even if the Λ-modules N, M are not finitely generated.

PROOF. First consider the case that $N, M \in$ fingen(Λ). By Lemma 18.8, both ϕ' and ϕ are Ω-separated covers for some ring Ω such that $\Lambda \subseteq \Omega \subseteq \Gamma$ and ${}_\Lambda\Omega$ is finitely generated. Therefore the almost functorial property for Ω-separated covers applies ([**KL2**, 4.12] and Lemma 18.5) and immediately yields the desired conclusions.

Note that [**KL2**, 4.12] holds without the assumption that N, M be finitely generated [**KL2**, Remarks 4.8(ii)]. And when ${}_\Lambda\Gamma$ is finitely generated, Lemma 18.5 allows us to apply [**KL2**, 4.12] to Γ-separated covers. □

The following uniqueness result shows that all isomorphisms of finitely generated Λ-modules come from isomorphisms of Γ-modules.

COROLLARY 18.11 (Uniqueness of Γ-separated covers). *Let* $T \twoheadrightarrow N$ *and* $S \twoheadrightarrow M$ *be* Γ*-separated covers of modules* $M, N \in \text{fingen}(\Lambda)$, *with associated* Γ*-modules* $Y = \Gamma \otimes_\Lambda T = \Gamma T$ *and* $X = \Gamma \otimes_\Lambda S = \Gamma S$ *respectively. Then:*

(i) *Every* Λ*-isomorphism* $f\colon N \cong M$ *can be lifted to a* Λ*-isomorphism* $\theta\colon T \cong S$.

(ii) θ *can be further extended to a* Γ*-isomorphism* $Y \cong X$.

(iii) T *and* S *are finitely generated.*

If ${}_\Lambda\Gamma$ *is finitely generated, then the* Λ*-modules* N, M *need not be finitely generated in parts (i) and (ii).*

PROOF. (i) and (ii) With or without the finite generation of ${}_\Lambda\Gamma$, the given isomorphism $f\colon N \cong M$ can be lifted to a Λ-isomorphism $\theta\colon T \cong S$ [Theorem 18.10]. Since $Y = \Gamma \otimes_\Lambda T = \Gamma T$ and $X = \Gamma \otimes_\Lambda S = \Gamma S$, the map $1 \otimes \theta\colon Y \to X$ is the desired Γ-isomorphism.

(iii) Since M is finitely generated in this part, it has a Γ-separated cover $\phi'\colon S' \twoheadrightarrow M$ with S' finitely generated, by (18.4.1). Then part (i) implies that $S' \cong S$, and hence S is finitely generated. □

COROLLARY 18.12. *Let* $\phi\colon S \twoheadrightarrow M$ *be a* Γ*-separated cover with* $S, M \in \text{fingen}(\Lambda)$. *If* $\ker(\phi) \neq 0$, *then* M *has a nonzero submodule of finite length.*

PROOF. First we prove a general claim. *If* $M \in \text{fingen}(\Lambda)$ *has no nonzero submodules of finite length, then* M *is* Γ*-separated.* Let $K = \ker(M \to M_Q)$. Then $K_Q = 0$ and K is the largest submodule of M with this property. But since every maximal ideal of Λ has height 1 and K is finitely generated, $K_Q = 0$ is equivalent to the fact that K has finite length [Lemma 4.5]. Therefore $K = 0$. Since M_Q is a Λ_Q-module, we conclude that M is Λ_Q-separated, and therefore Γ-separated, as desired.

Now let M be as in the statement of the corollary. If M has no finite-length submodules, then M is separated, and hence the identity map on M is a separated cover. Therefore uniqueness of separated covers [Corollary 18.11] shows that $\ker(\phi) = 0$, a contradiction. □

THEOREM 18.13 (Γ- vs. $\Gamma_\mathfrak{m}$- vs. $\hat{\Gamma}_\mathfrak{m}$-separated covers). *Let* $\phi\colon S \twoheadrightarrow M$ *be a surjective homomorphism of* Λ*-modules, with* $M \in \text{fingen}(\Lambda)$.

(i) ϕ *is a* Γ*-separated cover if and only if* $\big(\forall \mathfrak{m} \in \text{maxspec}(\Lambda)\big)$ $\phi_\mathfrak{m}$ *is a* $\Gamma_\mathfrak{m}$*-separated cover of the* $\Lambda_\mathfrak{m}$*-module* $M_\mathfrak{m}$.

(ii) *Suppose also that* $S \in$ *fingen*(Λ), *and choose* $\mathfrak{m} \in \text{maxspec}(\Lambda)$. *Then* $\phi_\mathfrak{m}$ *is a* $\Gamma_\mathfrak{m}$*-separated cover of the* $\Lambda_\mathfrak{m}$*-module* $M_\mathfrak{m}$ *if and only if* $\hat{\phi}_\mathfrak{m}$ *is a* $\hat{\Gamma}_\mathfrak{m}$*-separated cover of the* $\hat{\Lambda}_\mathfrak{m}$*-module* $\hat{M}_\mathfrak{m}$.

PROOF. (i)(⇒) Suppose ϕ is a Γ-separated cover. We apply Lemma 18.8: Let $\mathcal{F}$ be the finite set of maximal ideals of Λ at which $M_\mathfrak{m}$ is not $\Lambda_\mathfrak{m}$-free, and let $\Omega = \Omega(\mathcal{F})$. Then that lemma states that ϕ is and Ω-separated cover. Therefore the localized Ω-separated cover $\phi_{\mathfrak{m}}$ is an $\Omega_{\mathfrak{m}}$-separated cover [Lemma 18.7]. We consider two cases.

If $\mathfrak{m} \in \mathcal{F}$, then $\Omega_{\mathfrak{m}} = \Gamma_{\mathfrak{m}}$ by the definition of $\Omega(\mathcal{F})$. Therefore $\phi_{\mathfrak{m}}$ is a $\Gamma_{\mathfrak{m}}$-separated cover. On the other hand, suppose that $\mathfrak{m} \notin \mathcal{F}$. Then $M_{\mathfrak{m}}$ is $\Lambda_{\mathfrak{m}}$-free; and hence the identity map on $M_{\mathfrak{m}}$ is (obviously) its $\Gamma_{\mathfrak{m}}$-separated cover as well as its $\Omega_{\mathfrak{m}}$-separated cover. As mentioned in the previous paragraph, $\phi_{\mathfrak{m}}$ is the $\Omega_{\mathfrak{m}}$-separated cover of $M_{\mathfrak{m}}$. Therefore, by uniqueness of $\Omega_{\mathfrak{m}}$-separated covers [Lemma 18.5] the map $\phi_{\mathfrak{m}}$ is a copy of the identity map on $M_{\mathfrak{m}}$, and hence is also the $\Gamma_{\mathfrak{m}}$-separated cover of $M_{\mathfrak{m}}$.

(i)($\Leftarrow$) Suppose that every $\phi_{\mathfrak{m}}$ is a $\Gamma_{\mathfrak{m}}$-separated cover. This implies that S is Γ-separated [Lemma 18.7]. Consider a factorization $\phi\colon S \twoheadrightarrow S' \twoheadrightarrow M$ of ϕ in which S' is Γ-separated. Then for each maximal ideal $\mathfrak{m}$ of Λ, localizing at $\mathfrak{m}$ yields a factorization of the $\Gamma_{\mathfrak{m}}$-separated cover $\phi_{\mathfrak{m}}$ in which $S'_{\mathfrak{m}}$ is $\Gamma_{\mathfrak{m}}$-separated, so that the map $S_{\mathfrak{m}} \to S'_{\mathfrak{m}}$ must be injective for every $\mathfrak{m}$. Therefore the map $S \to S'$ is injective, and hence ϕ is a Γ-separated cover.

(ii) We have, after $\mathfrak{m}$-localizing and changing notation, that Λ is a local Dedekind-like ring with maximal ideal $\mathfrak{m}$ and normalization Γ [see Corollary 10.7 and its proof]. In this situation, ${}_{\Lambda}\Gamma$ is finitely generated and $\mathfrak{m} = \operatorname{rad}(\Lambda) = \operatorname{rad}(\Gamma)$ [Proposition 10.6]. Moreover, in the notation of (18.3.1), we have $C = \mathfrak{m}$, the conductor ideal for Λ and Γ.

Write $\hat{\phi} = \hat{\phi}_{\mathfrak{m}}$, $\hat{M} = \hat{M}_{\mathfrak{m}}$, and so on. We make use of the standard identification $\hat{M} = \hat{\Lambda} \otimes_{\Lambda} M$. We also have that $\hat{\Lambda}$ is Dedekind-like with normalization $\hat{\Gamma}$ [Lemma 11.8]. Since $\mathfrak{m}$ is the conductor ideal for Λ and Γ, it follows that the maximal ideal $\hat{\mathfrak{m}}$ of $\hat{\Lambda}$ is the conductor ideal for $\hat{\Lambda}$ and $\hat{\Gamma}$.

We claim that S is Γ-separated if and only if $\hat{S}$ is $\hat{\Gamma}$-separated. Recall that S is Γ-separated if and only if the natural map $\tau\colon S \to \Gamma \otimes_{\Lambda} S$ is an injection [Lemma 17.2]. Similarly, $\hat{S}$ is $\hat{\Gamma}$-separated if and only if $\hat{\tau}\colon \hat{S} \to \hat{\Gamma} \otimes_{\hat{\Lambda}} \hat{S}$ is an injection. Therefore the claim follows from the facts that $\hat{\Lambda}$ is a faithfully flat Λ-module [**M**, Theorem 8.14].

(ii)($\Leftarrow$) Suppose that $\hat{\phi}$ is a $\hat{\Gamma}$-separated cover, and consider a factorization $\phi\colon S \twoheadrightarrow T \twoheadrightarrow M$ with T a Γ-separated Λ-module. Since $\hat{T}$ is $\hat{\Gamma}$-separated, the map $\hat{S} \twoheadrightarrow \hat{T}$ is an injection. Then faithful flatness of ${}_{\Lambda}\hat{\Lambda}$ shows that $S \twoheadrightarrow T$ is also an injection.

(ii)($\Rightarrow$) Suppose that ϕ is a Γ-separated cover. The necessary and sufficient conditions for this are (a) $\ker(\phi)$ has no nonzero Γ-submodules, and (b) $\ker(\phi) \subseteq \mathfrak{m}S$ [Lemma 18.6]. Thus it suffices to establish the analogs (a)$'$ and (b)$'$ of these two conditions for $\hat{\phi}$. Condition (b)$'$ is immediate since $\ker(\hat{\phi}) \subseteq (\ker\phi)\hat{} = \hat{\mathfrak{m}}\hat{S}$.

For (a)$'$, note that $\ker(\phi)$ has finite length [Corollary 18.9] and therefore is unchanged by completion. Since any $\hat{\Gamma}$-submodule of $(\ker\phi)\hat{} = \ker(\phi)$ is also a Γ-submodule, condition (a)$'$ follows from condition (a). □

THEOREM 18.14. *A finite direct sum of Γ-separated covers of finitely generated Λ-modules is again a Γ-separated cover.*

PROOF. It suffices to prove the Theorem for the direct sum of two separated covers. Let $\phi_i\colon S_i \twoheadrightarrow M_i$ be Γ-separated covers with associated Γ-modules $X_i = \Gamma S_i$ $(i = 1, 2)$. We need to show that $\phi = \phi_1 \oplus \phi_2$ is also a Γ-separated cover. Let $\mathcal{F}$ be the finite set of maximal ideals $\mathfrak{m}$ of Λ such that at least one of $(M_1)_{\mathfrak{m}}$ and

$(M_2)_{\mathfrak{m}}$ is not $\Lambda_{\mathfrak{m}}$-free, and apply Lemma 18.8 with $\Omega = \Omega(\mathcal{F})$. Then each ϕ_i is an Ω-separated cover, and it suffices to show that ϕ is an Ω-separated cover.

Each ϕ_i satisfies the necessary and sufficient conditions (a) and (b) of Lemma 18.6 for Ω-separated covers. We check that $\phi = \phi_1 \oplus \phi_2$ does as well. Let C be the conductor ideal for Λ and Ω. Condition (b), that $\ker(\phi_i) \subseteq CS$, is obviously preserved by direct sums.

Consider condition (a), that $K_i = \ker(\phi_i)$ contains no nonzero Γ-submodules. Let Y be any Γ-submodule of $K = K_1 \oplus K_2$, and let Y_i be the projection of Y in K_i $(i = 1, 2)$. It suffices to prove that each $Y_i = 0$, and for this it suffices to show that each Y_i is a Γ-submodule of K_i. But each $K_i \subseteq X_i$, so each Y_i is also the projection of Y in the Γ-module X_i with respect to the direct sum $X_1 \oplus X_2$, and this makes each Y_i a Γ-submodule of K_i. □

We conclude this section with two minimality properties of separated covers.

THEOREM 18.15. *Let ϕ: $S \to M$ be a Γ-separated cover, with $M \in \mathrm{fingen}(\Lambda)$. Then:*

(i) *Each Λ-module surjection ϕ': $S' \twoheadrightarrow M$ with $S' \in \mathrm{fingen}(\Lambda)$ and Γ-separated has a factorization $S' \overset{\theta}{\twoheadrightarrow} S \overset{\phi}{\twoheadrightarrow} M$.*

(ii) *ϕ is a "minimal epimorphism" (no submodule properly smaller than S is mapped by ϕ onto M).*

If ${}_\Lambda\Gamma$ is finitely generated then the Λ-modules S', M need not be finitely generated.

PROOF. First consider the case that $S', M \in \mathrm{fingen}(\Lambda)$.

(i) Apply the almost functorial property [Theorem 18.10] to the following diagram.

$$\begin{array}{ccc} S' & \overset{\theta}{\dashrightarrow} & S \\ \big\downarrow{\scriptstyle 1} & & \big\downarrow{\scriptstyle \phi} \\ S' & \overset{\phi'}{\longrightarrow} & M \end{array}$$

(ii) By Lemma 18.8, ϕ is an Ω-separated cover for some ring Ω such that $\Lambda \subseteq \Omega \subseteq \Gamma$ and ${}_\Lambda\Omega$ is finitely generated. Then, by Lemma 18.5 we can apply [**KL2**, Lemma 4.10], which states the desired conclusion.

Now suppose that ${}_\Lambda\Gamma$ is finitely generated. Then Lemma 18.5, together with [**KL2**, Remarks 4.8(ii)] allow us to apply the almost functorial property [**KL2**, Theorem 4.12] and minimal epimorphism property [**KL2**, Lemma 4.10], as done above, but with no finite generation hypotheses on the modules involved. □

CHAPTER 5

Isomorphism Classes in a Genus, Idèle Group Action

Let $M \in \text{fingen}(\Lambda)$, and suppose that we already know genus(M), that is, the package of isomorphism classes of completions $\hat{M}_{\mathfrak{m}}$ at maximal ideals of Λ. What additional information is needed, to specify the isomorphism class of M?

By our point-of-view that fingen(Γ) is an approximation to fingen(Λ), the first invariant is the Γ-isomorphism class of $\Gamma \otimes_\Lambda M$. This partitions genus(M) into "restricted genera" [Definition 19.1].

For what comes next, there is no loss of generality in working with $\text{fingen}_\infty(\Lambda)$ instead of fingen(Λ), as explained in Section 7. We do this whenever convenient. [But we cannot restrict our attention to $\text{fingen}_\infty(\Gamma)$!]

The main content of this chapter is to describe the action of a group — the group of "residue unit idèles" — on the set of isomorphism classes of an arbitrary $M \in \text{fingen}_\infty(\Lambda)$ [Section 22]. The set of orbits of this action turns out to be the collection of isomorphism classes in the restricted genus of M. This type of action has been used extensively, for example in [**G1, GL2, W1**], but always under the assumption that ${}_\Lambda\Gamma$ is finitely generated — and hence a conductor ideal exists for Λ and Γ. The present chapter drops this finiteness assumption, taking advantage of the fact that Dedekind-like rings always have local conductor ideals; more precisely, $\mathfrak{m}_{\mathfrak{m}}$ is always an ideal of $\Gamma_{\mathfrak{m}}$.

Unfortunately, the definition of our action is quite complicated, and requires a preliminary action by a group of "residue endo-idèles" that occupies Section 21; and this, in turn, requires an extension of a lifting theorem of Wiegand [Section 20]. We determine the stabilizer of this action in sections 23 and 24. The latter section requires some very complicated matrix calculations, and is the only part of this memoir that requires technical details of the matrix-theoretical results in [**KL2**]. This difficulty should not be surprising since, even in the case of torsionfree modules over well-studied rings, very few of these stabilizers are known in detail.

19. Restricted Genus and Separated Covers

DEFINITION 19.1. Let $M \in \text{fingen}(\Lambda)$. The *restricted genus* rgen(M) is the collection of isomorphism classes of modules $N \in \text{genus}(M)$ such that $\Gamma \otimes_\Lambda N \cong \Gamma \otimes_\Lambda M$ as Γ-modules.

This brief section relates rgen(M) to the Γ-separated cover of M.

NOTATION 19.2 (Torsion submodule $t(X)$). The *torsion submodule* $t(X)$ of a module $X \in \text{fingen}(\Gamma)$ (respectively Λ) is the largest Γ-submodule (respectively Λ-submodule) of X such that $X_Q = 0$. Since all maximal ideals of Dedekind-like rings have height 1 [Lemma 10.5], this condition is equivalent to the condition that $t(X)$

is the largest finite-length Γ-submodule (respectively Λ-submodule) of X [Lemma 4.5].

LEMMA 19.3. *Let* $\phi\colon S \twoheadrightarrow M$ *be a* Γ*-separated cover of* $M \in \text{fingen}(\Lambda)$, *with associated* Γ*-module* $X = \Gamma \otimes_\Lambda S = \Gamma S$. *Then*

$$(19.3.1) \qquad (\Gamma \otimes_\Lambda M)/t(\Gamma \otimes_\Lambda M) \cong X/t(X) \qquad (\textit{as } \Gamma\textit{-modules.})$$

PROOF. Let $K = \ker(\phi)$. Tensoring the exact sequence $K \hookrightarrow S \twoheadrightarrow M$ over Λ with Γ, and using right exactness of the tensor product, shows that $X/\operatorname{im}(\Gamma \otimes_\Lambda K) \cong \Gamma \otimes_\Lambda M$. Since $\Gamma \otimes_\Lambda K$ has finite length over Γ [Corollary 18.9], we have $\operatorname{im}(\Gamma \otimes_\Lambda K) \subseteq t(X)$. Reducing the previous isomorphism modulo torsion therefore completes the proof. □

LEMMA 19.4. *Let* $\phi\colon S \twoheadrightarrow M$ *and* $\phi'\colon S' \twoheadrightarrow M'$ *be* Γ*-separated covers of modules in* $\text{fingen}(\Lambda)$, *with associated* Γ*-modules* $X = \Gamma \otimes_\Lambda S$ *and* $X' = \Gamma \otimes_\Lambda S'$, *respectively. If* $\text{genus}(M) = \text{genus}(M')$, *then the following statements are equivalent.*

(i) $\text{rgen}(M) = \text{rgen}(M')$.
(ii) $X/t(X) \cong X'/t(X')$ *(as* Γ*-modules).*
(iii) $X \cong X'$ *(as* Γ*-modules).*

Moreover, when these conditions hold, we have $\text{rgen}(S) = \text{rgen}(S')$.

PROOF. (i)⇒(ii). Given a Γ-isomorphism $\Gamma \otimes_\Lambda M \cong \Gamma \otimes_\Lambda M'$, it must take $t(\Gamma \otimes_\Lambda M)$ onto $t(\Gamma \otimes_\Lambda M')$. Therefore by (19.3.1), it induces an isomorphism as in (ii).

(ii)⇒(i). Given a Γ-isomorphism $X/t(X) \cong X'/t(X')$, (19.3.1) yields a Γ-isomorphism $(\Gamma \otimes M)/t(\Gamma \otimes M) \cong (\Gamma \otimes M')/t(\Gamma \otimes M')$. Since Γ is a direct sum of Dedekind domains, $t(\Gamma \otimes M)$ and $t(\Gamma \otimes M')$ are direct summands of $\Gamma \otimes M$ and $\Gamma \otimes M'$, respectively. Since modules of finite length cancel from direct sums [**F**, 4.5], it suffices to prove that $t(\Gamma \otimes M) \cong t(\Gamma \otimes M')$.

For this, note that $t(\Gamma \otimes_\Lambda M)$ and $t(\Gamma \otimes_\Lambda M')$ have finite length as Γ-modules, and hence as Λ-modules [Lemma 10.10]. Therefore, each is the direct sum of its localizations at the maximal ideals of Λ [Corollary 6.4]. Moreover, corresponding terms in these two local decompositions are isomorphic, since $\text{genus}(M) = \text{genus}(M')$, completing the proof of (i).

(ii)⇒(iii). As noted above, $t(X)$ and $t(X')$ are direct summands of X and X', respectively, so it suffices to prove that $t(X) \cong t(X')$. Also as above, it suffices to prove that $t(X_\mathfrak{m}) \cong t(X'_\mathfrak{m})$ for each $\mathfrak{m} \in \text{maxspec}(\Lambda)$.

Fix a maximal ideal $\mathfrak{m} \in \text{maxspec}(\Lambda)$. The maps $\phi_\mathfrak{m}$ and $\phi'_\mathfrak{m}$ are $\Gamma_\mathfrak{m}$-separated covers of $M_\mathfrak{m}$ and $M'_\mathfrak{m}$, respectively [Theorem 18.13]. Since $\text{genus}(M) = \text{genus}(M')$, we have $M_\mathfrak{m} \cong M'_\mathfrak{m}$, so by uniqueness of Γ-separated covers [Corollary 18.11], we get that $S_\mathfrak{m} \cong S'_\mathfrak{m}$. Tensoring with $\Gamma_\mathfrak{m}$ over $\Lambda_\mathfrak{m}$ yields $X_\mathfrak{m} \cong X'_\mathfrak{m}$, and therefore $t(X_\mathfrak{m}) \cong t(X'_\mathfrak{m})$, completing the proof of (iii).

(iii)⇒(ii) is obvious.

Supplementary statement. The proof of (ii)⇒(iii) shows that $\text{genus}(S) = \text{genus}(S')$. Note that the identity maps on S and S' are separated covers of these modules. Therefore (iii)⇒(i), applied to the modules S and S', proves that $\text{rgen}(S) = \text{rgen}(S')$. □

20. Wiegand Lifting Theorem

Let $(S \subseteq X)$ be a $(\Lambda \subseteq \Gamma)$-inclusion [Definition 17.1]. We know that S is determined by X and the residue inclusions of $(S \subseteq X)$ [Theorem 17.4]. In this section we prove a theorem for lifting an automorphism of determinant 1, of some $\bar{X}(\mathfrak{n})$, to an automorphism of X — when ${}_{\Gamma}X$ is projective — in a way that leaves all residue inclusions except $\big(\bar{S}(\mathfrak{n}) \subseteq \bar{X}(\mathfrak{n})\big)$ undisturbed [Theorem 20.6]. As with Wiegand's original result, this theorem is a critical step in the proof that our group action that constructs global isomorphism classes [Section 22] is well defined. Unfortunately, in our case, the use of the lifting theorem is deeply buried in preliminary considerations [proof of Theorem 21.8, just after (21.8.3)].

Notes. This section makes use of a lot of previously defined notation. See the notation index at the end of this memoir for help. We make frequent use of the fact that only finitely many maximal ideals of Λ can be strictly split [Proposition 11.4].

REMARKS 20.1 (Determinant). Recall Goldman's definition [**G**] of the *determinant* of an endomorphism θ of a finitely generated projective module X over a commutative ring: Choose P such that $X \oplus P$ is free of finite rank. Then the endomorphism $\theta \oplus 1_P$ of $X \oplus P$ has an ordinary determinant, and one defines $\det(\theta)$ to be $\det(\theta \oplus 1_P)$. This definition of $\det(\theta)$ is independent of the choice of P, is a unit of the ring if and only if θ is an automorphism, and agrees with the ordinary definition of $\det(\theta)$ if X is free.

Let $X \in \operatorname{fingen}(\Gamma)$ and, for $\mathfrak{m} \in \operatorname{maxspec}(\Lambda)$, let θ be a $\bar{\Gamma}$-endomorphism of $\bar{X}(\mathfrak{m})$. If $\mathfrak{m}$ is split, then $\theta = (\theta_1, \theta_2)$ where θ_1 and θ_2 are $k(\mathfrak{m})$-endomorphisms of $\big(k(\mathfrak{m})_1, 0\big)\cdot \bar{X}(\mathfrak{m})$ and $\big(0, k(\mathfrak{m})_2\big)\cdot \bar{X}(\mathfrak{m})$, respectively. Note that these two $k(\mathfrak{m})$-vector spaces can have different dimensions. However, $\det(\theta) = \big(\det(\theta_1), \det(\theta_2)\big) \in k(\mathfrak{m})_1 \times k(\mathfrak{m})_2$.

We adopt the usual convention that *the identity automorphism of the zero vector space has determinant 1.* For example, this situation occurs when, in the notation of the previous paragraph, $\big(k(\mathfrak{m})_1, 0\big)\cdot \bar{X}(\mathfrak{m}) = 0$ or $\big(0, k(\mathfrak{m})_2\big)\cdot \bar{X}(\mathfrak{m}) = 0$.

LEMMA 20.2. *Let $\mathfrak{m} \in \operatorname{singspec}(\Lambda)$ be unsplit or nonstrictly split, Γ_h the unique coordinate ring of Γ supported by $\mathfrak{m}$, and H a nonzero ideal of Γ_h. Then $\bar{H}(\mathfrak{m}) \cong \bar{\Gamma}(\mathfrak{m})$ ($\bar{\Gamma}$-isomorphism).*

PROOF. This proof uses the fact that $\bar{H}(\mathfrak{m})$ can be interpreted as either $H/\mathfrak{m}H$ or $H_{\mathfrak{m}}/(\mathfrak{m}_{\mathfrak{m}}H_{\mathfrak{m}})$ [see (11.2.1)].

We have $H_{\mathfrak{m}} \neq 0$ because $(\Gamma_h)_{\mathfrak{m}} \neq 0$. Since Γ_h is the unique coordinate ring supported by $\mathfrak{m}$ we have $(\Gamma_h)_{\mathfrak{m}} = \Gamma_{\mathfrak{m}}$ in $\Gamma_{\mathfrak{m}}$, and therefore $H_{\mathfrak{m}}$ is a nonzero ideal of the principal ideal ring $\Gamma_{\mathfrak{m}}$. Therefore $H_{\mathfrak{m}} \cong \Gamma_{\mathfrak{m}}$ ($\Gamma_{\mathfrak{m}}$-isomorphism). Reducing modulo $\mathfrak{m}_{\mathfrak{m}}$ therefore yields $\bar{H}_{\mathfrak{m}} \cong \bar{\Gamma}(\mathfrak{m})$. □

LEMMA 20.3. *Let $\gamma \in \Gamma$. Then its natural image $\bar{\gamma}(\mathfrak{m})$ in $\bar{\Gamma}(\mathfrak{m})$ satisfies $\bar{\gamma}(\mathfrak{m}) \in k(\mathfrak{m})$ $\big(\text{a}\forall \mathfrak{m} \in \operatorname{singspec}(\Lambda)\big)$.*

PROOF. $\big(\text{a}\forall \mathfrak{m} \in \operatorname{singspec}(\Lambda)\big)$ the element $\gamma/1$ of $\Gamma_{\mathfrak{m}}$ satisfies $\gamma/1 \in \Lambda_{\mathfrak{m}}$ [Corollary 8.3]. Reducing modulo $\mathfrak{m}_{\mathfrak{m}}$ gives the claimed result. □

Converting problems about linear transformations of vector spaces to matrix problems requires making arbitrary (and convenient) choices of bases of those spaces. We now introduce the definitions of the choices we need, for converting problems about endomorphisms of projective Γ-modules and their residue inclusions to matrix problems. These definitions are used in the proof of the Lifting Theorem. "Global matrix setups" are quite different from the matrix setups used in the complete local case (see Remarks 24.10 and [**KL2**, 5.6]) because the present setups are designed to deal with the difficulties that arise when singspec(Λ) is infinite.

DEFINITION 20.4. A *global matrix setup* $\mathcal{X}$ is a collection $\{X_{ih}\}$ of Γ-modules, where each X_{ih} is isomorphic to a nonzero ideal of coordinate ring Γ_h of Γ (and each i is an element of some index set), together with the following identifications (Γ-module isomorphisms which we usually write as equality) and Λ-coherence conditions.

Identifications. For each pair of indices i, h and each singular maximal ideal $\mathfrak{m} \in \mathrm{maxsupp}_\Lambda(\Gamma_h)$, the setup $\mathcal{X}$ contains an identification:

$$(20.4.1)\qquad \bar{X}_{ih}(\mathfrak{m}) = \begin{cases} \bar{\Gamma}(\mathfrak{m}) & \text{if } \mathfrak{m} \text{ is unsplit or nonstrictly split;} \\ k(\mathfrak{m})_1 \text{ or } k(\mathfrak{m})_2 & \text{if } \mathfrak{m} \text{ is strictly split (see details below).} \end{cases}$$

If $\mathfrak{m}$ is unsplit or nonstrictly split, there exist Γ-isomorphisms $\bar{X}_{ih} \cong \bar{\Gamma}(\mathfrak{m})$ by Lemma 20.2. The coherence conditions in the present definition restrict which of these isomorphisms can appear in $\mathcal{X}$.

If $\mathfrak{m}$ is strictly split, then $\mathfrak{m}$ is in the support of two coordinate rings $\Gamma_{h_1}, \Gamma_{h_2}$ and we have $\bar{\Gamma}(\mathfrak{m}) = k(\mathfrak{m})_1 \times k(\mathfrak{m})_2$ [Notation 11.5]. It follows easily that $\bar{X}_{ih}(\mathfrak{m})$ is Γ-isomorphic to only one of $k(\mathfrak{m})_1$ and $k(\mathfrak{m})_2$, and this determines which must be chosen in (20.4.1). Our definition of $\mathcal{X}$ places no further restrictions on the choice of identifications, in the strictly split situation.

Λ*-coherence conditions.* We require the identifications (20.4.1) — when $\mathfrak{m}$ is unsplit or nonstrictly split — to be made in such a way that the following two conditions hold for each i, h. Such identifications always exist by Lemma 20.5 below.

(i) There exists a ($\Lambda \subseteq \Gamma$)-inclusion ($T_{ih} \subseteq X_{ih}$) such that for almost all unsplit $\mathfrak{m}$ and almost all nonstrictly split $\mathfrak{m}$ ($\mathfrak{m} \in \mathrm{maxsupp}_\Lambda(\Gamma_h)$) we have

$$(20.4.2)\qquad \big(\bar{T}_{ih}(\mathfrak{m}) \subseteq \bar{X}_{ih}(\mathfrak{m})\big) = \big(k(\mathfrak{m}) \subseteq \bar{\Gamma}(\mathfrak{m})\big)$$

with respect to the identification $\bar{X}_{ih}(\mathfrak{m}) = \bar{\Gamma}(\mathfrak{m})$ in (20.4.1).

(ii) For every Γ-homomorphism of the form $\sigma\colon X_{ih} \to X_{jh}$, and for *almost all* unsplit and strictly split $\mathfrak{m} \in \mathrm{maxsupp}_\Lambda(\Gamma_h)$, the multiplication map

$$(20.4.3)\qquad \bar{\sigma}(\mathfrak{m})\colon \bar{\Gamma}(\mathfrak{m}) \to \bar{\Gamma}(\mathfrak{m}) \qquad \big(\bar{\sigma}(\mathfrak{m}) \in \bar{\Gamma}(\mathfrak{m})\big)$$

induced by σ with respect to the identifications $\bar{X}_{ih}(\mathfrak{m}) = \bar{\Gamma}(\mathfrak{m})$ and $\bar{X}_{jh}(\mathfrak{m}) = \bar{\Gamma}(\mathfrak{m})$ in (20.4.1) satisfies $\bar{\sigma}(\mathfrak{m}) \in k(\mathfrak{m})$.

Recall that, in the nonstrictly split case, the requirement $\bar{\sigma}(\mathfrak{m}) \in k(\mathfrak{m})$ means that $\bar{\sigma}(\mathfrak{m})$ is in the diagonal copy of $k(\mathfrak{m})$ in $k(\mathfrak{m})_1 \times k(\mathfrak{m})_2$.

LEMMA 20.5 (Global matrix setups exist). *Every collection $\{X_{ih}\}$ of Γ-modules, where each X_{ih} is isomorphic to a nonzero ideal of coordinate ring Γ_h of Γ (and each i is an element of some index set), is the collection of modules in some global matrix setup $\mathcal{X}$.*

PROOF. Let $\mathcal{X}$ be the given collection of modules. To make $\mathcal{X}$ into a global matrix setup, we need to define the identifications in such a way that the Λ-coherence conditions hold. First identify each X_{ih} with an ideal of Γ_h that is isomorphic to it.

In this proof *we deal only with unsplit and nonstrictly split maximal ideals* $\mathfrak{m}$, because these are the only maximal ideals involved in the Λ-coherence conditions.

Identifications. Let $\rho^{\mathfrak{m}}$ denote the natural map $\Gamma \twoheadrightarrow \bar{\Gamma}(\mathfrak{m})$. Each $\mathfrak{m}$ under consideration supports a unique coordinate ring Γ_h of Γ. Therefore $\rho^{\mathfrak{m}}(\Gamma_h) = \rho^{\mathfrak{m}}(\Gamma) = \bar{\Gamma}(\mathfrak{m})$. Choose some X_{ih}.

(20.5.1) Make the identification $\bar{X}_{ih} = \bar{\Gamma}(\mathfrak{m})$ via (the map induced by) $\rho^{\mathfrak{m}}$ whenever $\rho^{\mathfrak{m}}(X_{ih}) = \bar{\Gamma}(\mathfrak{m})$; otherwise via any Γ-isomorphism $\bar{X}_{ih} \cong \bar{\Gamma}(\mathfrak{m})$ [see Lemma 20.2].

To show that these identifications satisfy the Λ-coherence conditions, we first claim that, for each fixed i and h, we have: $\rho^{\mathfrak{m}}(X_{ih}) = \bar{\Gamma}(\mathfrak{m})$ *for almost all unsplit and strictly split* $\mathfrak{m} \in \operatorname{maxsupp}_\Lambda(\Gamma_h)$.

In order to find a contradiction, suppose that the set $\mathcal{S}$ of unsplit and nonstrictly split $\mathfrak{m} \in \operatorname{maxsupp}_\Lambda(\Gamma_h)$ such that $\rho^{\mathfrak{m}}(X_{ih}) \subset \bar{\Gamma}(\mathfrak{m})$ is infinite. Since i and h are now fixed, write $X = X_{ih}$, and let $\tilde{X}$ be the direct sum of X with all coordinate rings Γ_k such that $k \neq h$. Since every $\mathfrak{m} \in \mathcal{S}$ supports only Γ_h, we have $\rho^{\mathfrak{m}}(\tilde{X}) = \rho^{\mathfrak{m}}(X)$. Therefore $\rho^{\mathfrak{m}}(\tilde{X}) \subset \bar{\Gamma}(\mathfrak{m})$ for every $\mathfrak{m} \in \mathcal{S}$. Equivalently: for each $\mathfrak{m} \in \mathcal{S}$, $\tilde{X}$ fails to be comaximal with $\ker(\rho^{\mathfrak{m}})$. Therefore, for all $\mathfrak{m} \in \mathcal{S}$, some maximal ideal, say $M(\mathfrak{m})$ of Γ contains both $\tilde{X}$ and $\ker(\rho^{\mathfrak{m}}) = \Gamma\mathfrak{m}$, and hence contains both $\tilde{X}$ and $\mathfrak{m}$.

We have $\Gamma/\tilde{X} \cong \Gamma_h/X$ as Γ-modules, and therefore $\Gamma/\tilde{X}$ has finite length. Consequently, $\tilde{X}$ is contained in only finitely many maximal ideals of Γ. Therefore the set $\mathcal{T}$ of maximal ideals $M(\mathfrak{m})$ ($\mathfrak{m} \in \mathcal{S}$) of Γ is a finite set, and each $M(\mathfrak{m})$ contains $\mathfrak{m}$. Therefore some $M \in \mathcal{T}$ contains infinitely many $\mathfrak{m} \in \mathcal{S}$. But this is impossible because each maximal ideal M of Γ contains a unique maximal ideal of Λ, namely the maximal ideal $M \cap \Lambda$.

Λ-coherence conditions. (i) Choose $X = X_{ih}$, and let d be any nonzero element of X. We show that $(T \subseteq X) = (\Lambda d \subseteq X)$ works. Except for $\mathfrak{m}$ in some finite set $\mathcal{E}$ of singular maximal ideals in $\operatorname{maxsupp}_\Lambda(\Gamma_h)$, we have $\bar{X}(\mathfrak{m}) = \rho^{\mathfrak{m}}(X) = \bar{\Gamma}(\mathfrak{m})$. Therefore, for $\mathfrak{m} \in \operatorname{maxsupp}_\Lambda(\Gamma_h) - \mathcal{E}$ we have $\bar{d}(\mathfrak{m}) = \rho^{\mathfrak{m}}(d)$ [with respect to the identification $\bar{X}(\mathfrak{m}) = \bar{\Gamma}(\mathfrak{m})$]. Therefore, after a finite enlargement of $\mathcal{E}$ we have $\bar{d}(\mathfrak{m}) \in k(\mathfrak{m})$ for all $\mathfrak{m} \in \operatorname{maxsupp}_\Lambda(\Gamma_h) - \mathcal{E}$ [Lemma 20.3]. For such $\mathfrak{m}$, $\rho^{\mathfrak{m}}(\Lambda d)$ equals either $k(\mathfrak{m})$ or 0. Thus it now suffices to show that $\rho^{\mathfrak{m}}(d) \neq 0$ for almost all unsplit and strictly split $\mathfrak{m} \in \operatorname{maxsupp}_\Lambda(\Gamma_h)$. This holds since the nonzero element d of the noetherian domain Γ_h of dimension 1 is contained in only finitely many maximal ideals of Γ_h [Lemma 8.1], because Γ_h is the unique coordinate ring supported by $\mathfrak{m}$, and $\Gamma_h\mathfrak{m}$ is the intersection of at most two maximal ideals of Γ_h [Lemma 11.6]. [For the full details here, see the contradiction argument following (20.5.1).]

(ii) Fix some Γ_h and consider any Γ-homomorphism $\sigma\colon X_{ih} \to X_{jh}$. We allow the possibility $i = j$. Since Γ_h is an integral domain and its two ideals X_{ih} and X_{jh} are both nonzero, Q-localization of σ yields the $(\Gamma_h)_Q$-homomorphism σ_Q of $(\Gamma_h)_Q$ to itself, and therefore σ_Q equals multiplication by some element of $(\Gamma_h)_Q$ that we call simply σ. We have $\sigma = x/c$ for some $x, c \in \Lambda$, since $\Gamma_Q = \Lambda_Q$. Moreover, the regular element c is contained in only finitely many maximal ideals of Λ [Lemma 8.1]. For any other maximal ideal $\mathfrak{m}$ of Λ we have $\bar{\sigma}(\mathfrak{m}) = \bar{x}(\mathfrak{m})/\bar{c}(\mathfrak{m}) \in k(\mathfrak{m})$, as desired. □

As already mentioned, when singspec(Λ) is finite the following lifting theorem becomes a very special case of what is proved in Wiegand's Lifting Theorem [**W1**, 1.1].

THEOREM 20.6 (Lifting theorem). *Let* $(S \subseteq X)$ *be a* $(\Lambda \subseteq \Gamma)$*-inclusion, where* ${}_\Gamma X$ *is projective. Let* $\mathfrak{n} \in \operatorname{singspec}(\Lambda)$*, and let* $\mathcal{E}$ *be a finite subset of* $\operatorname{singspec}(\Lambda) - \{\mathfrak{n}\}$*. Also let* $\psi \in \operatorname{Aut}_{\bar\Gamma(\mathfrak{n})} \bar X(\mathfrak{n})$ *be such that* $\det(\psi) = 1$*.*

Then there is an automorphism $\alpha \in \operatorname{Aut}_\Gamma(X)$ *such that* $\det(\alpha) = 1$ *and the following conditions hold.*

(i) $\bar\alpha(\mathfrak{n}) = \psi$.
(ii) $\bar\alpha(\mathfrak{m}) = 1$ $(\forall \mathfrak{m} \in \mathcal{E})$.
(iii) $\bar S(\mathfrak{m})\,\bar\alpha(\mathfrak{m}) = \bar S(\mathfrak{m})$ *for all* $\mathfrak{m} \in \operatorname{singspec}(\Lambda) - \{\mathfrak{n}\}$.

Here $\bar\alpha(\mathfrak{m})$ *denotes the automorphism of* $\bar X(\mathfrak{m})$ *induced by* α*.*

PROOF. Let $Y = \oplus_\nu Y_\nu$ be a decomposition of a module over some ring. An *elementary transvection automorphism* of Y (with respect to the given decomposition) is an automorphism of the form $1 + \sigma$ where σ maps some Y_μ to some Y_ν such that $\mu \neq \nu$, and $\sigma(Y_\rho) = 0$ for all $\rho \neq \mu$. We call $Y_\mu \to Y_\nu$ the *active part* of σ. Since X is projective there is a decomposition

$$X = \oplus_h \oplus_{i=1}^{n(h)} X_{ih} \tag{20.6.1}$$

where each X_{ih} is isomorphic to a nonzero ideal of coordinate ring Γ_h (a Dedekind domain) of Γ, and the summation runs over all possible h. Since we allow $n(h) = 0$, we are not requiring every h to actually occur in the summation. Choose a global matrix setup $\mathcal{X}$ whose set of modules consists of the modules X_{ih} [Lemma 20.5]. Then for each $\mathfrak{m} \in \operatorname{singspec}(\Lambda)$ one of the following two decompositions holds, according to the nature of $\mathfrak{m}$.

(20.6.2)
$$\bar X(\mathfrak{m}) = \begin{cases} \oplus_{i=1}^{n(h)} \bar X_{ih}(\mathfrak{m}) = \bar\Gamma(\mathfrak{m})^{n(h)} & (\mathfrak{m} \text{ unsplit or nonstrictly split}) \\ (\oplus_{i=1}^{n(h_1)} \bar X_{ih_1}(\mathfrak{m})) \oplus (\oplus_{i=1}^{n(h_2)} \bar X_{ih_2}(\mathfrak{m})) & \\ \qquad = k(\mathfrak{m})_1^{n(h_1)} \oplus k(\mathfrak{m})_2^{n(h_2)} & (\mathfrak{m} \text{ strictly split}) \end{cases}$$

In the first decomposition, h is the unique index such that $\mathfrak{m}$ is in the support of Γ_h, while in the second decomposition (corresponding to the strictly split case), h_1 and h_2 arise from the two coordinate rings Γ_{h_1} and Γ_{h_2} whose support contains $\mathfrak{m}$ [and one or both of $n(h_1)$ and $n(h_2)$ can equal zero]. The final equality in each of the two decompositions results from the identifications (20.4.1) in the global matrix setup $\mathcal{X}$. Thus, in the unsplit and nonstrictly split cases, each summand $\bar X_{ih}(\mathfrak{m})$ is identified (by $\mathcal{X}$) with the copy of $\bar\Gamma(\mathfrak{m})$ in coordinate i of $\bar\Gamma(\mathfrak{m})^{n(h)}$. An analogous statement holds for the strictly split case.

Now consider the given ψ. We claim that, in order to lift ψ as described, *we may assume that* ψ *is an elementary transvection automorphism of* $\bar X(\mathfrak{n})$, with respect to the appropriate decomposition in (20.6.2).

To prove this, suppose first that $\mathfrak{n}$ is unsplit, so that $\bar\Gamma(\mathfrak{m})$ is a field. Then use the well-known (and easily proved) fact that every matrix of determinant 1 over a field is a product of elementary transvection matrices.

Suppose next that $\mathfrak{n}$ is nonstrictly or strictly split. Then $\bar\Gamma(\mathfrak{m}) = k(\mathfrak{m})_1 \times k(\mathfrak{m})_2$, and so $\psi = (\psi_1, \psi_2)$, can be identified with an ordered pair of matrices over fields.

Therefore $\psi = (\psi_1, 1)\cdot(1, \psi_2)$; and so factoring each ψ_i as in the previous paragraph proves the claim.

By the claim, we start with $\psi = 1 + \bar{\sigma}(\mathfrak{n})$, an elementary transvection automorphism. To keep the notation simple, we may assume that the active part of $\bar{\sigma}(\mathfrak{n})$ is $\bar{X}_{11}(\mathfrak{n}) \to \bar{X}_{21}(\mathfrak{n})$. We lift ψ in two approximations. Each lifted map will again be an elementary transvection automorphism of X, and hence will have determinant 1 as desired. Therefore we ignore this determinant from now on.

For the first approximation, we ignore all maximal ideals except $\mathfrak{n}$. Since X_{11} and X_{21} are projective we can lift $\bar{\sigma}(\mathfrak{n})$ to a homomorphism $\sigma\colon X_{11} \to X_{21}$. Then the elementary transvection automorphism $\alpha' = 1 + \sigma$ of X lifts σ as specified in statement (i) of the theorem.

We claim that there is a finite subset $\mathcal{E}' \subseteq \operatorname{singspec}(\Lambda)$ such that for every $\mathfrak{m} \in \operatorname{singspec}(\Lambda) - \mathcal{E}'$, $\mathfrak{m}$ is unsplit or nonstrictly split, and there is a decomposition

$$(20.6.3) \qquad \left(\bar{S}(\mathfrak{m}) \subseteq \bar{X}(\mathfrak{m})\right) = \oplus_{i=1}^{n(h)} \left(V_{ih}(\mathfrak{m}) \subseteq \bar{X}_{ih}(\mathfrak{m})\right) = \left(k(\mathfrak{m}) \subseteq \bar{\Gamma}(\mathfrak{m})\right)^{n(\mathfrak{m})}$$

in which each $\bar{X}_{ih}(\mathfrak{m})$ is identified (by $\mathcal{X}$) with the copy of $\bar{\Gamma}(\mathfrak{m})$ in coordinate i of $\bar{\Gamma}(\mathfrak{m})^{n(h)}$, and $V_{ih}(\mathfrak{m}) = k(\mathfrak{m})$ via this identification.

To prove this, begin with $\mathcal{E}'$ equal to the finite set of strictly split maximal ideals. (We shall soon enlarge $\mathcal{E}'$, keeping it finite.) Then each $\bar{X}(\mathfrak{m})$ ($\mathfrak{m} \in \operatorname{singspec}(\Lambda) - \mathcal{E}'$) is the direct sum of the $n(h)$ summands $\bar{X}_{ih}$, as shown by the first equality in (20.6.3). In addition, each $\bar{X}_{ih}$ is identified with the coordinate-i copy of $\bar{\Gamma}(\mathfrak{m})$, as shown by the second equality of (20.6.3).

Temporarily fix i, h. By Λ-coherence condition (20.4.2) in the definition of "global matrix setup", there is a $(\Lambda \subseteq \Gamma)$-inclusion $(T_{ih} \subseteq X_{ih})$ such that, for *almost all* unsplit and nonstrictly split $\mathfrak{m} \in \operatorname{maxsupp}_\Lambda(X_{ih})$, we have $\left(\bar{T}_{ih}(\mathfrak{m}) \subseteq \bar{X}_{ih}(\mathfrak{m})\right) = \left(k(\mathfrak{m}) \subseteq \bar{\Gamma}(\mathfrak{m})\right)$. Direct-summing over i and h yields a $(\Lambda \subseteq \Gamma)$-inclusion $(T \subseteq X)$ whose residue inclusions are $\left(k(\mathfrak{m}) \subseteq \bar{\Gamma}(\mathfrak{m})\right)^{n(h)}$ for almost all $\mathfrak{m}$ in $\operatorname{singspec}(\Lambda)$. [Note: This holds for all $\mathfrak{m}$ not in the support of X because we allow $n(h) = 0$.] Therefore, after a finite enlargement of $\mathcal{E}'$ we get $V_{ih}(\mathfrak{m}) = k(\mathfrak{m})$ for all i, h and all $\mathfrak{m} \in \operatorname{singspec}(\Lambda) - \mathcal{E}'$, as shown. This completes the proof of (20.6.3).

Return to our first approximation $\alpha' = 1 + \sigma$, and recall that the active part of σ is $X_{11} \to X_{21}$. For each unsplit or nonstrictly split $\mathfrak{m} \in \operatorname{singspec}(\Lambda)$, the identifications $X_{11} = \bar{\Gamma}(\mathfrak{m})$ and $X_{21} = \bar{\Gamma}(\mathfrak{m})$ provided by $\mathcal{X}$ allow us to consider $\bar{\sigma}(\mathfrak{m})$ to be a multiplication map on $\bar{\Gamma}(\mathfrak{m})$, by an element of $\bar{\Gamma}(\mathfrak{m})$ that we again call $\bar{\sigma}(\mathfrak{m})$. Moreover, by Λ-coherence condition (ii), we have $\bar{\sigma}(\mathfrak{m}) \in k(\mathfrak{m})$ *for almost all unsplit or nonstrictly split* $\mathfrak{m}$.

Let $\mathcal{F}$ be the finite subset of $\operatorname{singspec}(\Lambda) - \{\mathfrak{n}\}$ consisting of the following maximal ideals.

(20.6.4) (i) The maximal ideals in the given set $\mathcal{E}$.
- (ii) All strictly split maximal ideals in $\operatorname{singspec}(\Lambda) - \{\mathfrak{n}\}$ [a finite set by Proposition 11.4].
- (iii) All unsplit and nonstrictly split maximal ideals in the finite set $\mathcal{E}'$ mentioned above (20.6.3).
- (iv) All unsplit and nonstrictly split maximal ideals $\mathfrak{m} \in \operatorname{singspec}(\Lambda) - \{\mathfrak{n}\}$ such that $\bar{\sigma}(\mathfrak{m}) \notin k(\mathfrak{m})$.

It suffices to prove the theorem with $\mathcal{F}$ in place of $\mathcal{E}$. Note that we have partitioned $\operatorname{singspec}(\Lambda)$ into three subsets: The 1-element set $\{\mathfrak{n}\}$, the finite set $\mathcal{F}$, and $\mathcal{F}' =$

$\text{singspec}(\Lambda) - \{\mathfrak{n}\} - \mathcal{F}$. By the Chinese Remainder Theorem, there exists $\lambda \in \Lambda$ such that $\bar{\lambda}(\mathfrak{n}) = 1$, and $\bar{\lambda}(\mathfrak{m}) = 0$ for all $\mathfrak{m} \in \mathcal{F}$. Since we are applying this theorem in Λ, we have $\bar{\lambda}(\mathfrak{m}) \in k(\mathfrak{m})$ for all maximal ideals $\mathfrak{m}$.

Let $\alpha = 1 + \lambda\sigma$. Since $\bar{\lambda}(\mathfrak{n}) = 1$, we have $\bar{\alpha}(\mathfrak{n}) = \bar{\alpha}'(\mathfrak{n}) = \psi$, as demanded by condition (i) of the theorem. Since $\bar{\lambda}(\mathfrak{m}) = 0$ for all $\mathfrak{m} \in \mathcal{F}$, we have $\bar{\alpha}(\mathfrak{m}) = 1$ for all $\mathfrak{m} \in \mathcal{F}$, as demanded by condition (ii).

To verify condition (iii) it suffices to consider the active part $\bar{X}_{11}(\mathfrak{m}) \to \bar{X}_{21}(\mathfrak{m})$ of $\bar{\alpha}(\mathfrak{m})$. Take $\mathfrak{m} \in \mathcal{F}'$. Then $\mathfrak{m}$ is unsplit or strictly split and we are in the situation of (20.6.3). In this situation, $\bar{X}_{11}(\mathfrak{m})$ and $\bar{X}_{21}(\mathfrak{m})$ are both identified with copies of $\bar{\Gamma}(\mathfrak{m})$, and $V_{11}(\mathfrak{m})$ and $V_{12}(\mathfrak{m})$ are both identified with the same $k(\mathfrak{m})$-subspace $k(\mathfrak{m})$ of $\bar{\Gamma}(\mathfrak{m})$.

With respect to this identification, we can view $\bar{\sigma}(\mathfrak{m})$ as either a map $\bar{X}_{11} \to \bar{X}_{21}$ or as multiplication on $\bar{\Gamma}(\mathfrak{m})$ by $\bar{\sigma}(\mathfrak{m})$, viewed as an element of $\bar{\Gamma}(\mathfrak{m})$. Moreover, by (20.6.4)(iv) and our choice of λ, both $\bar{\sigma}(\mathfrak{m})$ and $\bar{\lambda}(\mathfrak{m})$ are elements of $k(\mathfrak{m})$, in the present situation.

The critical observation is that multiplication by the element $\bar{\lambda}(\mathfrak{m})\bar{\sigma}(\mathfrak{m})$ of $k(\mathfrak{m})$ takes $k(\mathfrak{m})$ to itself. This means that the *map* $\bar{\lambda}(\mathfrak{m})\bar{\sigma}(\mathfrak{m})$ takes $V_{11}(\mathfrak{m}) \to V_{21}(\mathfrak{m})$. Since $\bar{\alpha}(\mathfrak{m})$ is the identity on all inactive coordinates, we see that $\bar{\alpha}(\mathfrak{m})$ takes $\bar{S}(\mathfrak{m})$ to itself, as demanded by condition (iii). □

21. Residue Endo-idèles on Γ-modules

This section develops a group action whose orbit is the set of isomorphism classes in $\text{rgen}(S)$ for an arbitrary Γ-separated Λ-module S. This is a crude preliminary version of the action of residue unit idèles developed in Section 22 and, as just mentioned, appies only to separated modules. The modules S considered in the present section will occur as separated covers $S \twoheadrightarrow M$ of arbitrary modules $M \in \text{fingen}_\infty(\Lambda)$, in the next section. However, we can have $S \notin \text{fingen}_\infty(\Lambda)$ when $M \in \text{fingen}(\Lambda)$; and so *we do not assume that* $S \in \text{fingen}_\infty(\Lambda)$ *in the present section.*

DEFINITION 21.1 (Residue endo-idèle on X). Let $X \in \text{fingen}(\Gamma)$, and assume that $\bar{X}(\mathfrak{m})$ is $\bar{\Gamma}(\mathfrak{m})$-free $\big(\forall \mathfrak{m} \in \text{maxspec}(\Lambda)\big)$. Recall that a module $X \in \text{fingen}(\Gamma)$ has this property if and only there exists a finitely generated Λ-submodule $S \subseteq X$ such that $X = \Gamma \otimes_\Lambda S = \Gamma S$ [Theorem 17.10]. In the terminology of Definition 17.1, we call $(S \subseteq X)$ a *standard* $(\Lambda \subseteq \Gamma)$-inclusion.

A *residue endo-idèle on* X is an indexed family of automorphisms

$$\boldsymbol{\psi} = \big\{\psi(\mathfrak{m}) \in \text{Aut}_\Gamma\big(\bar{X}(\mathfrak{m})\big) \bigm| \mathfrak{m} \in \text{singspec}(\Lambda)\big\}$$

such that the following two conditions hold.

(21.1.1) (i) For some standard $(\Lambda \subseteq \Gamma)$-inclusion $(S \subseteq X)$, we have $\bar{S}(\mathfrak{m})\psi(\mathfrak{m}) = \bar{S}(\mathfrak{m})$ $\big(\text{a}\forall \mathfrak{m} \in \text{singspec}(\Lambda)\big)$. (By Theorem 17.6, this then holds for every such standard inclusion.)

(ii) Each $\psi(\mathfrak{m})$ is *locally liftable;* that is, there is a $\Gamma_\mathfrak{m}$-automorphism $\widetilde{\psi}(\mathfrak{m})$ of $X_\mathfrak{m}$ that induces $\psi(\mathfrak{m})$ modulo $\mathfrak{m}_\mathfrak{m}$. (Note that $\widetilde{\psi}(\mathfrak{m})$ is not uniquely determined by $\psi(\mathfrak{m})$.)

We say that the residue endo-idèle $\boldsymbol{\psi}$ on X is *globally liftable* if there is a Γ-automorphism θ of X such that, for every $\mathfrak{m} \in \text{singspec}(\Lambda)$, $\bar{\theta}(\mathfrak{m}) = \psi(\mathfrak{m})$.

Note that the set of residue endo-idèles on X is a group under the pointwise product: $\boldsymbol{\psi}\boldsymbol{\phi}$ is the indexed family of automorphisms $\psi(\mathfrak{m})\phi(\mathfrak{m})$.

Since we do matrix-like manipulations with these idèles, they act on the right-hand side of their arguments.

NOTATION 21.2 (Triangular form). Let X be as in Definition 21.1. Since Γ is a direct sum of Dedekind domains, the torsion submodule $t(X)$ is a direct summand of X. Choose a decomposition

$$X = X_\infty \oplus t(X) \tag{21.2.1}$$

With respect to this decomposition, every residue endo-idèle $\boldsymbol{\psi}$ on X has, for each $\mathfrak{m} \in \text{singspec}(\Lambda)$, an upper triangular form(acting on the right!):

$$\psi(\mathfrak{m}) = \begin{pmatrix} \psi_\infty(\mathfrak{m}) & \psi_{12}(\mathfrak{m}) \\ 0 & \psi_t(\mathfrak{m}) \end{pmatrix}$$

where $\psi_\infty(\mathfrak{m})$ and $\psi_t(\mathfrak{m})$ are Γ-automorphisms of $\bar{X}_\infty(\mathfrak{m})$ and $\overline{t(X)}(\mathfrak{m})$, respectively, and $\psi_{12}(\mathfrak{m})$ is a Γ-homomorphism $\bar{X}_\infty(\mathfrak{m}) \to \overline{t(X)}(\mathfrak{m})$. We abbreviate this by the notation:

$$\boldsymbol{\psi} = \begin{pmatrix} \boldsymbol{\psi}_\infty & \boldsymbol{\psi}_{12} \\ 0 & \boldsymbol{\psi}_t \end{pmatrix} \tag{21.2.2}$$

If $\boldsymbol{\psi}_{12} = 0$, we use the following more compact notation in place of (21.2.2).

$$\boldsymbol{\psi} = \boldsymbol{\psi}_\infty \oplus \boldsymbol{\psi}_t \tag{21.2.3}$$

Although $\bar{X}(\mathfrak{m})$ is a free $\bar{\Gamma}(\mathfrak{m})$-module, its direct summand $\bar{X}_\infty(\mathfrak{m})$ need not be free if $\mathfrak{m}$ is split. However, $\psi_\infty(\mathfrak{m})$ does have a determinant, as defined in Remarks 20.1, and $\det(\psi_\infty(\mathfrak{m})) \in \bar{\Gamma}(\mathfrak{m})^\times$. Let $\det(\boldsymbol{\psi}_\infty)$ denote the indexed family $\{\det(\psi_\infty(\mathfrak{m})) \mid \mathfrak{m} \in \text{singspec}(\Lambda)\}$. We note:

(21.2.4) $\det(\boldsymbol{\psi}_\infty)$ is independent of the choice of summand X_∞ complementing $t(X)$, in the decomposition (21.2.1).

This holds because $\det(\boldsymbol{\psi}_\infty) = \det(\boldsymbol{\psi})/\det(\boldsymbol{\psi}_t)$.

LEMMA 21.3. *Let $X \in \text{fingen}(\Gamma)$ be such that $\bar{X}(\mathfrak{m})$ is $\bar{\Gamma}(\mathfrak{m})$-free for every $\mathfrak{m} \in \text{singspec}(\Lambda)$, and let $\theta \in \text{Aut}_\Gamma(X)$. Then the indexed family $\boldsymbol{\theta} = \{\bar{\theta}(\mathfrak{m}) \mid \mathfrak{m} \in \text{singspec}(\Lambda)\}$ is a globally liftable residue endo-idèle on X.*

PROOF. Since condition (21.1.1)(ii) obviously holds, we need only check condition (21.1.1)(i). Let $(S \subseteq X)$ be a standard $(\Lambda \subseteq \Gamma)$-inclusion. Since $S' = (S)\theta$ is a finitely generated Λ-submodule of X satisfying $\Gamma S' = X$, we have $\bar{S}'(\mathfrak{m}) = \bar{S}(\mathfrak{m})$ (a$\forall\mathfrak{m}$) [Theorem 17.6]. Thus, $\bar{S}(\mathfrak{m})\bar{\theta}(\mathfrak{m}) = \bar{S}(\mathfrak{m})$ (a$\forall\mathfrak{m}$), as desired. □

DEFINITION 21.4 (Action of endo-idèles on standard $(\Lambda \subseteq \Gamma)$-inclusions). Let $(S \subseteq X)$ be a standard $(\Lambda \subseteq \Gamma)$-inclusion, and let $\boldsymbol{\psi}$ be a residue endo-idèle on X. We define $S^{\boldsymbol{\psi}}$ in such a way that

$$(S^{\boldsymbol{\psi}} \subseteq X) \text{ is a standard } (\Lambda \subseteq \Gamma)\text{-inclusion.} \tag{21.4.1}$$

By (21.1.1)(i), $\bar{S}(\mathfrak{m})\psi(\mathfrak{m}) = \bar{S}(\mathfrak{m})$ (a$\forall\mathfrak{m}$). Moreover, since $\Gamma S = X$, we have $\bar{\Gamma}(\mathfrak{m})(\bar{S}(\mathfrak{m})\psi(\mathfrak{m})) = \bar{X}(\mathfrak{m})\psi(\mathfrak{m}) = \bar{X}(\mathfrak{m})$ ($\forall\mathfrak{m}$). Therefore, by Theorem 17.7, there is a unique standard $(\Lambda \subseteq \Gamma)$-inclusion, which we denote by $(S^{\boldsymbol{\psi}} \subseteq X)$, such that $\overline{S^{\boldsymbol{\psi}}}(\mathfrak{m}) = \bar{S}(\mathfrak{m})\psi(\mathfrak{m})$ ($\forall\mathfrak{m}$).

Thus, given X, $\boldsymbol{\psi}$ acts on the set of submodules S for which $(S \subseteq X)$ is a standard $(\Lambda \subseteq \Gamma)$-inclusion, and this is a group action in the sense that $S^{(\boldsymbol{\psi}\boldsymbol{\phi})} = (S^{\boldsymbol{\psi}})^{\boldsymbol{\phi}}$ and $S^{\mathbf{1}} = S$. In Corollary 21.11 we show that the isomorphism class of $S^{\boldsymbol{\psi}}$ is determined by X, $\boldsymbol{\psi}$, and the isomorphism class of S.

LEMMA 21.5 (Double cosets). *Let $(S \subseteq X)$ be a standard $(\Lambda \subseteq \Gamma)$-inclusion, and let $\boldsymbol{\psi}$ and $\boldsymbol{\phi}$ be residue endo-idèles on X. Then $S^{\boldsymbol{\psi}} \cong S^{\boldsymbol{\phi}}$ if and only if there are residue endo-idèles $\boldsymbol{\alpha}$ and $\boldsymbol{\theta}$ on X such that $S^{\boldsymbol{\alpha}} = S$, $\boldsymbol{\theta}$ is globally liftable, and $\boldsymbol{\psi} = \boldsymbol{\alpha}\boldsymbol{\phi}\boldsymbol{\theta}$.*

PROOF. Suppose first that $S^{\boldsymbol{\phi}} \cong S^{\boldsymbol{\psi}}$ via some isomorphism θ. Then $1 \otimes \theta$ induces an isomorphism between $X \cong \Gamma \otimes_\Lambda S^{\boldsymbol{\phi}}$ and $X \cong \Gamma \otimes_\Lambda S^{\boldsymbol{\psi}}$, extending θ; call this extended map θ also. For every $\mathfrak{m} \in \operatorname{singspec}(\Lambda)$, θ induces a Γ-automorphism $\bar{\theta}(\mathfrak{m})$ of $\bar{X}(\mathfrak{m})$, and the indexed family $\boldsymbol{\theta} = \{\bar{\theta}(\mathfrak{m}) \mid \mathfrak{m} \in \operatorname{singspec}(\Lambda)\}$ is a globally liftable residue endo-idèle on X [Lemma 21.3]. Since $S^{\boldsymbol{\psi}} = S^{\boldsymbol{\phi}}\theta$, we have $\bar{S}(\mathfrak{m})\psi(\mathfrak{m}) = \bar{S}(\mathfrak{m})\phi(\mathfrak{m})\bar{\theta}(\mathfrak{m})$, and hence $\bar{S}(\mathfrak{m})\psi(\mathfrak{m})\bar{\theta}(\mathfrak{m})^{-1}\phi(\mathfrak{m})^{-1} = \bar{S}(\mathfrak{m})$ $(\forall \mathfrak{m})$. Therefore, the residue endo-idèle $\boldsymbol{\alpha} = \boldsymbol{\psi}\boldsymbol{\theta}^{-1}\boldsymbol{\phi}^{-1}$ satisfies $S^{\boldsymbol{\alpha}} = S$, as desired.

Conversely, suppose that $\boldsymbol{\psi} = \boldsymbol{\alpha}\boldsymbol{\phi}\boldsymbol{\theta}$, where $S^{\boldsymbol{\alpha}} = S$ and $\boldsymbol{\theta}$ is globally liftable, and let θ be an automorphism of X lifting $\boldsymbol{\theta}$. Then $S^{\boldsymbol{\psi}} = S^{\boldsymbol{\alpha}\boldsymbol{\phi}\boldsymbol{\theta}} = S^{\boldsymbol{\phi}\boldsymbol{\theta}} = S^{\boldsymbol{\phi}}\theta \cong S^{\boldsymbol{\phi}}$. □

LEMMA 21.6. *Let $X \in \operatorname{fingen}(\Gamma)$ be such that $\bar{X}(\mathfrak{m})$ is $\bar{\Gamma}(\mathfrak{m})$-free for every $\mathfrak{m} \in \operatorname{singspec}(\Lambda)$, and let $\boldsymbol{\psi}$ be a residue endo-idèle on X. Fix a decomposition $X = X_\infty \oplus t(X)$, and view $\boldsymbol{\psi}$ in the upper triangular form displayed in (21.2.2). Define the indexed family $\boldsymbol{\sigma}$ by:*

$$\sigma(\mathfrak{m}) = \begin{pmatrix} 1_\infty(\mathfrak{m}) & \sigma_{12}(\mathfrak{m}) \\ 0 & \boldsymbol{\psi}_t(\mathfrak{m}) \end{pmatrix}$$

where $1_\infty(\mathfrak{m})$ is the identity map on $\bar{X}_\infty(\mathfrak{m})$, and $\sigma_{12}(\mathfrak{m})$ is any Γ-linear map from $\bar{X}_\infty(\mathfrak{m})$ to $\overline{t(X)}(\mathfrak{m})$. Then $\boldsymbol{\sigma}$ is a globally liftable endo-idèle on X.

Moreover, $\boldsymbol{\sigma}$ can be lifted to a Γ-automorphism θ of X whose restriction θ_t to $t(X)$ is any pre-assigned lifting of the family $\boldsymbol{\psi}_t$, and such that $\theta_\infty = 1$.

PROOF. (i) By the definition of a residue endo-idèle, each $\psi(\mathfrak{m})$ is locally liftable to an automorphism $\widetilde{\psi}(\mathfrak{m})$ of $X_\mathfrak{m}$. Identify each $t(X)_\mathfrak{m}$ with the $\mathfrak{m}$-primary component of the Γ-module $t(X)$ of finite length [Lemma 6.2]. Since the Γ-module $t(X)$ of finite length is the direct sum of its finitely many nonzero primary components [Lemma 6.3], the direct sum of the maps $\widetilde{\psi}_t(\mathfrak{m})$ is a Γ-automorphism θ_t of $t(X)$, lifting the map $\boldsymbol{\psi}_t$. Let θ_t be any Γ-automorphism of $t(X)$ that lifts $\boldsymbol{\psi}_t$.

Now consider the following commutative square.

$$\begin{array}{ccc} X_\infty & \xrightarrow{\ \theta_{12}\ } & t(X) \\ \downarrow & & \downarrow \\ \oplus_\mathfrak{m} \bar{X}_\infty(\mathfrak{m}) & \xrightarrow{\ \oplus\sigma_{12}(\mathfrak{m})\ } & \oplus_\mathfrak{m} \overline{t(X)}(\mathfrak{m}) \end{array}$$

The direct sums in the bottom row are taken over the (finitely many) maximal ideals $\mathfrak{m}$ in $\operatorname{singspec}(\Lambda)$ such that $\overline{t(X)}(\mathfrak{m}) \neq 0$. The vertical arrows are the natural maps, surjective by the Chinese Remainder Theorem. The lower horizontal arrow is as labeled, and the upper (dashed) horizontal arrow exists because the Γ-module X_∞ is projective.

Then the desired Γ-automorphism θ of X is given in matrix form by

$$\theta = \begin{pmatrix} 1 & \theta_{12} \\ 0 & \theta_t \end{pmatrix} \tag{21.6.1}$$

with respect to the decomposition $X = X_\infty \oplus t(X)$. □

THEOREM 21.7. *Let* $(S \subseteq X)$ *be a standard* $(\Lambda \subseteq \Gamma)$*-inclusion. Then the collection of isomorphism classes of* Λ*-modules* $S^{\boldsymbol{\psi}}$*, in which* $\boldsymbol{\psi}$ *ranges over all residue endo-idèles on* X*, is the restricted genus* $\mathrm{rgen}(S)$.

PROOF. Since $(S \subseteq X)$ is a standard $(\Lambda \subseteq \Gamma)$-inclusion, we conclude that $(S_\mathfrak{m} \subseteq X_\mathfrak{m})$ is a standard $(\Lambda_\mathfrak{m} \subseteq \Gamma_\mathfrak{m})$-inclusion $\big(\forall \mathfrak{m} \in \mathrm{maxspec}(\Lambda)\big)$. Now choose $T \in \mathrm{rgen}(S)$. We claim that we may assume that $(T \subseteq X)$ is a standard $(\Lambda \subseteq \Gamma)$ inclusion.

Since $T \in \mathrm{rgen}(S)$ we have $\Gamma \otimes_\Lambda T \cong \Gamma \otimes_\Lambda S = X$. Thus we have a natural map $T \to X$, namely $t \to 1 \otimes t$. Moreover, the map $T \to \Gamma \otimes_\Lambda T$ is an injection because each of its localizations at a maximal ideal $\mathfrak{m}$ of Λ becomes the injection $S_\mathfrak{m} \to \Gamma_\mathfrak{m} \otimes_{\Lambda_\mathfrak{m}} S_\mathfrak{m}$. Thus, after replacing T by its isomorphic copy in X, the claim is proved.

By Lemma 8.2(iii), we have $S_\mathfrak{m} = T_\mathfrak{m}$ in $X_\mathfrak{m}$ for almost all $\mathfrak{m} \in \mathrm{singspec}(\Lambda)$. For each such $\mathfrak{m}$, let $\widetilde{\psi}(\mathfrak{m})$ be the identity isomorphism on $S_\mathfrak{m} = T_\mathfrak{m}$. For each of the finitely many remaining $\mathfrak{m} \in \mathrm{singspec}(\Lambda)$, since $T \in \mathrm{genus}(S)$, we can choose a $\Lambda_\mathfrak{m}$-isomorphism $\widetilde{\psi}(\mathfrak{m})\colon S_\mathfrak{m} \cong T_\mathfrak{m}$. Then for all $\mathfrak{m} \in \mathrm{singspec}(\Lambda)$, we can extend $\widetilde{\psi}(\mathfrak{m})$ to the $\Gamma_\mathfrak{m}$-automorphism $1 \otimes \widetilde{\psi}(\mathfrak{m})$ of $\Gamma_\mathfrak{m} \otimes_{\Lambda_\mathfrak{m}} S_\mathfrak{m} = X_\mathfrak{m} = \Gamma_\mathfrak{m} \otimes_{\Lambda_\mathfrak{m}} T_\mathfrak{m}$. Call this extended map $\widetilde{\psi}(\mathfrak{m})$ also. Reducing each $\widetilde{\psi}(\mathfrak{m})$ modulo $\mathfrak{m}_\mathfrak{m}$ yields a residue endo-idèle $\boldsymbol{\psi} = \{\psi(\mathfrak{m})\}$ on X, and $T = S^{\boldsymbol{\psi}}$ by Theorem 17.4.

Conversely, suppose that $T = S^{\boldsymbol{\psi}}$. Then part of the meaning of "standard" in (21.4.1) is that $\Gamma \otimes_\Lambda T = X = \Gamma \otimes_\Lambda S$. Thus, it suffices to show that $T \in \mathrm{genus}(S)$.

Each $\psi(\mathfrak{m})$ is liftable to a $\Gamma_\mathfrak{m}$-automorphism $\widetilde{\psi}(\mathfrak{m})$ of $X_\mathfrak{m}$. We have $\bar{S}(\mathfrak{m})\psi(\mathfrak{m}) = \bar{T}(\mathfrak{m})$ in $\bar{X}(\mathfrak{m})$ for each $\mathfrak{m} \in \mathrm{singspec}(\Lambda)$, which implies that $S_\mathfrak{m}\widetilde{\psi}(\mathfrak{m}) = T_\mathfrak{m}$ in $X_\mathfrak{m}$, because each of $S_\mathfrak{m}$ and $T_\mathfrak{m}$ is the pullback of its unique residue inclusion. (This is the well-known local case of Theorem 17.4.) Thus, $S_\mathfrak{m} \cong T_\mathfrak{m}$ for all $\mathfrak{m} \in \mathrm{singspec}(\Lambda)$. On the other hand, if $\mathfrak{m}$ is a nonsingular maximal ideal of Λ, then $\Lambda_\mathfrak{m} = \Gamma_\mathfrak{m}$ in $\Gamma_\mathfrak{m}$, and therefore localizing the relation $\Gamma \otimes_\Lambda S = \Gamma \otimes_\Lambda T$ at $\mathfrak{m}$ again shows that $S_\mathfrak{m} \cong T_\mathfrak{m}$. Therefore, $T \in \mathrm{genus}(S)$. □

THEOREM 21.8. *Let* $(S \subseteq X)$ *be a standard* $(\Lambda \subseteq \Gamma)$*-inclusion and* $\boldsymbol{\psi}, \boldsymbol{\phi}$ *residue endo-idèles on* X*, expressed in triangular form with respect to some decomposition* $X = X_\infty \oplus t(X)$*, as in* (21.2.2)*. If* $\det(\boldsymbol{\psi}_\infty) = \det(\boldsymbol{\phi}_\infty)$*, then* $S^{\boldsymbol{\psi}} \cong S^{\boldsymbol{\phi}}$ *via a* Γ*-automorphism of* X*. Moreover, if* $\boldsymbol{\psi}_t = \boldsymbol{\phi}_t = \mathbf{1}_t$ *(the identity map on* $\overline{t(X)}(\mathfrak{m})$ *for all* $\mathfrak{m}$*), then this isomorphism can be chosen to equal the identity map on* $t(X)$.

PROOF. Fix a decomposition $X = X_\infty \oplus t(X)$, and let $\boldsymbol{\sigma}$ be the indexed family defined for $\mathfrak{m} \in \mathrm{singspec}(\Lambda)$ by:

$$\sigma(\mathfrak{m}) = \begin{pmatrix} 1_\infty(\mathfrak{m}) & -\psi_\infty(\mathfrak{m})^{-1}\psi_{12}(\mathfrak{m})\psi_t(\mathfrak{m})^{-1} \\ 0 & \psi_t(\mathfrak{m})^{-1} \end{pmatrix} \tag{21.8.1}$$

By Lemma 21.6, the indexed family $\boldsymbol{\sigma}$ is a globally liftable residue endo-idèle on X, and hence $S^{\boldsymbol{\psi}} \cong S^{\boldsymbol{\psi\sigma}}$. If we let $\boldsymbol{\psi}' = \boldsymbol{\psi\sigma}$, then for all $\mathfrak{m}$ we have:

$$(21.8.2) \qquad \psi'(\mathfrak{m}) = \begin{pmatrix} \psi_\infty(\mathfrak{m}) & 0 \\ 0 & 1_t(\mathfrak{m}) \end{pmatrix} = \psi_\infty(\mathfrak{m}) \oplus 1_t(\mathfrak{m})$$

where $1_t(\mathfrak{m})$ is the identity map on $\overline{t(X)}(\mathfrak{m})$. Moreover, if $\psi_t(\mathfrak{m}) = 1_t(\mathfrak{m})$ on $\overline{t(X)}(\mathfrak{m})$ for all $\mathfrak{m}$, then we can choose the Γ-automorphism lifting $\boldsymbol{\sigma}$ to equal the identity map 1_t on $t(X)$ [Lemma 21.6].

Similarly, we can define a globally liftable residue endo-idèle $\boldsymbol{\tau}$ on X such that $\boldsymbol{\phi\tau} = \boldsymbol{\phi}_\infty \oplus \mathbf{1}_t$. Moreover, if $\phi_t(\mathfrak{m}) = 1_t(\mathfrak{m})$ on $\overline{t(X)}(\mathfrak{m})$ for all $\mathfrak{m}$, then we can choose the Γ-automorphism lifting $\boldsymbol{\tau}$ to equal the identity map 1_t on $t(X)$. Changing notation, we can assume that $\boldsymbol{\psi} = \boldsymbol{\psi}_\infty \oplus \mathbf{1}_t$ and $\boldsymbol{\phi} = \boldsymbol{\phi}_\infty \oplus \mathbf{1}_t$.

The rest of the proof is more complicated than the preceding because — if singspec(Λ) is an infinite set — there might be no globally liftable endo-idèle $\boldsymbol{\theta}$ such that $\boldsymbol{\phi} = \boldsymbol{\psi\theta}$.

Let $\boldsymbol{\varepsilon} = \boldsymbol{\phi}^{-1}\boldsymbol{\psi}$. Then, because of our new forms for $\boldsymbol{\phi}$ and $\boldsymbol{\psi}$, we have $\boldsymbol{\varepsilon} = \boldsymbol{\varepsilon}_\infty \oplus \mathbf{1}_t$, and $\det(\boldsymbol{\varepsilon}_\infty) = 1$. Since $\boldsymbol{\phi}$ and $\boldsymbol{\psi}$ are residue endo-idèles on X, we have

$$(21.8.3) \qquad \bar{S}(\mathfrak{m})\phi(\mathfrak{m}) = \bar{S}(\mathfrak{m}) = \bar{S}(\mathfrak{m})\psi(\mathfrak{m}) \qquad (\text{a}\forall\mathfrak{m})$$

Let $\mathcal{E}$ be the finite set of "exceptional" maximal ideals consisting of those $\mathfrak{m}$ for which (21.8.3) fails, together with those $\mathfrak{m}$ for which $\overline{t(X)}(\mathfrak{m}) \neq 0$.

By Lifting Theorem 20.6, for each $\mathfrak{n} \in \mathcal{E}$ we can lift the Γ-automorphism $\varepsilon_\infty(\mathfrak{n})$ of determinant 1 to a Γ-automorphism $\theta_\infty(\mathfrak{n})$ of X_∞ which induces $\varepsilon_\infty(\mathfrak{n})$ modulo $\mathfrak{n}$, induces the identity map modulo every other maximal ideal in $\mathcal{E}$, and takes $\overline{S^\phi}(\mathfrak{m})$ onto itself for every $\mathfrak{m} \in \text{maxspec}(\Lambda) - \mathcal{E}$. Let θ_∞ be the product (in some order) of these lifted automorphisms $\theta_\infty(\mathfrak{n})$, and let $\theta = \theta_\infty \oplus 1_t$, a Γ-automorphism of X. Then $\boldsymbol{\theta} = \{\bar{\theta}(\mathfrak{m}) \mid \mathfrak{m} \in \text{singspec}(\Lambda)\}$ is a globally liftable residue endo-idèle on X, by Lemma 21.3.

We claim that $S^{\boldsymbol{\phi\theta}} = S^{\boldsymbol{\psi}}$. Since these modules are the pullbacks of their residue inclusions [Theorem 17.4], it suffices to check this relation modulo each $\mathfrak{m} \in \text{singspec}(\Lambda)$. That is, it suffices to show that $\bar{S}(\mathfrak{m})\phi(\mathfrak{m})\bar{\theta}(\mathfrak{m}) = \bar{S}(\mathfrak{m})\psi(\mathfrak{m})$ for each $\mathfrak{m} \in \text{singspec}(\Lambda)$. Since $\phi(\mathfrak{m}) = \phi_\infty(\mathfrak{m}) \oplus 1_t$, and $\bar{\theta}(\mathfrak{m})$ and $\psi(\mathfrak{m})$ have analogous decompositions, it suffices to check that

$$(21.8.4) \qquad \bar{S}_\infty(\mathfrak{m})\phi_\infty(\mathfrak{m})\bar{\theta}_\infty(\mathfrak{m}) = \bar{S}_\infty(\mathfrak{m})\psi_\infty(\mathfrak{m}) \qquad (\forall\mathfrak{m} \in \text{singspec}(\Lambda))$$

If $\mathfrak{m} \in \mathcal{E}$, then (21.8.4) holds because $\bar{\theta}(\mathfrak{m}) = \varepsilon_\infty(\mathfrak{m})$ and $\boldsymbol{\varepsilon} = \boldsymbol{\phi}^{-1}\boldsymbol{\psi}$. If $\mathfrak{m} \notin \mathcal{E}$, then $\overline{t(X)}(\mathfrak{m}) = 0$, and hence $\bar{S}(\mathfrak{m}) = \bar{S}_\infty(\mathfrak{m})$; therefore, (21.8.4) follows from (21.8.3) together with the fact that $\bar{\theta}_\infty(\mathfrak{m})$ takes $\overline{S^\phi}(\mathfrak{m}) = \bar{S}(\mathfrak{m})\phi(\mathfrak{m})$ onto itself for every such $\mathfrak{m}$. Thus, $S^{\boldsymbol{\phi\theta}} = S^{\boldsymbol{\psi}}$, as claimed.

Now $S^{\boldsymbol{\psi}} = S^{\boldsymbol{\phi\theta}} = S^{\boldsymbol{\phi}}\theta \cong S^{\boldsymbol{\phi}}$, as desired.

Suppose, in addition, that $\boldsymbol{\psi}_t = \boldsymbol{\phi}_t = \mathbf{1}_t$, the family of identity maps on the modules $\overline{t(X)}(\mathfrak{m})$. Then as noted above, for the globally liftable residue endo-idèles $\boldsymbol{\sigma}$ and $\boldsymbol{\tau}$ such that $\boldsymbol{\psi\sigma} = \boldsymbol{\psi}_\infty \oplus \mathbf{1}_t$ and $\boldsymbol{\phi\tau} = \boldsymbol{\phi}_\infty \oplus \mathbf{1}_t$, we can choose the Γ-automorphisms lifting $\boldsymbol{\sigma}$ and $\boldsymbol{\tau}$ to equal the identity map on $t(X)$. Since also $\theta = \theta_\infty \oplus 1_t$, it follows from the above proof that the isomorphism $S^{\boldsymbol{\psi}} \cong S^{\boldsymbol{\phi}}$ can be chosen to be the identity map on $t(X)$. □

Recall that the direct-sum notation $\boldsymbol{\psi} = \boldsymbol{\psi}_\infty \oplus \mathbf{1}_t$ was defined in (21.2.3).

COROLLARY 21.9. *Let* $(S \subseteq X)$ *be a standard* $(\Lambda \subseteq \Gamma)$*-inclusion, and fix a decomposition* $X = X_\infty \oplus t(X)$. *Then*

(i) *The collection of isomorphism classes of* Λ*-modules of the form* $S^{\boldsymbol{\psi}}$, *in which* $\boldsymbol{\psi} = \boldsymbol{\psi}_\infty \oplus \mathbf{1}_t$ *with respect to this decomposition, is the restricted genus* $\mathrm{rgen}(S)$.

(ii) *Consider any decomposition* $X = Y_\infty \oplus t(X)$ *and a residue endo-idèle* $\boldsymbol{\phi} = \boldsymbol{\phi}_\infty \oplus \mathbf{1}_t$ *with respect to this decomposition. If* $\det(\boldsymbol{\psi}_\infty) = \det(\boldsymbol{\phi}_\infty)$, *then* $S^{\boldsymbol{\psi}} \cong S^{\boldsymbol{\phi}}$ *by a* Γ*-automorphism of* X *that equals the identity on* $t(X)$.

PROOF. (i) The set of isomorphism classes of modules $S^{\boldsymbol{\psi}}$, where $\boldsymbol{\psi}$ ranges over all residue endo-idèles on X, concides with $\mathrm{rgen}(S)$, by Theorem 21.7. Write $\boldsymbol{\psi}$ in upper triangular form (21.2.2) with respect to the specified decomposition, and then replace the (1,2)- and (2,2)-entries of this triangular form by 0 and 1_t, respectively. This again yields a residue endo-idèle, which by Theorem 21.8 is still isomorphic to $S^{\boldsymbol{\psi}}$.

(ii) $\boldsymbol{\phi}$ has a triangular form (21.2.2) with respect to the decomposition $X = X_\infty \oplus t(X)$, say:

$$\boldsymbol{\phi} = \begin{pmatrix} \boldsymbol{\phi}'_\infty & \boldsymbol{\phi}'_{12} \\ 0 & \mathbf{1}_t \end{pmatrix}$$

Moreover $\det(\boldsymbol{\phi}'_\infty) = \det(\boldsymbol{\phi}_\infty)$ because this determinant does not depend on the complement of $t(X)$ in X [see (21.2.4)]. Since the idèle $\boldsymbol{\psi} = \psi_\infty \oplus \mathbf{1}_t$ is also expressed with respect to the decomposition $X = X_\infty \oplus t(X)$, the desired conclusion is given in Theorem 21.8. □

COROLLARY 21.10. *Let* $(S \subseteq X)$ *be a standard* $(\Lambda \subseteq \Gamma)$*-inclusion and* $\boldsymbol{\psi}, \boldsymbol{\phi}, \boldsymbol{\chi}$ *residue endo-idèles on* X. *If* $S^{\boldsymbol{\psi}} \cong S^{\boldsymbol{\phi}}$, *then* $S^{\boldsymbol{\psi\chi}} \cong S^{\boldsymbol{\phi\chi}}$.

PROOF. If $S^{\boldsymbol{\psi}} \cong S^{\boldsymbol{\phi}}$, then we have a relation $\boldsymbol{\psi} = \boldsymbol{\alpha\phi\theta}$, where $S^{\alpha} = S$ and $\boldsymbol{\theta}$ is globally liftable [Lemma 21.5]. Therefore, $\boldsymbol{\psi\chi} = \boldsymbol{\alpha\phi\theta\chi} = \boldsymbol{\alpha\phi}(\boldsymbol{\theta\chi\theta}^{-1})\boldsymbol{\theta}$, and hence $S^{\boldsymbol{\psi\chi}} \cong S^{\boldsymbol{\phi}(\boldsymbol{\theta\chi\theta}^{-1})}$. Since conjugation does not change determinants, it follows from Theorem 21.8 that $S^{\boldsymbol{\phi}(\boldsymbol{\theta\chi\theta}^{-1})} \cong S^{\boldsymbol{\phi\chi}}$. □

COROLLARY 21.11. *Let* $(S \subseteq X)$ *and* $(T \subseteq X)$ *be standard* $(\Lambda \subseteq \Gamma)$*-inclusions such that* $S \cong T$. *Then for every residue endo-idèle* $\boldsymbol{\mathcal{X}}$ *on* X *we have* $S^{\boldsymbol{\mathcal{X}}} \cong T^{\boldsymbol{\mathcal{X}}}$.

PROOF. Choose a Λ-isomorphism $f\colon S \cong T$. Then, by standardness, $\psi = 1 \otimes f$ is a Γ-automorphism of $X = \Gamma \otimes_\Lambda S$ that extends f. Moreover, the indexed family $\boldsymbol{\psi} = \{\bar{\psi}(\mathfrak{m}) \mid \mathfrak{m} \in \mathrm{singspec}(\Lambda)\}$ is a globally liftable residue endo-idèle on X [Lemma 21.3], and $S^{\boldsymbol{\psi}} = T$.

We have $S^{\boldsymbol{\psi}} = T \cong S = S^{\mathbf{1}}$. Therefore $S^{\boldsymbol{\psi\mathcal{X}}} \cong S^{\mathbf{1}\boldsymbol{\mathcal{X}}} = S^{\boldsymbol{\mathcal{X}}}$ [Corollary 21.10], and hence $T^{\boldsymbol{\mathcal{X}}} = S^{\boldsymbol{\psi\mathcal{X}}} \cong S^{\boldsymbol{\psi}}$. □

22. Residue Unit Idèles

This section contains our construction of the set of isomorphism classes in the restricted genus of an arbitrary $M \in \mathrm{fingen}_\infty(\Lambda)$, in terms of a group action $M \to M^{\boldsymbol{u}}$ where $\boldsymbol{u}$ is an element of the group of residue unit idèles $\mathcal{I}$ [see (22.1.2)]. The construction is complicated by the fact that the full group of idèles only acts on faithful modules. For the unfaithful case, a direct summand $\mathcal{I}_{C(M)}$ of $\mathcal{I}$ is what actually acts. We define this action in Definition 22.4, and prove that its orbit is $\mathrm{rgen}(M)$ [Theorem 22.6]. Then we show how the isomorphism class of $(M \oplus N)^{\boldsymbol{u}}$ is obtained from M, N, and $\boldsymbol{u}$ [Theorem 22.7].

DEFINITIONS 22.1 ($\mathcal{I}$, $\mathcal{C}(M)$, $\Gamma_{\mathcal{C}(M)}$, $\mathcal{I}_{\mathcal{C}(M)}$). Recall [Notation 3.1] that $\Gamma = \oplus_{h\in\mathcal{H}}\Gamma_h$. We define the *torsionfree support set* of a Λ-module or Γ-module M to be

(22.1.1) $\mathcal{C}(M) = \{h \in \mathcal{H} \mid \Gamma_h M_Q \neq 0\}$; equivalently: $\{\Gamma_h \mid \Gamma_h M_Q \neq 0\}$

In terms of previously used terminology: *$\mathcal{C}(M)$ is the set of all Γ_h such that the Γ_h-rank of M is nonzero, and is therefore complete-locally determined.* [Definition 15.1, Lemma 15.2]

Let $\Gamma_{\mathcal{C}(M)} = \oplus_{h\in\mathcal{C}(M)}\Gamma_h$.

The group $\mathcal{I}$ of *residue unit idèles* is defined to be the set of indexed families $\boldsymbol{u}$ of the form:

(22.1.2) $\boldsymbol{u} = \{u(\mathfrak{m}) \in \bar{\Gamma}(\mathfrak{m})^\times \mid \mathfrak{m} \in \text{singspec}(\Lambda), \text{ and } u(\mathfrak{m}) \in k(\mathfrak{m})^\times \ (\text{a}\forall\mathfrak{m})\}$

under the operation or pointwise multiplication.

For Λ-module or Γ-module M, we define the group $\mathcal{I}_{\mathcal{C}(M)}$ of *$\mathcal{C}(M)$-idèles* as follows.

(22.1.3) $\mathcal{I}_{\mathcal{C}(M)}$ is the set of $\boldsymbol{u} \in \mathcal{I}$ such that the following conditions hold, for all $\mathfrak{m} \in \text{singspec}(\Lambda)$ and coordinate rings Γ_h of Γ.

- (i) $u(\mathfrak{m}) = 1$ if $\mathfrak{m}$ is in the support of only Γ_h, and $h \notin \mathcal{C}(M)$.
- (ii) $u(\mathfrak{m})_1 = 1$ if $\mathfrak{m}$ is in the support of two distinct coordinate rings Γ_{h_1} and Γ_{h_2} and $h_1 \notin \mathcal{C}(M)$. Similarly, $u(\mathfrak{m})_2 = 1$ if $\mathfrak{m}$ is in the support of Γ_{h_1} and Γ_{h_2} and $h_2 \notin \mathcal{C}(M)$.

The following remarks may help clarify this definition. Condition (i) applies when $\mathfrak{m}$ is unsplit or nonstrictly split, and condition (ii) applies when $\mathfrak{m}$ is strictly split [Lemma 11.6]. When $\mathfrak{m}$ is split (strictly or not) the assertion that $u(\mathfrak{m}) \in k(\mathfrak{m})$ in (22.1.2) means that $u(\mathfrak{m})$ is in the diagonal copy of $k(\mathfrak{m})$ in $k(\mathfrak{m})_1 \times k(\mathfrak{m})_2$ [see (11.5.1)].

Finally, we note the two extreme cases of this definition. *If M is faithful, then $\mathcal{C}(M) = \mathcal{H}$, so that $\mathcal{I}_{\mathcal{C}(M)} = \mathcal{I}$.* And, *If M has finite length, then $\mathcal{I}_{\mathcal{C}(M)} = \{1\}$.*

The next lemma gives the way that we usually view $\mathcal{C}(M)$.

LEMMA 22.2. *Let $S \twoheadrightarrow M$ be a Γ-separated cover of a finitely generated Λ-module, with associated Γ-module $X = \Gamma \otimes_\Lambda S = \Gamma S$, and choose a Γ-module decomposition $X = X_\infty \oplus t(X)$. Then $\mathcal{C}(M) = \mathcal{C}(X) = \mathcal{C}(X_\infty)$, and hence $\mathcal{I}_{\mathcal{C}(M)} = \mathcal{I}_{\mathcal{C}(X)}$.*

PROOF. We have $S/K \cong M$ where K has finite length [Corollary 18.9]. Since all maximal ideals of Dedekind-like rings have height 1 [Lemma 10.5], finite length of K implies that $K_Q = 0$ [Lemma 4.5], so that $S_Q \cong M_Q$. Moreover, we have $\Gamma S = X$ and $\Gamma_Q = \Lambda_Q$, and hence $X_Q \cong S_Q$. Since $t(X)$ has finite length, we have $(X_\infty)_Q \cong X_Q \cong M_Q$. □

The next lemma gives the basic relationship between residue unit idèles and residue endo-idèles. For the notation $\det(\psi_\infty)$, and the fact that it does not depend upon the particular decomposition $X = X_\infty \oplus t(X)$, see (21.2.4).

LEMMA 22.3. *Let $X \in \text{fingen}(\Gamma)$, suppose that $\bar{X}(\mathfrak{m})$ is $\bar{\Gamma}(\mathfrak{m})$-free $(\forall\mathfrak{m} \in \text{maxspec}(\Lambda))$ [This holds if and only if there is a standard $(\Lambda \subseteq \Gamma)$-inclusion $(S \subseteq X)$.], and fix a decomposition $X = X_\infty \oplus t(X)$. Then:*

(i) *For every residue endo-idèle* $\boldsymbol{\psi}$ *on* X, *the indexed family* $\boldsymbol{u} = \det(\boldsymbol{\psi}_\infty)$ *is a residue unit idèle in* $\mathcal{I}_{\mathcal{C}(X)}$.

(ii) *For every residue unit idèle* $\boldsymbol{u} = \det(\boldsymbol{\psi}_\infty)$ *in* $\mathcal{I}_{\mathcal{C}(X)}$, *there is a residue endo-idèle on* X *of the form* $\boldsymbol{\psi} = \boldsymbol{\psi}_\infty \oplus \mathbf{1}_t$ *(with respect to the given decomposition of* X*) such that* $\det(\boldsymbol{\psi}_\infty) = \boldsymbol{u}$.

PROOF. The equivalence of the freeness hypothesis with the existence of a standard inclusion is part of Theorem 17.10.

(i) First we prove that $\boldsymbol{u} = \det(\boldsymbol{\psi}_\infty)$ is an element of $\mathcal{I}_{\mathcal{C}(X)}$. If the maximal ideal $\mathfrak{m}$ is in the support of the single coordinate ring Γ_h, and $h \notin \mathcal{C}(X)$, then $\bar{X}_\infty(\mathfrak{m}) = 0$. Therefore, by convention, $\det\big(\psi_\infty(\mathfrak{m})\big) = 1$ [Remarks 20.1]. It follows that (22.1.3)(i) is satisfied; (22.1.3)(ii) follows similarly.

Recall, from Definition 21.1 of "residue endo-idèle", that there exists a standard $(\Lambda \subseteq \Gamma)$-inclusion $(S \subseteq X)$, such that $\bar{S}(\mathfrak{m})\psi(\mathfrak{m}) = \bar{S}(\mathfrak{m})$ $\big(\text{a}\forall\mathfrak{m} \in \operatorname{singspec}(\Lambda)\big)$.

It remains to show that $\det\big(\psi_\infty(\mathfrak{m})\big) \in k(\mathfrak{m})^\times$ (a$\forall\mathfrak{m}$). For this, we may disregard the finite number of striclty split $\mathfrak{m}$ and the finite number of $\mathfrak{m}$ such that $t(X)(\mathfrak{m}) \neq 0$. For all remaining $\mathfrak{m} \in \operatorname{singspec}(\Lambda)$, we have $\det\big(\psi_\infty(\mathfrak{m})\big) = \det\big(\psi(\mathfrak{m})\big)$. By (21.1.1)(i), $\psi(\mathfrak{m})$ acts as an automorphism on the $k(\mathfrak{m})$-vector space $\bar{S}(\mathfrak{m})$ (a$\forall\mathfrak{m}$), and for these $\mathfrak{m}$, the determinant of $\psi(\mathfrak{m})$ restricted to the $k(\mathfrak{m})$-vector space $\bar{S}(\mathfrak{m})$ is an element $\alpha \in k(\mathfrak{m})^\times$. To show that $\det\big(\psi(\mathfrak{m})\big) = \alpha$ it suffices to show that $\bar{S}(\mathfrak{m})$ contains a basis of the free $\bar{\Gamma}(\mathfrak{m})$-module $\bar{X}(\mathfrak{m})$. This holds because $\Gamma S = X$.

(ii) Fix a standard $(\Lambda \subseteq \Gamma)$-inclusion $(S \subseteq X)$. It exists by the freeness hypothesis, as mentioned above.

Before using the given $\boldsymbol{u}$ to construct the desired $\boldsymbol{\psi}$, we obtain a useful decomposition of $\bar{X}(\mathfrak{m})$ for every $\mathfrak{m} \in \operatorname{singspec}(\Lambda)$. Let $\mathcal{E}$ be the set of $\mathfrak{m} \in \operatorname{singspec}(\Lambda)$ such that either $t(X)_\mathfrak{m} \neq 0$ or $\mathfrak{m}$ is strictly split. We claim that $\mathcal{E}$ *is a finite set.*

The Γ-module $t(X)$ of finite length has finite length as a Λ-module [Lemma 10.10], and therefore $t(X)_\mathfrak{m} = 0$ for almost all $\mathfrak{m}$ [Lemma 6.3]. Since the number of strictly split $\mathfrak{m}$ is finite [Proposition 11.4] the claim is proved.

Next we choose a decomposition of each $\bar{X}(\mathfrak{m})$. Since reduction modulo $\mathfrak{m}$ distributes over direct sums, we have:

$$\bar{X}(\mathfrak{m}) = \overline{X_\infty}(\mathfrak{m}) \oplus \overline{t(X)}(\mathfrak{m}) \qquad \big(\forall\mathfrak{m} \in \operatorname{singspec}(\Lambda)\big) \tag{22.3.1}$$

For $\mathfrak{m} \in \operatorname{singspec}(\Lambda) - \mathcal{E}$ we refine this decomposition as follows. Standardness of $(S \subseteq X)$ as a $(\Lambda \subseteq \Gamma)$-inclusion implies that the residue inclusion $\big(\bar{S}(\mathfrak{m}) \subseteq \bar{X}(\mathfrak{m})\big)$ is a standard $\big(\bar{\Lambda}(\mathfrak{m}) \subseteq \bar{\Gamma}(\mathfrak{m})\big)$-inclusion. Since ${}_\Lambda S$ is finitely generated, the $k(\mathfrak{m})$-vector space $\bar{S}(\mathfrak{m})$ is finite-dimensional. Choose a decomposition into a direct sum of 1-dimensional subspaces V_i, as displayed in (22.3.2).

$$\bar{S}(\mathfrak{m}) = \oplus_{i=1}^{n(\mathfrak{m})} V_i(\mathfrak{m}) \qquad\qquad \bar{X}(\mathfrak{m}) = \overline{X_\infty}(\mathfrak{m}) = \oplus_{i=1}^{n(\mathfrak{m})} W_i(\mathfrak{m}) \tag{22.3.2}$$

Since the $(\Lambda \subseteq \Gamma)$-inclusion $(S \subseteq X)$ is standard, the $\big(\bar{\Lambda}(\mathfrak{m}) \subseteq \bar{\Gamma}(\mathfrak{m})\big)$-inclusion $\big(\bar{S}(\mathfrak{m}) \subseteq \bar{X}(\mathfrak{m})\big)$ is also standard. Therefore, as displayed in (22.3.2), $\bar{X}(\mathfrak{m})$ has a decomposition where each $W_i = \bar{\Gamma}(\mathfrak{m}) \otimes_{k(\mathfrak{m})} V_i(\mathfrak{m}) = \bar{\Gamma}(\mathfrak{m}) \cdot V_i(\mathfrak{m}) \cong \bar{\Gamma}(\mathfrak{m})$. Moreover, $\bar{X}(\mathfrak{m}) = \overline{X_\infty}(\mathfrak{m})$ by the definition of $\mathcal{E}$.

Now let $\boldsymbol{u} \in \mathcal{I}_{\mathcal{C}(X)}$ be given. We define $\boldsymbol{\psi} = \boldsymbol{\psi}_\infty \oplus \mathbf{1}_t$. For each $\mathfrak{m}$, let $1_t(\mathfrak{m})$ be the identity automorphism of $\overline{t(X)}(\mathfrak{m})$. We define each $\psi_\infty(\mathfrak{m}) \in \operatorname{Aut}(\,\overline{X_\infty}\,)(\mathfrak{m})$ in three cases, according to the nature of $\mathfrak{m}$.

Case 1: $\mathfrak{m} \in \mathcal{E}$ and $\mathfrak{m}$ is either unsplit or nonstrictly split. We claim that *the* $\bar{\Gamma}(\mathfrak{m})$*-module* $\overline{X_\infty}(\mathfrak{m})$ *is free.*

If $\mathfrak{m}$ is unsplit, then $\bar{\Gamma}(\mathfrak{m})$ is a field, and hence all of its modules are free. Suppose next that $\mathfrak{m}$ is nonstrictly split, and let $H = X_\infty$. Then $\Gamma_\mathfrak{m}$ is a principal ideal domain with exactly two maximal ideals [Theorem 11.3], and hence the projective $\Gamma_\mathfrak{m}$-module $H_\mathfrak{m}$ is free. Since $\bar{H}(\mathfrak{m}) = \overline{(H_\mathfrak{m})}(\mathfrak{m}_\mathfrak{m})$, this module is therefore $\bar{\Gamma}(\mathfrak{m}) = \overline{\Gamma_\mathfrak{m}}(\mathfrak{m}_\mathfrak{m})$-free, as claimed.

To define $\psi_\infty(\mathfrak{m})$, write the free $\bar{\Gamma}(\mathfrak{m})$-module $\bar{X}_\infty(\mathfrak{m})$ as a direct sum of free modules of rank 1, and let $\psi_\infty(\mathfrak{m})$ be the automorphism of $\overline{X_\infty}(\mathfrak{m})$ that equals multiplication by $u(\mathfrak{m})$ on the first summand, and equals the identity automorphism on the remaining summands.

Case 2: $\mathfrak{m} \in \mathcal{E}$ and $\mathfrak{m}$ is strictly split. Here $\bar{\Gamma}(\mathfrak{m})$ is the direct sum of two fields, say $\bar{\Gamma}(\mathfrak{m})_1$ and $\bar{\Gamma}(\mathfrak{m})_2$ [Theorem 11.3], and hence every $\bar{\Gamma}(\mathfrak{m})$-module is a direct sum of a free $\bar{\Gamma}(\mathfrak{m})_1$-module and a free $\bar{\Gamma}(\mathfrak{m})_2$-module.

Take such a decomposition, $\overline{X_\infty}(\mathfrak{m}) = A_1(\mathfrak{m}) \oplus A_2(\mathfrak{m})$, and recall that $u(\mathfrak{m}) = \big(u(\mathfrak{m})_1, u(\mathfrak{m})_2\big)$ in this situation. Therefore we can work separately with each $A_i(\mathfrak{m})$, proceeding as in Case 1. If either $A_1(\mathfrak{m})$ or $A_2(\mathfrak{m})$ is zero, the definition of $\mathcal{I}_{C(X)}$ assures us that the corresponding coordinate of $u(\mathfrak{m})$ equals 1, so there is nothing that we need to assign to that $A_i(\mathfrak{m})$.

Case 3: $\mathfrak{m} \notin \mathcal{E}$. Here we make use of (22.3.2). Let $\psi(\mathfrak{m})$ be the automorphism of $\overline{X_\infty}(\mathfrak{m})$ that equals multiplication by $u(\mathfrak{m})$ on $W_1(\mathfrak{m})$ and equals the identity on each other $W_i(\mathfrak{m})$.

We now have an indexed family $\boldsymbol{\psi} = \boldsymbol{\psi}_\infty \oplus \mathbf{1}_t$ of automorphisms, and we need to show that $\boldsymbol{\psi}$ is a residue endo-idèle on X. There are two things to be checked: (a) and (b) below [see (21.1.1)].

(a) Local liftability. Here $\Gamma_\mathfrak{m}$ is either a DVR, a principal ideal ring with exactly 2 maximal ideals, or the direct sum of two DVRs. In any of these cases, every automorphism of the projective $\bar{\Gamma}(\mathfrak{m})$-module $\overline{X_\infty}(\mathfrak{m})$ can be lifted to an automorphism of the projective $\Gamma_\mathfrak{m}$-module $(X_\infty)_\mathfrak{m}$.

(b) $\bar{S}(\mathfrak{m})\psi(\mathfrak{m}) = \bar{S}(\mathfrak{m})$ for almost all $\mathfrak{m}$. In proving this, we may disregard the finitely many $\mathfrak{m} \in \mathcal{E}$. We may also disregard the finitely many $\mathfrak{m}$ such that (by the definition of "residue unit idèle") $u(\mathfrak{m})$ is not a unit of $k(\mathfrak{m})$. Then, in the notation of (22.3.2), it suffices to verify that each $V_i(\mathfrak{m})u(\mathfrak{m}) = V_i$, and this holds because $u(\mathfrak{m})$ is a unit of $k(\mathfrak{m})$. □

DEFINITION 22.4 ($M^{\boldsymbol{u}}$). Let $M \in \text{fingen}_\infty(\Lambda)$ and $\boldsymbol{u} \in \mathcal{I}_{C(M)}$. In Theorem 22.6 below, we shall show that the following series of steps defines a module $M^{\boldsymbol{u}} \in \text{fingen}_\infty(\Lambda)$, up to isomorphism.

(22.4.1) (i) Choose a Γ-separated cover $S \twoheadrightarrow M$, with kernel K and associated Γ-module $X = \Gamma \otimes_\Lambda S = \Gamma S$. To keep the notation simple, identify $S/K = M$, so that the separated cover becomes the natural homomorphism $S \twoheadrightarrow S/K$.
(ii) Choose a residue endo-idèle on X of the form $\boldsymbol{\psi} = \boldsymbol{\psi}_\infty \oplus \mathbf{1}_t$ with respect to some decomposition $X = X_\infty \oplus t(X)$, as defined in (21.2.3), such that $\det(\boldsymbol{\psi}_\infty) = \boldsymbol{u}$.
(iii) Let $M^{\boldsymbol{u}} = S^{\boldsymbol{\psi}}/K$ [Definition 21.4].

LEMMA 22.5. *With notation as above, the natural map* $S^{\boldsymbol{\psi}} \twoheadrightarrow S^{\boldsymbol{\psi}}/K = M^{\boldsymbol{u}}$ *is a* Γ*-separated cover with associated* Γ*-module* $X = \Gamma \otimes_\Lambda S^{\boldsymbol{\psi}} = \Gamma S^{\boldsymbol{\psi}}$*, and* $M^{\boldsymbol{u}} \in \text{rgen}(M)$*, the restricted genus of* M*.*

PROOF. *Separated cover.* Since $S \twoheadrightarrow S/K = M$ is a Γ-separated cover, its localization $S_{\mathfrak{m}} \twoheadrightarrow S_{\mathfrak{m}}/K_{\mathfrak{m}} = M_{\mathfrak{m}}$ is a $\Gamma_{\mathfrak{m}}$-separated cover for every $\mathfrak{m} \in \operatorname{maxspec}(\Lambda)$ [Theorem 18.13], and the associated $\Gamma_{\mathfrak{m}}$-module is $X_{\mathfrak{m}}$.

By the definition of a residue endo-idèle, for each $\mathfrak{m} \in \operatorname{singspec}(\Lambda)$ the $\bar{\Gamma}(\mathfrak{m})$-automorphism $\psi(\mathfrak{m})$ of $\bar{X}(\mathfrak{m})$ can be lifted to a Γ-automorphism $\widetilde{\psi}(\mathfrak{m})$ of $X_{\mathfrak{m}}$, and since $\psi_t(\mathfrak{m}) = 1$ [(22.4.1)(ii)], by Lemma 21.6 we can choose the lifting so that $\widetilde{\psi}_t(\mathfrak{m}) = 1$. In particular, $(K_{\mathfrak{m}})\widetilde{\psi}(\mathfrak{m}) = K_{\mathfrak{m}}$, so that the automorphism $\widetilde{\psi}(\mathfrak{m})$ transforms the $\Gamma_{\mathfrak{m}}$-separated cover $S_{\mathfrak{m}} \twoheadrightarrow S_{\mathfrak{m}}/K_{\mathfrak{m}}$ to a map $S_{\mathfrak{m}}\widetilde{\psi}(\mathfrak{m}) \twoheadrightarrow S_{\mathfrak{m}}\widetilde{\psi}(\mathfrak{m})/K_{\mathfrak{m}}$, and hence this latter map is also a $\Gamma_{\mathfrak{m}}$-separated cover. Since $\widetilde{\psi}(\mathfrak{m})$ lifts $\psi(\mathfrak{m})$, we have $S_{\mathfrak{m}}\widetilde{\psi}(\mathfrak{m}) = (S^{\boldsymbol{\psi}})_{\mathfrak{m}}$, and therefore the map $(S^{\boldsymbol{\psi}})_{\mathfrak{m}} \twoheadrightarrow (S^{\boldsymbol{\psi}})_{\mathfrak{m}}/K_{\mathfrak{m}} = (M^{\boldsymbol{u}})_{\mathfrak{m}}$ is a $\Gamma_{\mathfrak{m}}$-separated cover.

Thus we have shown that, for every $\mathfrak{m} \in \operatorname{singspec}(\Lambda)$, the $\mathfrak{m}$-localization of $S^{\boldsymbol{\psi}} \twoheadrightarrow S^{\boldsymbol{\psi}}/K = M^{\boldsymbol{u}}$ is a $\Gamma_{\mathfrak{m}}$-separated cover. If we can show that this conclusion remains true for every nonsingular maximal ideal, it will follow that $S^{\boldsymbol{\psi}} \twoheadrightarrow S^{\boldsymbol{\psi}}/K$ is a separated cover [Theorem 18.13]. But if $\mathfrak{m}$ is a nonsingular maximal ideal of Λ, the needed facts are obvious since $\Lambda_{\mathfrak{m}} = \Gamma_{\mathfrak{m}}$, $S_{\mathfrak{m}} = X_{\mathfrak{m}}$, and $K_{\mathfrak{m}} = 0$.

Associated Γ*-module.* The relation $X = \Gamma \otimes_{\Lambda} S^{\boldsymbol{\psi}} = \Gamma S^{\boldsymbol{\psi}}$ can be checked locally, at the maximal ideals $\mathfrak{m}$ of Λ. For nonsingular $\mathfrak{m}$, the localized relation follows from the fact that $\Lambda_{\mathfrak{m}} = \Gamma_{\mathfrak{m}}$, while for singular $\mathfrak{m}$, the localized relation follows from the fact that $\widetilde{\psi}(\mathfrak{m})$, as defined above, is a $\Gamma_{\mathfrak{m}}$-automorphism of $X_{\mathfrak{m}}$.

Genus. To see that $M^{\boldsymbol{u}} \in \operatorname{genus}(M)$, we need to show that its $\mathfrak{m}$-localization $S_{\mathfrak{m}}\widetilde{\psi}(\mathfrak{m})/K_{\mathfrak{m}}$ is isomorphic to the $\mathfrak{m}$-localization $S_{\mathfrak{m}}/K_{\mathfrak{m}}$ of M, for all maximal ideals $\mathfrak{m}$ of Λ. Again, this is clear for nonsingular $\mathfrak{m}$, while for singular $\mathfrak{m}$, the map $\widetilde{\psi}(\mathfrak{m})$ is an automorphism of $X_{\mathfrak{m}}$ that equals the identity on $K_{\mathfrak{m}}$.

Restricted genus. We already know that $M^{\boldsymbol{u}} \in \operatorname{genus}(M)$, and $S \twoheadrightarrow M$ and $S^{\boldsymbol{\psi}} \twoheadrightarrow M^{\boldsymbol{u}}$ are Γ-separated covers, both with associated Γ-module X. Therefore by Lemma 19.4, $M^{\boldsymbol{u}} \in \operatorname{rgen}(M)$. □

THEOREM 22.6. *Let* $M \in \operatorname{fingen}_{\infty}(\Lambda)$ *and* $\boldsymbol{u} \in \mathcal{I}_{C(M)}$. *Then:*

(i) $M^{\boldsymbol{u}}$ *is determined up to isomorphism by* $\boldsymbol{u}$ *and the isomorphism class of* M.

(ii) *For fixed* M, *the set of isomorphism classes* $M^{\boldsymbol{u}}$, *as* $\boldsymbol{u}$ *ranges over* $\mathcal{I}_{C(M)}$, *is the restricted genus* $\operatorname{rgen}(M)$.

(iii) $M^{(\boldsymbol{uv})} \cong (M^{\boldsymbol{u}})^{\boldsymbol{v}}$ *and* $M^{\boldsymbol{1}} \cong M$.

PROOF. (i) Let the natural homomorphism $S \twoheadrightarrow S/K = M$ be a separated cover of M, with associated Γ-module $X = X_{\infty} \oplus t(X)$, and let $\boldsymbol{\psi} = \boldsymbol{\psi}_{\infty} \oplus \mathbf{1}_t$ be a residue endo-idèle on X such that $\det(\boldsymbol{\psi}_{\infty}) = \boldsymbol{u}$. Similarly, for $N \cong M$, let $T \twoheadrightarrow T/K' = N$ be a separated cover of N, with associated Γ-module $Y = Y_{\infty} \oplus t(Y)$, and let $\boldsymbol{\phi} = \boldsymbol{\phi}_{\infty} \oplus \mathbf{1}_t$ be a residue endo-idèle on Y such that $\det(\boldsymbol{\phi}_{\infty}) = \boldsymbol{u}$. We need to show that $T^{\boldsymbol{\phi}}/K' \cong S^{\boldsymbol{\psi}}/K$.

We claim that we may suppose that $Y = X$ and $K' = K$ (but possibly, $Y_{\infty} \neq X_{\infty}$).

By the uniqueness of separated covers, there is a Γ-isomorphism $\theta : Y \cong X$ that takes T onto S and K' onto K [Corollary 18.11]. We now have $X = X'_{\infty} \oplus t(X)$ where $X'_{\infty} = (Y_{\infty})\theta$ Also, θ transforms $\boldsymbol{\phi}$ into the residue endo-idèle $\boldsymbol{\phi}' = \theta^{-1}\boldsymbol{\phi}\theta$ on X and $\boldsymbol{\phi}' = \boldsymbol{\phi}'_{\infty} \oplus \mathbf{1}_t$ where $\boldsymbol{\phi}'_{\infty}$ acts on X'_{∞} and satisfies $\det(\boldsymbol{\phi}'_{\infty}) = \det(\boldsymbol{\phi}_{\infty})$. This completes the proof of the claim, and allows us to change notation, after which we have $\boldsymbol{\phi}' = \boldsymbol{\phi}$.

In view of the claim we have two decompositions $X = X_\infty \oplus t(X)$ and $X = X'_\infty \oplus t(X)$ and residue endo-idèles $\boldsymbol{\psi} = \boldsymbol{\psi}_\infty \oplus \mathbf{1}_t$ and $\boldsymbol{\phi} = \boldsymbol{\phi}_\infty$ — with respect to these respective decompositions of X — such that $\det(\boldsymbol{\psi}_\infty) = \det(\boldsymbol{\phi}_\infty)$. Therefore $S^{\boldsymbol{\phi}} \cong S^{\boldsymbol{\psi}}$ by an automorphism of X which restricts to the identity map on $t(X)$ [Corollary 21.9], and hence takes K to K. This proves that $S^{\boldsymbol{\phi}}/K \cong S^{\boldsymbol{\psi}}/K$, as desired.

(ii) In view of Lemma 22.5, it suffices to show that every $N \in \mathrm{rgen}(M)$ is isomorphic to $M^{\boldsymbol{u}}$ for some $\boldsymbol{u}$. Choose a Γ-separated cover $S \twoheadrightarrow M$ of M, with associated Γ-module $X = \Gamma \otimes_\Lambda S = \Gamma S$, and choose a Γ-separated cover $T \twoheadrightarrow N$ of N. By Lemma 19.4, we may assume that the associated Γ-module of $T \twoheadrightarrow N$ is again X, and we also have that $T \in \mathrm{rgen}(S)$. Therefore, by Corollary 21.9, there is a residue endo-idèle $\boldsymbol{\psi} = \boldsymbol{\psi}_\infty \oplus \mathbf{1}_t$ on X such that $S^{\boldsymbol{\psi}} \cong T$. By Lemma 22.3, $\boldsymbol{u} = \det(\boldsymbol{\psi}_\infty)$ is a residue unit idèle, and therefore $N \cong M^{\boldsymbol{u}}$, as desired.

(iii) The first isomorphism follows from the product theorem for determinants and the formula $S^{(\boldsymbol{\psi}\boldsymbol{\phi})} = (S^{\boldsymbol{\psi}})^{\boldsymbol{\phi}}$; the second is obvious. □

THEOREM 22.7. *Let $M, N \in \mathrm{fingen}_\infty(\Lambda)$, $\boldsymbol{u} \in \mathcal{I}_{\mathcal{C}(M)}$, and $\boldsymbol{v} \in \mathcal{I}_{\mathcal{C}(N)}$. Then $\boldsymbol{uv} \in \mathcal{I}_{\mathcal{C}(M \oplus N)}$, and :*

$$M^{\boldsymbol{u}} \oplus N^{\boldsymbol{v}} \cong (M \oplus N)^{\boldsymbol{uv}}$$

PROOF. Clearly $\boldsymbol{uv} \in \mathcal{I}_{\mathcal{C}(M \oplus N)}$, because $\mathcal{C}(M \oplus N)$ contains both $\mathcal{C}(M)$ and $\mathcal{C}(N)$. Take Γ-separated covers $S \twoheadrightarrow M$ and $T \twoheadrightarrow N$ with associated Γ-modules $X = X_\infty \oplus t(X)$ and $Y = Y_\infty \oplus t(Y)$, and kernels K and L, respectively. Let $\boldsymbol{\psi} = \boldsymbol{\psi}_\infty \oplus \mathbf{1}_t$ and $\boldsymbol{\phi} = \boldsymbol{\phi}_\infty \oplus \mathbf{1}_t$ be residue endo-idèles on X and Y, respectively, such that $\det(\boldsymbol{\psi}_\infty) = \boldsymbol{u}$ and $\det(\boldsymbol{\phi}_\infty) = \boldsymbol{v}$ [Lemma 22.3].

Then $M^{\boldsymbol{u}} \cong S^{\boldsymbol{\psi}}/K$ and $N^{\boldsymbol{v}} \cong T^{\boldsymbol{\phi}}/L$. If we let $\boldsymbol{\psi} \oplus \boldsymbol{\phi}$ be the indexed family $\{\psi(\mathfrak{m}) \oplus \phi(\mathfrak{m}) \mid \mathfrak{m} \in \mathrm{singspec}(\Lambda)\}$, then $\boldsymbol{\psi} \oplus \boldsymbol{\phi}$ is a residue endo-idèle on $X \oplus Y$, and $(S \oplus T)^{\boldsymbol{\psi} \oplus \boldsymbol{\phi}}/(K \oplus L) \cong M^{\boldsymbol{u}} \oplus N^{\boldsymbol{v}}$. The theorem is therefore an immediate consequence of the product theorem for determinants, together with the fact that a direct sum of Γ-separated covers is again a Γ-separated cover [Theorem 18.14]. □

23. Stabilizers: Basic properties

This section studies the basic properties of the stabilizer $\mathrm{istab}(M)$ of the isomorphism class of M under the idèle action $M \to M^{\boldsymbol{u}}$. One main result [Theorem 23.5] is that $\mathrm{istab}(M)$ is the product of two factors: $\mathrm{eqstab}(M) \cdot \mathrm{im}\big((\Gamma_{\mathcal{C}(M)})^\times\big)$, the first of which ("equality stabilizer") is computable locally, and the second of which does not depend on any property of M other than its torsionfree support set.

The other main result is that the stabilizer of a direct sum is the product of the stabilizers of the summands. However, only half of the proof is in this section [Lemma 23.2(ii)–(iii)]. Completion of the proof [in Corollary 24.7] requires detailed computation of $\mathrm{eqstab}(M)$ of individual modules M which, in turn, requires some very complicated matrix calculations. This last computation is the subject of Section 24.

DEFINITIONS 23.1 (Stabilizer). Let $M \in \mathrm{fingen}_\infty(\Lambda)$. We define the *isomorphism stabilizer* $\mathrm{istab}(M)$ of M to be the subgroup of $\mathcal{I}_{\mathcal{C}(M)}$ consisting of those residue unit idèles $\boldsymbol{u}$ such that $M^{\boldsymbol{u}} \cong M$.

Let $S \twoheadrightarrow M$ be a Γ-separated cover with associated Γ-module $X = X_\infty \oplus t(X)$ and kernel K. We define the *equality stabilizer* $\mathrm{eqstab}(M)$ of M to be the subgroup of $\mathcal{I}_{\mathcal{C}(M)}$ consisting of those residue unit idèles $\boldsymbol{u}$ such that $\boldsymbol{u} = \det(\boldsymbol{\alpha}_\infty)$ for some residue endo-idèle $\boldsymbol{\alpha}$ on X for which both of the following conditions hold.

(23.1.1) (i) $S^{\boldsymbol{\alpha}} = S$ (equality, not just isomorphism).

(ii) $(K)\widetilde{\alpha}_t = K$ for some Γ-automorphism $\widetilde{\alpha}_t$ of $t(X)$ lifting $\boldsymbol{\alpha}_t$.

Caution. It does not seem possible to require that $\boldsymbol{\alpha} = \boldsymbol{\alpha}_\infty \oplus \mathbf{1}_1$ in this definition. See Lemma 23.4.

The definition of eqstab(M) is determined by the isomorphism class of M — that is, is independent of the particular Γ-separated cover chosen — because of uniqueness of separated covers [Corollary 18.11].

We prove in Lemma 23.3 below that eqstab(M) is locally determined; that is, eqstab$(M)(\mathfrak{m})$ = eqstab$(M_\mathfrak{m})$ $\big(\forall \mathfrak{m} \in \operatorname{maxspec}(\Lambda)\big)$. Then we prove, in Lemma 24.3, that each eqstab$(M_\mathfrak{m})$ = eqstab$(\hat{M}_\mathfrak{m})$.

LEMMA 23.2. *Let* $M, N \in \operatorname{fingen}(\Lambda)$.

(i) *If* genus(M) = genus(N), *then* $\mathcal{I}_{\mathcal{C}(M)} = \mathcal{I}_{\mathcal{C}(N)}$ *and* istab(M) = istab(N).

(ii) istab$(M \oplus N) \supseteq$ istab$(M)\cdot$ istab(N).

(iii) eqstab$(M \oplus N) \supseteq$ eqstab$(M)\cdot$ eqstab(N).

PROOF. (i) Suppose that genus(M) = genus(N). To show that $\mathcal{I}_{\mathcal{C}(M)} = \mathcal{I}_{\mathcal{C}(N)}$, it suffices to show that $\mathcal{C}(M) = \mathcal{C}(N)$. Recall that $M_Q = \oplus_\mathfrak{p} M_\mathfrak{p}$, where $\mathfrak{p}$ ranges over the finite set of minimal prime ideals of Λ [Corollary 6.4]. Therefore, it suffices to show that, for every coordinate Dedekind domain Γ_h and every minimal prime $\mathfrak{p}$, we have $\Gamma_h M_\mathfrak{p} = 0$ if and only if $\Gamma_h N_\mathfrak{p} = 0$. For any such Γ_h and $\mathfrak{p}$, if $\mathfrak{m}$ is a maximal ideal containing $\mathfrak{p}$, then $M_\mathfrak{m} \cong N_\mathfrak{m}$ by hypothesis, and hence $M_\mathfrak{p} \cong N_\mathfrak{p}$ as $\Lambda_\mathfrak{p}$-modules. In particular, these modules have the same annihilator in Γ, as desired.

Next we show that istab(M) = istab(N), the most interesting part of this proof. Take $\boldsymbol{u} \in$ istab$(M) \subseteq \mathcal{I}_{\mathcal{C}(M)}$. As we have just proved, this implies that $\boldsymbol{u} \in \mathcal{I}_{\mathcal{C}(N)}$. Therefore, Theorem 22.7 shows that

$$M^{\boldsymbol{u}} \oplus N \cong (M \oplus N)^{\boldsymbol{u}} \cong M \oplus N^{\boldsymbol{u}} \tag{23.2.1}$$

But $M^{\boldsymbol{u}} \cong M$, by assumption, and direct-sum cancellation holds in every genus of Λ-modules [Lemma 9.2]. Therefore (23.2.1) implies that $N \cong N^{\boldsymbol{u}}$, and hence $\boldsymbol{u} \in$ istab(N). Reversing the roles of M and N completes the proof of (i).

(ii) The inclusion istab$(M) \subseteq$ istab$(M \oplus N)$ follows from the first isomorphism in (23.2.1), and the inclusion istab$(N) \subseteq$ istab$(M \oplus N)$ follows from the second; the stated conclusion is then immediate.

(iii) This statement follows easily from the fact that the direct sum of separated covers of M and N is a separated cover of $(M \oplus N)$ [Theorem 18.14]. □

As already mentioned, equality holds in Lemma 23.2(ii) and (iii), but the proof requires the detailed matrix computation of equality stabilizers found in Section 24. See Corollary 24.7. Without these matrix computations, we can still show that eqstab(M) is "locally determined" by the various eqstab$(M_\mathfrak{m})$, as $\mathfrak{m}$ ranges over singspec(Λ), which we do in the next lemma. (In Lemma 24.3, we shall show that eqstab$(M_\mathfrak{m})$ = eqstab$(\hat{M}_\mathfrak{m})$.)

LEMMA 23.3 (eqstab is locally determined). *For* $M \in \operatorname{fingen}(\Lambda)$, eqstab$(M)$ $= \prod\{$eqstab$(M)(\mathfrak{m}) \mid \mathfrak{m} \in$ singspec$(\Lambda)\}$ *(cartesian product) and* eqstab$(M)(\mathfrak{m})$ = eqstab$(M_\mathfrak{m})$ *for each* $\mathfrak{m}$.

PROOF. For each $\mathfrak{m} \in \text{singspec}(\Lambda)$, let $\omega(\mathfrak{m})\colon T(\mathfrak{m}) \twoheadrightarrow M_\mathfrak{m}$ be a $\Gamma_\mathfrak{m}$-separated cover with associated $\Gamma_\mathfrak{m}$-module $Y(\mathfrak{m})$ and kernel $K(\mathfrak{m})$. Let G be the group of indexed families of the form $\boldsymbol{u} = \{\det(\alpha_\infty(\mathfrak{m})) \mid \mathfrak{m} \in \text{singspec}(\Lambda)\}$, where each $\alpha_\infty(\mathfrak{m})$ arises from a $\bar{\Gamma}(\mathfrak{m})$-automorphism $\alpha(\mathfrak{m})$ of $\bar{Y}(\mathfrak{m})$ that is liftable to a $\Gamma_\mathfrak{m}$-automorphism $\widetilde{\alpha}(\mathfrak{m})$ of $Y(\mathfrak{m})$ such that $(T(\mathfrak{m}))\widetilde{\alpha}(\mathfrak{m}) = T(\mathfrak{m})$ and $(K(\mathfrak{m}))\widetilde{\alpha}(\mathfrak{m}) = K(\mathfrak{m})$. To prove the lemma, it suffices to show that $G = \text{eqstab}(M)$.

Let $\phi\colon S \twoheadrightarrow M$ be the Γ-separated cover, with associated Γ-module X, used to define eqstab(M). Then for each $\mathfrak{m} \in \text{singspec}(\Lambda)$, the localization $\phi_\mathfrak{m}$ is a $\Gamma_\mathfrak{m}$-separated cover of $M_\mathfrak{m}$ with associated $\Gamma_\mathfrak{m}$-module $X_\mathfrak{m}$ [Theorem 18.13]. By uniqueness of separated covers [Corollary 18.11], there is a $\Gamma_\mathfrak{m}$-isomorphism $\theta(\mathfrak{m})\colon Y(\mathfrak{m}) \cong X_\mathfrak{m}$ that takes $T(\mathfrak{m})$ onto $S_\mathfrak{m}$ and $K(\mathfrak{m})$ onto $\ker(\phi_\mathfrak{m})$. Moreover, for each maximal ideal $\mathfrak{m}$ and each indexed family $\{\alpha(\mathfrak{m})\}$ as above, the determinant of the infinite part of the induced automorphism $\bar{\theta}(\mathfrak{m})^{-1}\alpha(\mathfrak{m})\bar{\theta}(\mathfrak{m})$ of $\bar{X}(\mathfrak{m})$ equals $\det(\alpha(\mathfrak{m})_\infty)$, because of the triangular form of both $\alpha(\mathfrak{m})$ and $\bar{\theta}(\mathfrak{m})$. Therefore, we may assume that $Y(\mathfrak{m}) = X_\mathfrak{m}$, $T(\mathfrak{m}) = S_\mathfrak{m}$, and $K(\mathfrak{m}) = \ker(\phi_\mathfrak{m})$.

Each $\alpha(\mathfrak{m})$ is now a $\bar{\Gamma}$-automorphism of $\bar{X}(\mathfrak{m})$ that is liftable to a $\Gamma_\mathfrak{m}$-automorphism $\widetilde{\alpha}(\mathfrak{m})$ of $X_\mathfrak{m}$, such that $\bar{S}(\mathfrak{m})\alpha(\mathfrak{m}) = \bar{S}(\mathfrak{m})$ and $(K(\mathfrak{m}))\widetilde{\alpha}(\mathfrak{m}) = K(\mathfrak{m})$ for all $\mathfrak{m}$. Therefore, the indexed families $\{\alpha(\mathfrak{m}) \mid \mathfrak{m} \in \text{singspec}(\Lambda)\}$ defining the group G are precisely the residue endo-idèles $\boldsymbol{\alpha}$ on X such that $S^{\boldsymbol{\alpha}} = S$ and $(K(\mathfrak{m}))\widetilde{\alpha}(\mathfrak{m}) = K(\mathfrak{m})$ for all $\mathfrak{m}$, that is, the residue endo-idèles defining eqstab(M). □

Our next lemma uses Goldman's definition of the determinant of an endomorphism of a projective module [Remarks 20.1].

LEMMA 23.4. *Let $S \twoheadrightarrow S/K$ be a Γ-separated cover, with associated Γ-module X, of some finitely generated Λ-module, and let $\boldsymbol{\psi}$ be a residue endo-idèle on X, such that $\boldsymbol{\psi}_t = \mathbf{1}_t$. Then $S^{\boldsymbol{\psi}}/K \cong S/K$ if and only if there is a factorization $\boldsymbol{\psi} = \boldsymbol{\alpha\theta}$, where $\boldsymbol{\alpha}$ and $\boldsymbol{\theta}$ are residue endo-idèles on X such that:*

(i) *$\boldsymbol{\theta}$ lifts to a Γ-automorphism θ of X such that $(K)\theta = K$;*
(ii) *$S^{\boldsymbol{\alpha}} = S$; and*
(iii) *$\boldsymbol{\alpha}_t$ lifts to a Γ-automorphism $\widetilde{\alpha}_t$ of $t(X)$ such that $(K)\widetilde{\alpha}_t = K$.*

PROOF. First, suppose that $S^{\boldsymbol{\psi}}/K \cong S/K$. Then by uniqueness of separated covers [Corollary 18.11] there is an automorphism θ of X such that $(S)\theta = S^{\boldsymbol{\psi}}$ and $(K)\theta = K$. The first of these equalities yields $S^{\boldsymbol{\theta}} = S^{\boldsymbol{\psi}}$, where $\boldsymbol{\theta}$ is the residue endo-idèle on X induced by θ [Lemma 21.3]. Then $\boldsymbol{\alpha} = \boldsymbol{\psi\theta}^{-1}$ satisfies $S^{\boldsymbol{\alpha}} = S$, as desired.

To see the effect of $\widetilde{\alpha}_t$ on K, choose a decomposition $X = X_\infty \oplus t(X)$, and write the relation $\boldsymbol{\psi} = \boldsymbol{\alpha\theta}$ in matrix form:

$$\begin{pmatrix} \boldsymbol{\psi}_\infty & \boldsymbol{\psi}_{12} \\ 0 & \mathbf{1}_t \end{pmatrix} = \begin{pmatrix} \boldsymbol{\alpha}_\infty & \boldsymbol{\alpha}_{12} \\ 0 & \boldsymbol{\alpha}_t \end{pmatrix} \begin{pmatrix} \boldsymbol{\theta}_\infty & \boldsymbol{\theta}_{12} \\ 0 & \boldsymbol{\theta}_t \end{pmatrix}$$

Then $\boldsymbol{\alpha}_t\boldsymbol{\theta}_t = \mathbf{1}_t$, and therefore we can choose the lifting $\widetilde{\alpha}_t$ of $\boldsymbol{\alpha}_t$ to be θ_t^{-1}. Since $(K)\theta = K$, it follows that $(K)\widetilde{\alpha}_t = K$ also.

Conversely, suppose that $\boldsymbol{\psi} = \boldsymbol{\alpha\theta}$, with $\boldsymbol{\alpha}$ and $\boldsymbol{\theta}$ as described. Then $S^{\boldsymbol{\psi}}/K = S^{\boldsymbol{\alpha\theta}}/K = S^{\boldsymbol{\theta}}/K = (S)\theta/(K)\theta \cong S/K$. □

The final theorem in this section gives a clear separation of the structure of the group istab(M) into a local factor and a factor that becomes trivial locally. Note that $\Gamma^\times$, the group of units in Γ, has a natural image in $\mathcal{I}$. [This is essentially

Corollary 8.3.] Moreover, the group of units $(\Gamma_{\mathcal{C}(M)})^\times$ in $\Gamma_{\mathcal{C}(M)}$ can be viewed in a natural way as a subgroup of $\Gamma^\times$, by extending $(d_h)_{h\in\mathcal{C}(M)} \in (\Gamma_{\mathcal{C}(M)})^\times$ to all of $\mathcal{H}$ by setting $d_h = 1$ whenever $h \notin \mathcal{C}(M)$. Composing these two maps, the group $(\Gamma_{\mathcal{C}(M)})^\times$ has a natural image in $\mathcal{I}$; in fact, by its definition, this image is contained in the subgroup $\mathcal{I}_{\mathcal{C}(M)}$ of $\mathcal{I}$.

THEOREM 23.5. *For every $M \in \mathrm{fingen}(\Lambda)$ we have*

$$\mathrm{istab}(M) = \mathrm{eqstab}(M)\cdot \mathrm{im}\big((\Gamma_{\mathcal{C}(M)})^\times\big) \tag{23.5.1}$$

where $\mathrm{im}((\Gamma_{\mathcal{C}(M)})^\times)$ denotes the natural image of $(\Gamma_{\mathcal{C}(M)})^\times$ in $\mathcal{I}_{\mathcal{C}(M)}$, as defined above.

PROOF. Let $\phi\colon S \twoheadrightarrow M$ be a Γ-separated cover, with associated Γ-module X and $\ker(\phi) = K$. Fix a decomposition $X = X_\infty \oplus t(X)$, and recall that $\mathcal{C}(X) = \mathcal{C}(M)$ [Lemma 22.2]. Then for every residue unit idèle $\boldsymbol{u} \in \mathcal{I}_{\mathcal{C}(M)}$, we have $M^{\boldsymbol{u}} \cong S^{\boldsymbol{\psi}}/K$ for any residue endo-idèle of the form $\boldsymbol{\psi} = \boldsymbol{\psi}_\infty \oplus \mathbf{1}_t$ on X such that $\det(\boldsymbol{\psi}_\infty) = \boldsymbol{u}$ [Definition 22.4]. In the proof that follows, we say that such a $\boldsymbol{\psi}$ *realizes* $M^{\boldsymbol{u}}$.

First, suppose that $\boldsymbol{u} \in \mathrm{istab}(M)$, and choose some $\boldsymbol{\psi}$ that realizes $M^{\boldsymbol{u}}$. Then we have a factorization $\boldsymbol{\psi} = \boldsymbol{\alpha\theta}$ such that $S^{\boldsymbol{\alpha}} = S$, and some Γ-automorphism θ of X lifts $\boldsymbol{\theta}$ and satisfies $(K)\theta = K$ [Lemma 23.4]; and $\boldsymbol{\alpha} \in \mathrm{eqstab}(M)$. If we let $d = \det(\theta_\infty)$, then $d \in (\Gamma_{\mathcal{C}(M)})^\times$. And if we view d "multiplicatively", in the form $d = (d_h)_{h\in\mathcal{H}}$ with $d_h = 1$ for all $h \notin \mathcal{C}(M)$, then the indexed family $\boldsymbol{d} = \{\bar{d}(\mathfrak{m}) \mid \mathfrak{m} \in \mathrm{singspec}(\Lambda)\}$ satisfies $\boldsymbol{d} \in \mathcal{I}_{\mathcal{C}(M)}$ [Lemma 22.3]. Therefore, the relation $\boldsymbol{\psi} = \boldsymbol{\alpha\theta}$, together with the triangular form of residue endo-idèles [Notation 21.2], yields the desired factorization $\boldsymbol{u} = \det(\boldsymbol{\psi}_\infty) = \det(\boldsymbol{\alpha}_\infty)\cdot\det(\boldsymbol{\theta}_\infty) = \det(\boldsymbol{\alpha}_\infty)\cdot \mathrm{im}(d)$.

Conversely, suppose that $\boldsymbol{u} = \det(\boldsymbol{\alpha}_\infty)\cdot \mathrm{im}(d)$ for some residue endo-idèle $\boldsymbol{\alpha} \in \mathrm{eqstab}(M)$ and some unit $d \in (\Gamma_{\mathcal{C}(M)})^\times$. As above, we view $d = (d_h)_{h\in\mathcal{H}} \in \Gamma^\times$, by taking $d_h = 1$ for each $h \notin \mathcal{C}(M)$. Let $\boldsymbol{d}$ be the indexed family $\{\bar{d}(\mathfrak{m}) \mid \mathfrak{m} \in \mathrm{singspec}(\Lambda)\}$ as in the previous paragraph. Then $\boldsymbol{d} = \mathrm{im}(d) \in \mathcal{I}_{\mathcal{C}(M)}$.

We have $X_\infty = \oplus_\nu X_\nu$, where each X_ν is a nonzero ideal of some coordinate Dedekind domain Γ_h of Γ. Moreover, an ideal of Γ_h actually occurs if and only if $h \in \mathcal{C}(M)$. Define a Γ-automorphism θ_∞ of X_∞ by letting θ_∞ be multiplication by d_h on one Γ_h-ideal X_ν for each $\Gamma_h \in \mathcal{C}(M)$, and letting θ_∞ be the identity on every other summand X_ν. Then $\det(\theta_\infty) = d$.

To define the finite part of the Γ-automorphism θ we are building, consider the residue endo-idèle $\boldsymbol{\alpha}$. By the definition of $\mathrm{eqstab}(M)$, we have a separated cover $S \twoheadrightarrow M$ with kernel K and associated Γ-module X, such that $S^{\boldsymbol{\alpha}} = S$ and $(K)\widetilde{\alpha}_t = K$, for some Γ-automorphism $\widetilde{\alpha}_t$ of $t(X)$ lifting $\boldsymbol{\alpha}_t$. Let $\theta = \theta_\infty \oplus (\widetilde{\alpha}_t)^{-1}$, a Γ-automorphism of X such that $(K)\theta = K$.

Let $\boldsymbol{\phi} = \boldsymbol{\alpha\theta}$, where $\boldsymbol{\theta}$ is the residue endo-idèle induced by θ [Lemma 21.3]. Then the triangular form of residue endo-idèles shows that $\det(\boldsymbol{\phi}_\infty) = \det(\boldsymbol{\alpha}_\infty)\cdot \mathrm{im}(d) = \boldsymbol{u}$. Moreover, $\boldsymbol{\phi}_t = \boldsymbol{\alpha}_t\boldsymbol{\theta}_t = \mathbf{1}_t$, and so $S^{\boldsymbol{\phi}}/K \cong S/K \cong M$ [Lemma 23.4].

Finally, let $\boldsymbol{\psi}$ be obtained from $\boldsymbol{\phi}$ by changing $\boldsymbol{\phi}_{12}$ to zero. Then $S^{\boldsymbol{\psi}} \cong S^{\boldsymbol{\phi}}$ by an automorphism of X that equals the identity on [$t(X)$, hence on] K [Theorem 21.8], and therefore $S^{\boldsymbol{\psi}}/K \cong S^{\boldsymbol{\phi}}/K \cong M$. Moreover, $\boldsymbol{u} = \det(\boldsymbol{\alpha}_\infty)\cdot \mathrm{im}(d) = \det(\boldsymbol{\psi}_\infty)$, and therefore the form $\boldsymbol{\psi} = \boldsymbol{\psi}_\infty \oplus \mathbf{1}_t$ shows that $M^{\boldsymbol{u}} \cong S^{\boldsymbol{\psi}}/K \cong S^{\boldsymbol{\phi}}/K \cong M$, as desired. □

24. Equality Stabilizers

This section establishes those properties of stabilizers whose proofs require the matrix-theoretic details of the structure of Λ-modules, developed in [**KL2**] in the case that Λ is a complete local ring. Almost all of the work involves equality stabilizers. We begin by showing that eqstab(M) is complete-locally determined [Lemma 24.3, in light of Lemma 23.3]. Then we state the main results: the explicit description of stabilizers of modules and their direct sums [Theorem 24.5 and its non-local consequence, Corollary 24.7]. Finally, we do the detailed matrix-theoretic proofs needed for the complete local case. Actually, most of the work in this section invokes the following slightly more general context (i.e. more general than completeness), assumed in almost all of [**KL2**].

HYPOTHESIS 24.1. *Let the local Dedekind-like ring* $(\Lambda, \mathfrak{m}, k)$ *be unsplit or strictly split, and let* $M = M_1 \oplus \ldots \oplus M_n \neq 0$*, where each* $M_i \in \operatorname{fingen}(\Lambda)$ *is indecomposable and has infinite (composition) length.*

We remark that every complete local Dedekind-like ring is either a DVR or one of the two types of rings mentioned in Hypothesis 24.1 [Proposition 11.8], and rings of this type satisfy the Krull-Schmidt theorem for finitely generated modules, even if they are not complete [**KL2**, 1.3].

NOTATION 24.2 (Simplifications in local case). Let $(\Lambda, \mathfrak{m}, k)$ be a local Dedekind-like ring.

(Already mentioned:) For any Λ-module H we set $\bar{H} = H/\mathfrak{m}H$, simplifying the notation $\bar{H}(\mathfrak{m})$ used in the non-local case. Since Λ is local, $\mathfrak{m}$ is an ideal of Γ, and therefore $\bar{\Gamma} = \Gamma/\mathfrak{m}$. Similarly, when Λ is unsplit we write $\bar{\Gamma} = F$ rather than $F(\mathfrak{m})$. We also write $\hat{M}$ in place of $\hat{M}_{\mathfrak{m}}$.

Let S be a Λ-submodule of a Γ-module X such that $\Gamma S = X$. Recall that the natural map $\bar{S} \to \bar{X}$ is an injection [Lemma 11.1] which we almost always consider to be inclusion. It therefore makes sense to refer to S as the inverse image of $\bar{S}$ in X.

Since Λ is local, a *residue endo-idèle* ψ *on* X [Definition 21.1] is merely a Γ-automorphism of $\bar{X}$ that can be lifted to a Γ-automorphism $\widetilde{\psi}$ of X, because condition 21.1.1(i) becomes vacuous when Λ is local. In particular, S^{ψ} [Definition 21.4] is the inverse image of $\bar{S}\psi$ in X.

Recall that what we call Γ-separated modules and Γ-separated covers are merely called a separated modules and separated covers in [**KL2**].

Finally, let $S \twoheadrightarrow M$ be a Γ-separated cover of M, with associated Γ-module $X = \Gamma S = \Gamma \otimes_{\Lambda} X$ and $\ker(S \twoheadrightarrow M) = K$. Then eqstab($M$) [Definition 23.1] is the set of all elements $\det(\psi_{\infty}) \in \bar{\Gamma}^{\times}$ [Notation 21.2], where ψ ranges over the residue endo-idèles on X such that:

$$\bar{S}\psi = \bar{S} \qquad \text{and} \qquad K\widetilde{\psi}_t = K \tag{24.2.1}$$

for some lifting $\widetilde{\psi}_t$ of ψ_t to an automorphism of $t(X)$.

LEMMA 24.3 (eqstab is complete-locally determined). *Let* $M \in \operatorname{fingen}_{\infty}(\Lambda)$.

(i) *Suppose that* Λ *is local and* $S \twoheadrightarrow M$ *is a separated cover of* M*, with associated* Γ*-module* $X = \Gamma S = \Gamma \otimes_{\Lambda} S$*. Then a* Γ*-automorphism* ψ *of* $\bar{X}$ *is a residue endo-idèle on* X *if and only it is a residue endo-idèle on* $\hat{X}$.

(ii) *(For non-local Λ:)* $\operatorname{eqstab}(M) = \prod\{\operatorname{eqstab}(M)(\mathfrak{m}) \mid \mathfrak{m} \in \operatorname{singspec}(\Lambda)\}$ *(cartesian product) and each* $\operatorname{eqstab}(M)(\mathfrak{m}) = \operatorname{eqstab}(\hat{M})$.

PROOF. (i) Implicit in both the statement and proof of this lemma is the fact that $\bar{\Gamma}$ and $\bar{X}$ do not change when we pass from X to $\hat{X}$.

In view of the simplifications spelled out in Notation 24.2, the only part of the definition of "residue endo-idèle" that we need to check is that ψ is liftable to an automorphism of X if and only if it is liftable to an automorphism of $\hat{X}$. To prove the nontrivial half of this, suppose that ψ is liftable to an automorphism of $\hat{X}$. Using the 2×2 upper triangular form of ψ defined in Notation 21.2, it suffices to lift the three entries ψ_∞, ψ_{12}, and ψ_t individually.

For ψ_∞, since $\mathfrak{m} = \operatorname{rad}\Gamma$ when Λ is local and X_∞ is Γ-projective, the natural map $X_\infty \twoheadrightarrow \bar{X}_\infty$ is a projective cover, and hence ψ_∞ lifts to an automorphism of X_∞ (regardless of the liftability hypothesis). Similarly, ψ_{12} lifts to a homomorphism $X_\infty \to t(X)$ because X_∞ is projective. (The proof again does not depend upon the liftability hypothesis.) Finally, since ψ lifts to an automorphism of $\hat{X}$, ψ_t lifts to an automorphism of $t(\hat{X}) = t(X)$, and this completes the proof.

(ii) Because of Lemma 23.3, we may assume that Λ is local with maximal ideal $\mathfrak{m}$. Both sides of the equation are subgroups of $(\Gamma/\mathfrak{m})^\times = (\hat{\Gamma}/\hat{\mathfrak{m}})^\times$. Therefore, statement (ii) follows immediately from (i). □

We now state the main result of this section. Recall that "rank" (i.e. torsionfree rank) was defined in Definition 15.1.

NOTATION 24.4 (N_k^F). For an unsplit local Dedekind-like ring Λ, we use the notation N_k^F for the norm map, from the set of *nonzero* elements of of F (the residue field of Γ) to those of k. Thus $\operatorname{im}(N_k^F) \subseteq k^\times$.

THEOREM 24.5. *Assume Hypothesis 24.1.*

(i) *If Λ is unsplit and M is indecomposable, then one of the following holds.*
 (a) *M has rank 2, and* $\operatorname{eqstab}(M) = \operatorname{im}(N_k^F)$.
 (b) *M has rank 1 and* $\operatorname{eqstab}(M) = k^\times$ *or* $F^\times$. *In particular,* $\operatorname{eqstab}(\Lambda) = k^\times$.

(ii) *If Λ is strictly split and M is indecomposable, then one of the following holds.*
 (a) *M has rank (1,1) and* $\operatorname{eqstab}(M) = k^\times$ *(the diagonal copy of $k^\times$ in $k_1^\times \times k_2^\times$). In particular,* $\operatorname{eqstab}(\Lambda) = k^\times$.
 (b) *M has rank* $(1,0)$ *or* $(0,1)$, *and* $\operatorname{eqstab}(M) = k_1^\times \times 1$ *or* $1 \times k_2^\times$ *respectively. [See* (11.5.1) *for the significance of the notation k, k_1, and k_2.]*

Moreover, all of the possibilities enumerated above actually occur for every Λ of the specified type (unsplit or strictly split). In addition, $\operatorname{eqstab}(\oplus_i M_i) = \prod_i \operatorname{eqstab}(M_i)$ *(product in $\bar{\Gamma}^\times$).*

REMARK 24.6. *Caution.* Hypothesis 24.1 does not allow the possibility $M = 0$. If we allow this possibility in Theorem 24.5 then we need to note our convention that $\operatorname{eqstab}(M) = \{1\}$ if $M = 0$ (otherwise eqstab does not behave properly in direct sums).

The proof of this theorem occupies most of the remainder of this section, beginning with an outline in subsection 24.9. But first we state and prove the main non-local consequences of this theorem, which are needed for the rest of the paper, and which seem to require the computation of the equality stabilizer of every module in $\mathrm{fingen}_\infty(\Lambda)$.

COROLLARY 24.7. *For the (not necessarily local) Dedekind-like ring* Λ, *let* $M, N \in \mathrm{fingen}(\Lambda)$. *Then:*

(i) $\mathrm{eqstab}(M \oplus N) = \mathrm{eqstab}(M)\,\mathrm{eqstab}(N)$;
(ii) $\mathrm{istab}(M \oplus N) = \mathrm{istab}(M)\,\mathrm{istab}(N)$; *and*
(iii) $\mathrm{istab}(M \oplus N) = \mathrm{eqstab}(M)\,\mathrm{eqstab}(N)\,\mathrm{im}((\Gamma_{\mathcal{C}(M\oplus N)})^\times)$.

PROOF. (i) follows from the final sentence of Theorem 24.5, together with Lemma 23.3. For (ii) and (iii), recall that $\mathrm{istab}(M) = \mathrm{eqstab}(M) \cdot \mathrm{im}((\Gamma_{\mathcal{C}(M)})^\times)$ [Theorem 23.5], so that it suffices to show that:

$$\mathrm{im}((\Gamma_{\mathcal{C}(M\oplus N)})^\times) = \mathrm{im}((\Gamma_{\mathcal{C}(M)})^\times) \cdot \mathrm{im}((\Gamma_{\mathcal{C}(N)})^\times)$$

where the three images are in $\mathcal{I}_{\mathcal{C}(M\oplus N)}$, $\mathcal{I}_{\mathcal{C}(M)}$, and $\mathcal{I}_{\mathcal{C}(N)}$, respectively. But this equation holds because $\mathcal{C}(M \oplus N) = \mathcal{C}(M) \cup \mathcal{C}(N)$, and because every unit of Γ is a tuple of the form $x = (x_h)_{h\in\mathcal{H}}$, where each x_h is a unit in the coordinate ring Γ_h of Γ. □

COROLLARY 24.8. *Let* $A = \oplus_{i=1}^n A_i$ *and* $B = \oplus_{i-1}^n B_i$ *be modules in* $\mathrm{fingen}_\infty(\Lambda)$ *such* $\mathcal{C}(A_i) = \mathcal{C}(B_i)$ *for each* i, *and* $\mathcal{C}(A_i) \cap \mathcal{C}(A_j) = \varnothing$ *when* $i \neq j$. *Then* $\mathrm{eqstab}(A) = \mathrm{eqstab}(B)$ *if and only if every* $\mathrm{eqstab}(A_i) = \mathrm{eqstab}(B_i)$.

PROOF. First we show that it suffices to prove the lemma when the Dedekind-like ring Λ is complete local. Each $\hat{\Lambda}_\mathfrak{m}$ remains Dedekind-like [Theorem 11.9], $\mathrm{eqstab}(A)$ determines and is determined by the set $\{\mathrm{eqstab}(\hat{A}_\mathfrak{m}) \mid \mathfrak{m} \in \mathrm{singspec}(\Lambda)\}$ [Lemma 23.3]; and the same relationship holds between $\mathcal{C}(A)$ and the set of $\mathcal{C}(\hat{A}_\mathfrak{m})$ such that $\mathfrak{m} \in \mathrm{maxspec}(\Lambda)$ [Lemma 15.2]. Also, since $A \in \mathrm{fingen}_\infty(\Lambda)$ we have $\hat{A}_\mathfrak{m} \in \mathrm{fingen}_\infty(\hat{\Lambda}_\mathfrak{m})$ $(\forall \mathfrak{m} \in \mathrm{maxspec}(\Lambda))$ [Proposition 7.3].

We now assume that Λ is a complete local ring with maximal ideal $\mathfrak{m}$, and break the proof into two cases two cases.

Case 1: $\mathfrak{m}$ is split — necessarily strictly split because of completeness [Proposition 11.10] — and the two coordinate rings Γ_1 and Γ_2 of Γ are in distinct sets, say $\mathcal{C}(A_1)$ and $\mathcal{C}(A_2)$. Then for each $i \geq 3$, both ranks of A_i are zero, and hence A_i has finite length [Lemma 15.3]. Since $A \in \mathrm{fingen}_\infty(\Lambda)$, this implies that $A_i = 0$ for $i \geq 3$; equivalently, $A = A_1 \oplus A_2$.

By Theorem 24.5 we have $\mathrm{eqstab}(A) = \mathrm{eqstab}(A_1)\cdot \mathrm{eqstab}(A_2)$. If we can prove that this product is a direct product, then — since everything said of about A is equally true of B, we will have $\mathrm{eqstab}(A) = \mathrm{eqstab}(B)$ if and only if each $\mathrm{eqstab}(A_i) = \mathrm{eqstab}(B_i)$.

Write A_1 and A_2 as direct sums of indecomposable modules. By assumption, A_1 has rank $(a, 0)$ with respect to the two coordinate rings of Γ, for some a; and similarly, A_2 has rank $(0, b)$ for some b. Therefore A has rank (a, b). By Theorem 24.5, $\mathrm{eqstab}(A_1) = G_1 \times 1$ where G_1 equals $k_1^\times$ or 1 according as a is not, or is 0, respectively. Similarly, $\mathrm{eqstab}(A_2) = 1 \times G_2$ where G_2 equals $k_2^\times$ or 1 according as b is not, or is 0. These facts prove directness of the product $\mathrm{eqstab}(A_1)\cdot \mathrm{eqstab}(A_2)$, and complete the proof of case 1.

Case 2: The 1 or 2 coordinate rings of Γ (both) belong to a single $\mathcal{C}(A_i)$, say $\mathcal{C}(A_1)$. Then $A_i = 0$ for $i \geq 2$ as in case 1, and so $A = A_1$. Similarly, $B = B_1$, and what we are proving is now trivial. □

Now that the computation of eqstab(M) is reduced to the case in which Λ is complete local, we restrict our attention in the rest of this section to the slightly more general situation described in Hypothesis 24.1. Not having to worry about actual completeness simplifies matters a bit (and we ignore the trivial case in which Λ is a DVR).

24.9. Proof-outline of Theorem 24.5. We begin by reviewing some of the terminology from [**KL2**] that we have not needed until now, and is needed in order to understand the statements of the various blocks of our proof. Let Λ be as in Hypothesis 24.1. In the description of indecomposable modules $M \in$ fingen(Λ) given in [**KL2**], the structure of M is given in the form $M \cong S/K$, or more completely, $M(\mathcal{D}) = S(\mathcal{D})/K(\mathcal{D})$ where $\mathcal{D}$ is a diagram showing the structure of M [**KL2**, unsplit case: (2.2.11); strictly split case: (3.2.9)]. Moreover, the natural map $S \twoheadrightarrow M = S/K$ is a Γ-separated cover [**KL2**, 9.6, 10.4]. We discuss the two two types of rings Λ separately.

Unsplit case [**KL2**, §2]. The section begins with a discussion of "gluing" and "reduction" operations on uniserial Γ-modules, followed by a description of how these operations are put together to form indecomposable modules $M \in$ fingen(Λ). These gluing and reduction operations are displayed by means of "standard diagrams" $\mathcal{D}$. There are seven types of standard diagrams, but only the three in [**KL2**, (2.4.1)] occur when $M \in \text{fingen}_\infty(\Lambda)$: $\mathcal{D}_{\text{Nrd}}$ ("nonreduced"), $\mathcal{D}_{\text{Brd}}$ ("bottom-reduced"), and $\mathcal{D}_{\text{Trd}}$ ("top-reduced"). The main results about indecomposablity, and isomorphism invariants of these modules are stated in [**KL2**, Theorems 2.7 and 2.8].

The modules with nonreduced diagrams can have rank 1 or 2, while the others have rank 1. The diagram for Λ itself is top-reduced [**KL2**, (2.5.1)], and the rank is 1 in this case. The term "rank" was unfortunately not used in [**KL2**], but the rank of $M(\mathcal{D})$ — in the sense of the present paper [Definition 15.1] — is easily seen to be the number of uniserial Γ-modules of infinite length (hence $\cong \Gamma$) whose "length label" ∞ occurs in $\mathcal{D}$. Proofs of these statements about rank follow from the proof of Lemma 16.2 of the present paper.

Strictly split case [**KL2**, §3]. This is similar to the unsplit case, but much simpler. Only one type of standard diagram occurs for $M(\mathcal{D}) \in \text{fingen}_\infty(\Lambda)$, namely $\mathcal{D}_{\text{DCy}}$, the "deleted cycle" diagram in [**KL2**, (3.3.1)].

In the strictly split case we have $\Gamma = \Gamma_1 \oplus \Gamma_2$, the direct sum of two DVRs. Therefore the rank of $M(\mathcal{D})$ is an ordered pair of integers (a_1, a_2). Here each a_i equals the number of times the uniserial Γ_i-module of infinite length, hence with length label ∞, (i.e. Γ_i itself) occurs in $\mathcal{D}$. By the definition of this type of diagram, the possible ranks are $(1, 0)$, $(0, 1)$. The diagram for Λ is displayed in [**KL2**, (3.4.1)], and the rank is $(1, 1)$. See the proof of the proof of Lemma 16.2 for more details.

In either case. We refer to these modules by the name of their associated diagram; for example, a "deleted cycle module" is a module with a deleted cycle diagram. These modules are known in the finite dimensional algebra community as "string modules". The related "band modules" do not occur in $\text{fingen}_\infty(\Lambda)$ because they always have finite length.

In view of these facts, the proof of Theorem 24.5 is completed by the following collection of results.

Proofs: unsplit case. The proof that eqstab(M) $=$ im(N_k^F) for (necessarily nonreduced) indecomposables M of rank 2 is given in Theorem 24.17; the proof that eqstabM= $k^\times$ for top-reduced and bottom-reduced modules M is given in Theorem 24.18; and the proof that $\text{eqstab}(M) = F^\times$ for nonreduced modules M of rank 1 is given in Theorem 24.19. The fact that $\text{eqstab}(\oplus_i M_i) = \prod_i \text{eqstab}(M_i)$ follows by comparing the statements of these three theorems.

Proofs: strictly split case. We prove all of the assertions about this situation (which is conceptually simpler than the unsplit case) in Theorem 24.20. □

A large portion of [**KL2**] is devoted to the description of finitely generated Λ-modules in terms of matrices. In Remarks 24.10 below, we review the needed portion of that description and its relation to matrix representation of residue idèles.

REMARKS 24.10 (Matrices). In these Remarks, Λ is assumed to be a local, unsplit or strictly split Dedekind-like ring, as in Hypothesis 24.1. The construction described in these Remarks starts with a module $X \in \text{fingen}(\Gamma)$ and two Λ-submodules S and K, described by matrices. Appropriate theorems in [**KL2**] state that the resulting module $M = S/K$ ranges through all isomorphism classes of finitely generated Λ-modules. The number of matrices involved in the construction is either two or four, according as Λ is respectively unsplit or strictly split. We consider these two situations separately in items (i) and (ii) below.

(i) Λ *unsplit.* The construction starts with three ingredients: (a) a Γ-module decomposition $X = \oplus_{\nu=1}^n X_\nu$ where each $X_\nu = \Gamma/\mathfrak{m}^{\lambda_\nu}$ for λ_ν either a positive integer or ∞ (where we define $\mathfrak{m}^\infty = 0$); (b) a "standard" Γ-linear identification of each $\bar{X}_\nu$ with F; and (c) and a "standard" Γ-linear identification of each $\text{soc}_\Gamma(X_\nu)$ with F when ΓX_ν has finite length. This collection $\mathcal{X}$ of ingredients (a)–(c) is called a *matrix setup (unsplit case)* in [**KL2**, 5.6]. [These matrix setups are quite different from the global matrix setups used in Definition 20.4.] We need not concern ourselves here with the details of the standard identifications except that each matrix setup provides Γ-linear identifications $\bar{X} = F^{(n)}$ and $\text{soc}_\Gamma(X) = F^{(n-e_\infty)}$, where e_∞ is the number of X_ν that equal Γ.

Let A be an invertible $n \times n$ matrix over F, and B an $(n - e_\infty)$-column matrix over F whose rows are F-linearly independent. (The matrix B must satisfy some additional conditions which need not concern us here [**KL2**, (5.7.1)].) Let $\bar{S}(A)$ be the k (not F!) subspace of $\bar{X} = F^{(n)}$ generated by the rows of A, and $S(A)$ the inverse image of $\bar{S}(A)$ in X (via the standard maps). Also, let $K(A, B)$ be the k-subspace of $\text{soc}_\Gamma(X) = F^{(n-e_\infty)}$ generated by the rows of B. We allow the possibility that B has no rows, in which case $K = 0$. (See [**KL2**, 5.7, 5.8] for details.) Finally, set $M(A, B) = S(A)/K(A, B)$. By [**KL2**, Theorem 5.10], every finitely generated Λ-module is isomorphic to some such $M(A, B)$.

Note that each column of A and each column of B arises from some X_ν. We attach the *length label* λ_ν (the Γ-length of the associated X_ν) to each such column.

Although we refer to the matrix pair as (A, B), we always view B as lying under A, with each column of B lying under the corresponding column of A (that is, under the column of A that corresponds to the same summand X_ν). Thus, the columns of A with length label ∞ have no column of B below them. (For an example, see (24.17.1).)

We call the sequence $\lambda_1, \ldots, \lambda_n$ of length labels the *label sequence* of (A, B), and we usually write it above the columns of A. *The rank of $M(A, B)$ equals the number of infinite labels in its associated label sequence.*

(ii) Λ *strictly split.* The construction is similar to the unsplit case, except that four matrices are involved, because $\Gamma = \Gamma_1 \oplus \Gamma_2$. Start with a *matrix setup (strictly split case)* $\mathcal{X}$ consisting of three ingredients: (a) a Γ-module decomposition $X = X_1 \oplus X_2$ together with decompositions $X_1 = \oplus_{\nu=1}^n X_{1\nu}$ and $X_2 = \oplus_{\nu=1}^n X_{2\nu}$ (note that both X_i decompositions have the same number n of terms), where each $X_{1\nu} = \Gamma_1/\mathfrak{m}_1^{\lambda_\nu}$, $X_{2\nu} = \Gamma_2/\mathfrak{m}_2^{\mu_\nu}$, $\mathfrak{m}_i$ is the maximal ideal of Γ_i, λ_ν and μ_ν are positive integers or ∞, and $\mathfrak{m}_i^\infty = 0$; (b) a standard Γ_i-linear identification $\bar{X}_{i\nu} = k$ for each i and ν; and (c) standard Γ_i-linear identifications $\mathrm{soc}_{\Gamma_1}(\Gamma_1/\mathfrak{m}_1^{\lambda_\nu}) = k$ and $\mathrm{soc}_{\Gamma_2}(\Gamma_2/\mathfrak{m}_i^{\mu_\nu}) = k$ for each finite λ_ν and μ_ν. (For details, see [**KL2**, 6.3].)

Recall the slight ambiguity here: There are three nonisomorphic Γ-modules whose additive group is that of k. When it is necessary to distinguish among them, we write k_i for the copy of k that arises as $\bar{\Gamma}_i = \Gamma_i/\mathfrak{m}_i$ $(i = 1, 2)$, and simply k for the diagonal copy of k in $k_1 \times k_2$ as in (11.5.1).

The matrix setup provides Γ-linear identifications of each $\bar{X}_i = k_i^{(n)}$, $\mathrm{soc}_{\Gamma_1}(X_1) = k_1^{(n-d_\infty)}$, and $\mathrm{soc}_{\Gamma_2}(X_2) = k_2^{(n-e_\infty)}$, where d_∞ and e_∞ are respectively the number of terms in the decomposition of X_1 and X_2 that equal Γ_1 and Γ_2.

Let A_1 and A_2 be invertible $n \times n$ matrices over k, and let B_1 and B_2 be respectively $(n - d_\infty)$- and $(n - e_\infty)$-column matrices over k, with the same number of rows in each B_i, and with k-linearly independent rows. As above, we allow the number of rows to be zero, (and we ignore the additional conditions that each B_i must satisfy).

Let $\bar{S}(A_1, A_2)$ be the k-subspace of $\bar{X}_1 \oplus \bar{X}_2$ generated by the rows of (A_1, A_2), and $S(A_1, A_2)$ the full inverse image of $\bar{S}(A_1, A_2)$ in X. Let $K(B_1, B_2)$ be the k-subspace of $\mathrm{soc}_\Gamma(X_1 \oplus X_2) = \mathrm{soc}_{\Gamma_1}(X_1) \oplus \mathrm{soc}_{\Gamma_2}(X_2)$ generated by the rows of (B_1, B_2). (See [**KL2**, 6.4] for details. Note that $K(B_1, B_2)$ is called $\mathrm{im}(B_1, B_2)$ in [**KL2**, (6.4.4)].) Finally, set $M(A_1, A_2, B_1, B_2) = S(A_1, A_2)/K(B_1, B_2)$. As before, every finitely generated Λ-module is isomorphic to some such M [**KL2**, Theorem 6.6].

As before, we always view each B_i as lying below A_i, with each column of B_i lying under the corresponding column of A_i, and the length label of A_i (λ_ν or μ_ν) written above each column of A_i. For an example, see [**KL2**, (7.1.1)]. Thus, in the split case, we associate *two* label sequences, $\{\lambda_\nu\}$ and $\{\mu_\nu\}$ with $M = M(A_1, A_2, B_1, B_2)$. Moreover, M *has rank* (a, b) *if and only if the number of infinite labels* λ_ν *and* μ_ν *equal* a *and* b, *respectively.*

(iii) *Canonical form of* (A, B) *and* (A_1, A_2, B_1, B_2). In the preceding discussion, we avoided discussing precisely what matrices can occur as A and B matrices. The reason is that the proofs which follow only need to use the canonical forms of these matrices (with respect to the isomorphism class of the corresponding M). When a canonical form is used in these proofs, we say where to find its description in [**KL2**].

(iv) *Matrices of residue endo-idèles.* Note first that the natural maps $S \twoheadrightarrow M$ with kernel K in (i) and (ii) are Γ-separated covers of M [**KL2**, Theorems 5.10, 6.6]. Let ψ be a residue endo-idèle on X [Notation 24.2].

Unsplit case. Since ψ is a $\bar{\Gamma} = F$-automorphism of $\bar{X} = F^{(n)}$, ψ equals right multiplication by a unique invertible matrix Q_1 over F. Moreover, any lifting $\widetilde{\psi}$ to a Γ-automorphism of X restricts to a Γ-automorphism of $\mathrm{soc}_\Gamma(X) = F^{(n-e_\infty)}$. Therefore, $\widetilde{\psi}$ restricted to $\mathrm{soc}_\Gamma(X)$ equals right multiplication by a (unique) invertible matrix Q_2 over F. It is easy to see that the matrix Q_2 is independent of the lifting $\widetilde{\psi}$ of ψ.

Only rather special matrices Q_1, Q_2 occur in this way, because of the requirement of liftability of ψ to a Γ-automorphism of X. In order to state which matrices can occur, we introduce the following system of *length-labels* for Q_1 and Q_2.

(24.10.1) (i) Label both the rows and the columns of Q_1 by the label sequence of A.
(ii) Label both the rows and the columns of Q_2 by the label sequence of B (which consists of the sequence of finite length labels of A).
(iii) For $i = 1, 2$, $Q_i[\lambda\,, \lambda']$ refers to the submatrix of Q_i consisting of the intersection of the rows with length label λ and columns with length label λ'.

The answer to the question of which matrices can Q_i occur is stated in Lemma 24.11 below.

Strictly split case. This is similar to the unsplit case, and so we describe it more briefly. The residue endo-idèle ψ equals right multiplication by invertible matrices Q_{11} and Q_{12} (over k) on $k_1^{(n)}$ and $k_2^{(n)}$, respectively. Any lifting $\widetilde{\psi}$ acts on $\mathrm{soc}_{\Gamma_1}(X_1) \oplus \mathrm{soc}_{\Gamma_2}(X_2)$ by another pair of invertible matrices (over k) Q_{21} and Q_{22}.

As in (24.10.1), we attach length-labels as follows.

(24.10.2) (i) Label both the rows and the columns of each Q_{1i} by the label sequence of A_i.
(ii) Label both the rows and the columns of each Q_{2i} by the label sequence of B_i.
(iii) Define $Q_{ji}[\lambda\,, \lambda']$ as before.

The answer to the question of which matrices Q_{ji} can occur is stated in Lemma 24.14 below.

LEMMA 24.11. *Suppose that Λ is local and unsplit, and keep the notation in Remarks 24.10. Let Q_1 and Q_2 be invertible matrices over F, with length labels as in* (24.10.1). *Then Q_1 and Q_2 are induced by a residue endo-idèle on X if and only if the following conditions hold.*
(i) *Q_1 is length upper triangular; that is, if $\lambda < \lambda'$, then $Q_1[\lambda\,, \lambda'] = 0$.*
(ii) *Q_2 is length lower triangular; that is, if $\lambda > \lambda'$, then $Q_2[\lambda\,, \lambda'] = 0$.*
(iii) *$Q_1[\lambda\,, \lambda] = Q_2[\lambda\,, \lambda]$ whenever $\lambda < \infty$ ("diagonal block equality").*

For an example of this numbering scheme, see [**KL2**, (5.12.1)].

Note that, if the length-blocks of rows and columns of Q_1 and Q_2, together with their length labels, are simultaneously re-ordered so that all length labels occur in *decreasing* order, then the above "triangular forms" become actual block triangular forms, and (iii) states that each diagonal block of Q_2 equals the corresponding diagonal block of Q_1 (except that $Q_1[\infty\,, \infty]$ has no corresponding block in Q_2.)

PROOF. The proof of this occupies most of the proof of [**KL2**, Theorem 5.13], after Claim 1. The following comments are intended to bridge the gap between the notation in that paper and this one. The term "$\mathcal{X}$-triangular in [**KL2**] is replaced, in the present paper, by the more intuitive notions of length upper and lower triangular, and diagonal-block equality.

The term "idèle" does not occur in [**KL2**], since that paper deals only with local rings. Thus, the proof of [**KL2**, Theorem 5.13] refers to an automorphism τ of X (which is $\widetilde{\psi}$ in our present notation) that induces automorphisms of $F^{(n)}$ and $\mathrm{soc}_{\Gamma}(X)$ (which are ψ and the restriction of $\widetilde{\psi}$, respectively, in our present notation). The relation between τ and right multiplication by Q_1 (the matrix form of ψ) is shown in the right-most commutative square in [**KL2**, (5.13.2)], and the relation between τ and right multiplication by Q_2 on $F^{(n-e_\infty)}$ (i.e., the restriction of $\widetilde{\psi}$) is shown in the right-hand square of [**KL2**, (5.13.3)]). □

The next lemma translates the problem of finding eqstab(M) to a matrix problem, in the unsplit case.

LEMMA 24.12. *Suppose that Λ is local and unsplit, and let M be defined by a matrix pair (A, B) and a Γ-module X, as in Remarks 24.10. Then* eqstab(M) *is the set of elements* $\det Q_1[\infty\,,\infty] \in F^\times$ *such that there exist relations of the form*

$$P_1A = AQ_1 \quad and \quad P_2B = BQ_2 \tag{24.12.1}$$

where each P_i is an invertible matrix over k, and each Q_i is an invertible matrix over F, having the form described in Lemma 24.11(i)–(iii).

PROOF. From Notation 24.2, eqstab(M) is the set of all elements $\det(\psi_\infty) \in F^\times$ where ψ ranges over the residue endo-idèles on X such that (24.2.1) holds.

By the definitions of A and B, we have $\bar{S}(A) = k^{(n)}A$ and $K(A, B) = k^{(n-e_\infty)}B$. Also, right multiplication by a pair of invertible matrices Q_1 and Q_2 over F yields the action of some residue endo-idèle on X if and only if the conditions of Lemma 24.11 hold. Therefore, (24.2.1) is equivalent to the existence of such a pair Q_1, Q_2 satisfying $k^{(n)}AQ_1 = k^{(n)}A$ and $k^{(n-e_\infty)}BQ_2 = k^{(n-e_\infty)}B$. Rewriting this in the form shown in (24.12.1) is a straightforward excercise in linear algebra. □

LEMMA 24.13. *Suppose that Λ is local and strictly split, and keep the notation in Remarks 24.10. Let Q_{11}, Q_{12}, Q_{21}, and Q_{22} be invertible matrices over k, with length labels as in* (24.10.2). *These matrices are induced by a residue endo-idèle on X if and only if the following conditions hold.*

(i) *Each Q_{1i} is length upper triangular; that is, if $\lambda < \lambda'$, then $Q_{1i}[\lambda\,, \lambda'] = 0$.*
(ii) *Each Q_{2i} is length lower triangular; that is, if $\lambda > \lambda'$, then $Q_{2i}[\lambda\,, \lambda'] = 0$.*
(iii) *$Q_{1i}[\lambda\,, \lambda] = Q_{2i}[\lambda\,, \lambda]$ whenever $\lambda < \infty$ ("diagonal block equality").*

PROOF. This follows from the proof of [**KL2**, Theorem 6.7] in the same way that Lemma 24.11 followed from the proof of [**KL2**, Theorem 5.13]. □

The next lemma translates the problem of finding eqstab(M) to a matrix problem, in the strictly split case.

LEMMA 24.14. *Suppose that Λ is local and strictly split, and let M be defined by a matrix 4-tuple (A_1, A_2, B_1, B_2) and a Γ-module X, as in Remarks 24.10. Then* eqstab(M) *is the set of elements* $\big(\det Q_{11}[\infty\,,\infty], \det Q_{12}[\infty\,,\infty]\big) \in k^\times \times k^\times$ *such that there exist relations of the form*

$$\begin{aligned} P_1A_1 &= A_1Q_{11} \qquad & P_1A_2 &= A_2Q_{12} \\ P_2B_2 &= B_2Q_{21} \qquad & P_2B_2 &= B_2Q_{22} \end{aligned} \tag{24.14.1}$$

where each P_i and each Q_{ji} is an invertible matrix over k, with each Q_{ji} having the form described in Lemma 24.13(i)–(iii).

PROOF. This lemma follows from Lemma 24.13 in the same way that Lemma 24.12 followed from Lemma 24.11. □

We turn now to the question of calculating eqstab(M) in the case where Λ is a local, unsplit Dedekind-like ring.

NOTATION 24.15. For the matrix computations that follow, fix an element $\varepsilon \in F - k$. Since F is separable of dimension 2 over k, we have

$$F = k[\varepsilon] \quad \text{and} \quad \varepsilon \neq \tilde{\varepsilon} \tag{24.15.1}$$

where $\tilde{\varepsilon}$ denotes the conjugate of ε. To denote the conjugate of a long expression, we write the tilde as an exponent. For example, $Q_1[\mu,\nu]^{\sim}$ denotes the matrix whose entries are the conjugates of the entries of $Q_1[\mu,\nu]$.

The simple matrix computation in the following lemma is used repeatedly in the proof of Theorem 24.17.

LEMMA 24.16. *Let $\alpha, \beta, \gamma, \delta$ be $m \times n$ matrices over the field F. Then there exist $m \times n$ matrices a, b, c, d over k such that the following block matrix equation holds*

$$\begin{bmatrix} a & b \\ c & d \end{bmatrix} \begin{bmatrix} I_n & I_n \\ I_n\varepsilon & I_n\tilde{\varepsilon} \end{bmatrix} = \begin{bmatrix} I_m & I_m \\ I_m\varepsilon & I_m\tilde{\varepsilon} \end{bmatrix} \begin{bmatrix} \alpha & \beta \\ \gamma & \delta \end{bmatrix} \tag{24.16.1}$$

if and only if $\alpha = \tilde{\delta}$ and $\beta = \tilde{\gamma}$.

PROOF. Multiplying out both sides yields the equivalent equation:

$$\begin{bmatrix} a + b\varepsilon & a + b\tilde{\varepsilon} \\ c + d\varepsilon & c + d\tilde{\varepsilon} \end{bmatrix} = \begin{bmatrix} \alpha + \gamma & \beta + \delta \\ \alpha\varepsilon + \gamma\tilde{\varepsilon} & \beta\varepsilon + \delta\tilde{\varepsilon} \end{bmatrix} \tag{24.16.2}$$

Assume first that a, b, c, d exist. Since a, b, c, d are matrices over k, we have $a + b\tilde{\varepsilon} = (a + b\varepsilon)^{\sim}$ and $c + d\tilde{\varepsilon} = (c + d\varepsilon)^{\sim}$. Therefore, the (1,1)-block of each matrix must equal the conjugate of the (1,2)-block, and the (2,1)-block of each matrix must equal the conjugate of the (2,2)-block. This leads to the pair of equations

$$\begin{aligned} \alpha + \gamma &= \tilde{\beta} + \tilde{\delta} \\ \alpha\varepsilon + \gamma\tilde{\varepsilon} &= \tilde{\beta}\tilde{\varepsilon} + \tilde{\delta}\varepsilon \end{aligned}$$

If we multiply the first equation by ε and subtract, we get $\gamma(\varepsilon - \tilde{\varepsilon}) = \tilde{\beta}(\varepsilon - \tilde{\varepsilon})$. Since $\tilde{\varepsilon} \neq \varepsilon$, we can cancel $\varepsilon - \tilde{\varepsilon}$ from both sides to obtain the equation $\gamma = \tilde{\beta}$. Similarly, if we multiply the first equation by $\tilde{\varepsilon}$ and subtract, we obtain $\alpha(\tilde{\varepsilon} - \varepsilon) = \tilde{\delta}(\tilde{\varepsilon} - \varepsilon)$, and so $\alpha = \tilde{\delta}$.

Conversely suppose that $\alpha, \beta, \gamma, \delta$ satisfy the given conjugacy conditions. Since $F = k[\varepsilon]$, there exist matrices a, b, c, d over k such that the (1,1)- and (2,1)-entries on the left-hand side of (24.16.2) equal the corresponding entries on the right-hand side. Then the given conjugacy relations establish equality of the remaining two pairs of entries of the matrices in (24.16.2). □

THEOREM 24.17. *Assume Hypothesis 24.1, and suppose that Λ is unsplit. If every summand M_i has rank 2 (and hence is nonreduced), then* eqstab$(M) = \operatorname{im}(N_k^F)$.

PROOF. The matrix pair (A, B) in canonical form, for a nonreduced module $M(A, B)$ of rank 2, is as shown in (24.17.1), with its label sequence above it. Since the rank is 2, we have $\lambda_1 = \infty = \lambda_m$, and all other length labels are finite. Moreover, $m \geq 4$; that is, A has at least 4 columns. (Other than ellipses, shown by dots, all entries not shown are zero. See [**KL2**, (8.7.3)].) Throughout this proof, ε is as in Notation 24.15.

$$
\begin{array}{c}
\begin{array}{cccccccc}
\lambda_1 & \lambda_2 & \lambda_3 & \lambda_4 & \lambda_5 & \cdots & \lambda_{m-1} & \lambda_m
\end{array} \\
A = \begin{bmatrix}
1 & 1 & & & & \cdots & & \\
\varepsilon & \tilde{\varepsilon} & & & & \cdots & & \\
 & & 1 & 1 & & \cdots & & \\
 & & \varepsilon & \tilde{\varepsilon} & & \cdots & & \\
 & & & & & \ddots & & \\
 & & & & & \cdots & 1 & 1 \\
 & & & & & \cdots & \varepsilon & \tilde{\varepsilon}
\end{bmatrix} \\
B = \begin{bmatrix}
1 & 1 & & & \cdots & & \\
\varepsilon & \tilde{\varepsilon} & & & \cdots & & \\
 & & 1 & 1 & \cdots & & \\
 & & \varepsilon & \tilde{\varepsilon} & \cdots & & \\
 & & & & \ddots & & \\
 & & & & \cdots & 1 & 1 \\
 & & & & \cdots & \varepsilon & \tilde{\varepsilon}
\end{bmatrix}
\end{array}
\qquad (24.17.1)
$$

Now consider a general $(A, B) = (A_1, B_1)^{(e_1)} \oplus ... \oplus (A_n, B_n)^{(e_n)}$, where each (A_i, B_i) has the form displayed in (24.17.1), occurs with multiplicity e_i, and has label sequence $\lambda^i_1, ..., \lambda^i_{m_i}$. We choose the notation so that $M(A_i, B_i) \not\cong M(A_j, B_j)$ for $i \neq j$. We make frequent use of the following consequence of this: *Each label sequence $\lambda^i_1, ..., \lambda^i_{m_i}$ is different from its mirror image $\lambda^i_{m_i}, ..., \lambda^i_1$, and is also different from each other label sequence $\lambda^j_1, ..., \lambda^j_{m_i}$ and the mirror image of that sequence.*

The fact that the label sequence of every nonreduced pair must be different from its mirror image is a necessary condition for indecomposability [KL2, (8.7.2)(ii), Theorem 8.9, and Proposition 8.19]. The fact that each label sequence is different from each other label sequence and the reversal of that sequence follows from the uniqueness theorem for this type of module [KL2, Theorem 8.18].

Assume that matrix relations $P_1 A = A Q_1$ and $P_2 B = B Q_2$ hold, as described Lemma 24.12. To prove $\operatorname{eqstab}\big(M(A, B)\big) \subseteq \operatorname{im}(N^F_k)$, it suffices to show that $\det Q_1[\infty, \infty] \in \operatorname{im}(N^F_k)$ [Lemma 24.12]. The first step is to permute the rows and columns in in each summand $(A_i, B_i)^{(e_i)}$ — together with the length labels of the columns — to obtain the matrix pair $(A^{(e_i)}_i, B^{(e_i)}_i)$ consisting of the following matrices:

(24.17.2)

$$
\begin{array}{c}
\begin{array}{cccccccc} \sigma_1^i & \sigma_2^i & \sigma_3^i & \sigma_4^i & \sigma_5^i & \cdots & \sigma_{m_i-1}^i & \sigma_{m_i}^i \end{array}\\
\begin{bmatrix}
I & I & & & \cdots & & \\
I\varepsilon & I\tilde{\varepsilon} & & & \cdots & & \\
& & I & I & \cdots & & \\
& & I\varepsilon & I\tilde{\varepsilon} & \cdots & & \\
& & & & \ddots & & \\
& & & & \cdots & I & I \\
& & & & \cdots & I\varepsilon & I\tilde{\varepsilon}
\end{bmatrix}\\
\begin{bmatrix}
I & I & & & \cdots & & \\
I\varepsilon & I\tilde{\varepsilon} & & & \cdots & & \\
& & I & I & \cdots & & \\
& & I\varepsilon & I\tilde{\varepsilon} & \cdots & & \\
& & & & \ddots & & \\
& & & & \cdots & I & I \\
& & & & \cdots & I\varepsilon & I\tilde{\varepsilon}
\end{bmatrix}
\end{array}
$$

where each I denotes an identity matrix of order e_i, and each σ_j^i is a *subblock symbol* indicating a block of e_i columns each of whose length labels we denote by $|\sigma_j^i| = \lambda_j^i$. In order to preserve the relations $P_1A = AQ_1$ and $P_2B = BQ_2$, we need to make the same permutations of the rows and columns of Q_1 and Q_2, together with their length labels. We then label each row subblock and column subblock of Q_1 and Q_2 by the subblock symbol σ_j^i used for the corresponding column of A and B.

These column permutations correspond to rearranging the summands X_ν in the decomposition $X = \oplus_\nu X_\nu$ and therefore have no effect on the isomorphism class of M.

Note that, by our hypothesis that each $M(A_i, B_i)$ has rank 2, and we have (see the first paragraph of this proof):

(24.17.3)
$$|\sigma_j^i| = \infty \quad \Longleftrightarrow \quad j = 1 \text{ or } j = m_i$$

In order to show that the determinant of the matrix $Q_1[\infty, \infty]$ is an element of $\mathrm{im}(N_k^F)$, we show that *this matrix is block triangular with respect to the subblocks labeled by* σ_j^i, *and the diagonal subblocks occur in conjugate pairs.*

First consider the subblock of Q_1 consisting of the intersection of the rows labeled by σ_j^i and σ_{j+1}^i, with the columns labeled by σ_k^h and σ_{k+1}^h — for all values of the superscripts and subscripts that make sense. We claim that this subblock has the "cross-conjugacy" relations displayed in (24.17.4) when j and k are both odd.

(24.17.4)
$$Q_1: \quad \begin{array}{c c} & \begin{array}{cc} \sigma_k^h & \sigma_{k+1}^h \end{array} \\ \begin{array}{c} \sigma_j^i \\ \sigma_{j+1}^i \end{array} & \begin{bmatrix} \alpha & \beta \\ \tilde{\beta} & \tilde{\alpha} \end{bmatrix} \end{array} \qquad (j, k \text{ odd.})$$

Moreover, we claim that the same holds for Q_2 when j and k are both odd.

It is helpful to understand (24.17.4) as follows.

(24.17.5) To obtain a conjugate subblock, do one of the following.

In Q_1: If the row subscript is odd, call it j and increase it by one. If it is even, call it $j+1$ and reduce it by 1. The same applies to the column subscript, but with k in place of j.
In Q_2: The same holds if all parities are reversed.

Now we prove the claim. Fix i, j, h, k with j, k odd. By the Δ-block of Q_1 — or any matrix of the same size as Q_1 — we mean the submatrix occupying the same collection of rows and columns as displayed in (24.17.4). In particular, let X be the Δ-block of Q_1. The block diagonal form of A displayed in (24.17.2) — together with the fact that j and k are odd — shows that the Δ-block of AQ_1 is TX, where $T = \begin{bmatrix} I & I \\ I\varepsilon & I\tilde{\varepsilon} \end{bmatrix}$. On the other hand, P_1 is a matrix with entries in the field k. Therefore — again by the block diagonal form of A — the Δ-block of P_1A is KT where K is a matrix with entries in the field k.

Since $P_1A = AQ_1$, we have $KT = TX$. The claim for Q_1 now follows from Lemma 24.16, and similar reasoning applies to Q_2.

Now we show that $Q_1[\infty, \infty]$ is block triangular with respect to the row and column subblocks σ_j^i that occur in it. For this, it suffices to define a total ordering of the subblock symbols σ_j^i in such a way that $Q_1[\sigma_j^i, \sigma_k^h] = 0$ whenever $\sigma_j^i < \sigma_k^h$.

To begin, we list the n subblock sequences of the (A_i, B_i) twice, once forward and once backward:

(24.17.6)
$$\begin{aligned} \sigma_1 &= \sigma_1^1, \ldots, \sigma_{m_1}^1 & \qquad \sigma_{n+1} &= \sigma_{m_1}^1, \ldots, \sigma_1^1 \\ \sigma_2 &= \sigma_1^2, \ldots, \sigma_{m_2}^2 & \sigma_{n+2} &= \sigma_{m_2}^2, \ldots, \sigma_1^2 \\ &\vdots & &\vdots \\ \sigma_n &= \sigma_1^n, \ldots, \sigma_{m_n}^n & \sigma_{2n} &= \sigma_{m_n}^n, \ldots, \sigma_1^n \end{aligned}$$

By (24.17.3), each σ_a begins and ends with a subblock of value ∞, and all other subblocks are finite. Therefore, the set of subblocks of $Q_1[\infty, \infty]$ that we wish to order is the set of first terms of the sequences $\sigma_1, \ldots, \sigma_{2n}$. Since no subblock symbol σ_j^i appears more than once in (24.17.6), ordering these subblocks of $Q_1[\infty, \infty]$ is equivalent to ordering the sequences in (24.17.6). Recall that each m_i is even, as is apparent from (24.17.2).

Given two sequences $\sigma_a = \mu_1, \mu_2, \ldots$ and $\sigma_b = \nu_1, \nu_2, \ldots$, we define $\sigma_a < \sigma_b$ to mean that, for some r, $|\mu_k| = |\nu_k|$ for all $k < r$, and:

(24.17.7)
$$|\mu_r| < |\mu_r| \quad \text{if } r \text{ is even, but} \quad |\mu_r| > |\mu_r| \quad \text{if } r \text{ is odd.}$$

As we shall see, this twofold definition comes from the facts that Q_1 is length upper triangular while Q_2 is length lower triangular. We now prove that this is indeed a total ordering.

Comparability. Given sequences $\sigma_a = \mu_1, \mu_2, \ldots$ and $\sigma_b = \nu_1, \nu_2, \ldots$, with $a \neq b$, we need to show that $\sigma_a < \sigma_b$ or $\sigma_b < \sigma_a$. Let $|\sigma_a|$ and $|\sigma_b|$ be the corresponding sequences of length labels. The only ways that σ_a and σ_b could fail to be comparable is if the sequences $|\sigma_a|$ and $|\sigma_b|$ of length labels were identical, or if one were a proper initial segment of the other. The former cannot happen because, as observed below (24.17.1), the label sequences and their mirror images are all distinct. The latter cannot happen because each label sequence begins and ends with ∞, while all other terms are finite.

Triangularity. Given sequences $\sigma_a = \mu_1, \mu_2, \ldots$ and $\sigma_b = \nu_1, \nu_2, \ldots$, suppose that $\sigma_a < \sigma_b$. We prove that $Q_1[\mu_1, \nu_1] = 0$, the most complicated part of this proof.

We claim that $Q_1[\mu_1, \nu_1] = Q_1[\mu_2, \nu_2]^{\sim}$. There are four possibilities, according to whether each of σ_a and σ_b occur in the first or second column of (24.17.6). If σ_a occurs in the first column, then $\mu_1 = \sigma_1^i$ for some i. Then the row subscript 1 is odd, so according to rule (24.17.5), we find a conjugate subblock by increasing the subscript by 1, that is, by moving to the second term in σ_a. On the other hand, if σ_a occurs in the second column of (24.17.6), then the row subscript is some m_i, which is even. Therefore, by rule (24.17.5) we find a conjugate subblock by decreasing the subscript by 1, and that again moves us to the second term in the sequence σ_a. An analogous discussion of σ_b completes the proof of the claim.

In view of the claim, it suffices to prove that $Q_1[\mu_2, \nu_2] = 0$. This subblock of Q_1 is no longer in $Q_1[\infty, \infty]$. If $|\mu_2| \neq |\nu_2|$, then since $\sigma_a < \sigma_b$ and 2 is even we have $|\mu_2| < |\nu_2|$. Length upper triangularity of Q_1 [Lemma 24.11(i)] then implies that $Q_1[\mu_2, \nu_2] = 0$, and we are done.

If instead $|\mu_2| = |\nu_2|$, then $Q_1[\mu_2, \nu_2]$ is a diagonal subblock of Q_1. By diagonal block equality [Lemma 24.11(iii)], this subblock is equal to $Q_2[\mu_2, \nu_2]$. It therefore suffices to prove that $Q_2[\mu_2, \nu_2] = 0$.

We claim that $Q_2[\mu_2, \nu_2] = Q_2[\mu_3, \nu_3]^{\sim}$. Apply the part of (24.17.5) that applies to Q_2 (j and k even). If σ_a is in the first column of (24.17.6), then $\mu_2 = \sigma_2^i$, whose row subscript 2 is even. Therefore, according to rule (24.17.5), we get a conjugate subblock by increasing the row subscript from 2 to 3, that is, by moving to the third term in σ_a. On the other hand, if σ_a occurs in the second column of (24.17.6), then we get a conjugate subblock by decreasing the odd row subscript m_{i-1} by 1, again moving to the third term in σ_a. The analogous discussion of σ_b completes the proof of the claim.

If $|\mu_3| \neq |\nu_3|$, then since $\sigma_a < \sigma_b$ and 3 is odd, we have $|\mu_3| > |\nu_3|$. Therefore, length lower triangularity of Q_2 [Lemma 24.11(ii)] implies that $Q_2[\mu_3, \nu_3] = 0$, and we are done. If instead $|\mu_3| = |\nu_3|$, then $Q_2[\mu_3, \nu_3]$ is a diagonal subblock of Q_3, and we can move back to the corresponding subblock of Q_1 [Lemma 24.11(iii)], and repeat what was done before.

The procedure reaches a zero subblock when we reach the terms of the sequences σ_a and σ_b which determine that $\sigma_a < \sigma_b$. This completes the proof of triangularity.

Diagonal-block conjugacy. Each $(A_i^{e(i)}, B_i^{e(i)})$ contributes two infinite subblocks, namely, its first subblock and its last subblock. Therefore, it suffices to prove that $Q_1[\sigma_1^i, \sigma_1^i] = Q_1[\sigma_{m_i}^i, \sigma_{m_i}^i]^{\sim}$, for $1 \leq i \leq n$. We have:

$$\begin{aligned} Q_1[\sigma_1^i, \sigma_1^i] &= Q_1[\sigma_2^i, \sigma_2^i]^{\sim} &= Q_2[\sigma_2^i, \sigma_2^i]^{\sim} &= Q_2[\sigma_3^i, \sigma_3^i] \\ &= Q_1[\sigma_3^i, \sigma_3^i] &= Q_1[\sigma_4^i, \sigma_4^i]^{\sim} &= \ldots \end{aligned}$$

The first equality holds by (24.17.4). The second equality holds by diagonal block equality [Lemma 24.11(iii)]. The third equality holds by the Q_2-version of (24.17.4). The fourth equality holds by diagonal block equality again. The fifth equality holds by (24.17.4) again; and so on. Clearly this procedure can be repeated until we reach the end of the sequence σ_i. Since m_i is always even, the iteration always finishes with $Q_1[\sigma_{m_i}^i, \sigma_{m_i}^i]^{\sim}$, as desired.

Conversely, we show that every element $\theta \in \operatorname{im}(N_k^F)$ occurs in eqstab(M). Since eqstab$(M) \supseteq$ eqstab(M_i) for each direct summand M_i of M [Lemma 23.2], we may assume that M is indecomposable. Therefore, we may assume that (A, B) is as in (24.17.1).

Now $\theta = \alpha\widetilde{\alpha}$ for some $\alpha \in F^{\times}$, so by Lemma 24.16 there exist elements $a, b, c, d \in k$ such that

$$\begin{bmatrix} a & b \\ c & d \end{bmatrix}\begin{bmatrix} 1 & 1 \\ \varepsilon & \tilde{\varepsilon} \end{bmatrix} = \begin{bmatrix} 1 & 1 \\ \varepsilon & \tilde{\varepsilon} \end{bmatrix}\begin{bmatrix} \alpha & 0 \\ 0 & \widetilde{\alpha} \end{bmatrix} \tag{24.17.8}$$

as well as elements $a', b', c', d' \in k$ such that

$$\begin{bmatrix} a' & b' \\ c' & d' \end{bmatrix}\begin{bmatrix} 1 & 1 \\ \varepsilon & \bar{\varepsilon} \end{bmatrix} = \begin{bmatrix} 1 & 1 \\ \varepsilon & \bar{\varepsilon} \end{bmatrix}\begin{bmatrix} \bar{\alpha} & 0 \\ 0 & \alpha \end{bmatrix} \tag{24.17.9}$$

Take P_1 and P_2 to be block diagonal matrices of appropriate sizes, whose diagonal blocks all equal $\begin{bmatrix} a & b \\ c & d \end{bmatrix}$ and $\begin{bmatrix} a' & b' \\ c' & d' \end{bmatrix}$, respectively. Also, let Q_1 and Q_2 be diagonal matrices of appropriate sizes, whose diagonal entries are respectively $\alpha, \widetilde{\alpha}, \alpha, \widetilde{\alpha}, \ldots, \alpha, \widetilde{\alpha}$ and $\widetilde{\alpha}, \alpha, \widetilde{\alpha}, \alpha, \ldots, \widetilde{\alpha}, \alpha$. Then it is straightforward to verify that $P_1 A = AQ_1$ and $P_2 B = BQ_2$, where P_i and Q_i have the required forms, and $\det Q_1[\infty, \infty] = \theta$. □

THEOREM 24.18. *Assume Hypothesis 24.1, and suppose that Λ is unsplit. If some direct summand M_i of M is top-reduced or bottom-reduced, and no direct summand of M is nonreduced with rank 1, then* $\operatorname{eqstab}(M) = k^{\times}$.

PROOF. By hypothesis and the review of terminology at the beginning of subsection 24.9, every direct summand M_i is either nonreduced of rank 2, or else bottom or top-reduced (necessarily of rank 1). Moreover, each direct summand M_i is isomorphic to some $M(A_i, B_i)$, with (A_i, B_i) in canonical form. Therefore, we can take $M = M(A, B)$ with $(A, B) = (A_1, B_1) \oplus ... \oplus (A_n, B_n)$.

Recall that the matrix A is always invertible. We claim that each matrix B_i, in the present situation, is invertible, so that B is also invertible. If M_i is nonreduced of rank 2, then the claim follows easily from (24.17.1). If M_i is bottom-reduced, then the claim follows from [**KL2**, (8.7.5)] and the sentence preceding it. Finally, if M_i is top-reduced, then the claim follows from [**KL2**, (8.7.6)] (where the column of Z's indicates a column not present in B).

Now consider the relations $P_1 A = AQ_1$ and $P_2 B = BQ_2$. We need to compute $\det Q_1[\infty, \infty]$ [Lemma 24.12]. By the triangular form and diagonal block equality [Lemma 24.11(iii)], we have

$$\det Q_1 = \big(\det Q_1[\infty, \infty]\big) \cdot \det Q_2 \tag{24.18.1}$$

Since all of the matrices in the relations $P_1 A = AQ_1$ and $P_2 B = BQ_2$ are invertible, and P_1 and P_2 are matrices over k, taking determinants and then canceling $\det A$ and $\det B$ shows that $\det Q_1 \in k^{\times}$ and $\det Q_2 \in k^{\times}$, whence (24.18.1) shows that $\det Q_1[\infty, \infty] \in k^{\times}$, as desired.

Conversely, suppose that $\theta \in k^{\times}$. It suffices to show that $\theta \in \operatorname{eqstab}(M_i)$ for some i [Lemma 23.2]. We may therefore assume that $M = M(A, B)$ is either bottom-reduced or top-reduced, with label sequence $\lambda_1, ..., \lambda_m$, in which $\lambda_1 = \infty$ but $\lambda_2, ..., \lambda_m$ are all finite [**KL2**, (8.7.5) and (8.7.6)]. In this case, we need only take all of the matrices P_1, P_2, Q_1, Q_2 to be scalar matrices of the appropriate sizes with θ on the diagonal, to obtain $\det Q_1[\infty, \infty] = \theta$. □

THEOREM 24.19. *Assume Hypothesis 24.1, and suppose that* Λ *is unsplit. If some* M_i *is nonreduced of rank 1, then* $\text{eqstab}(M) = F^\times$.

PROOF. Given $\alpha \in F^\times$, it suffices to prove that $\alpha \in \text{eqstab}(M)$, and for this, it suffices to prove that $\alpha \in \text{eqstab}(M_i)$ for some direct summand M_i of M [Lemma 23.2]. Therefore, we may assume that M is itself nonreduced of rank 1, and we can take $M = M(A, B)$, where (A, B) is in canonical form. Let the label sequence of (A, B) be $\lambda_1, ..., \lambda_m$. By [**KL2**, (8.7.2)], we may assume that $\lambda_1 = \infty$ but $\lambda_2, ..., \lambda_m$ are all finite, and (A, B) is as in (24.17.1), with one change: B has one more column, a column of zeros. Moreover, this column lies under the last column of A, and the length label λ_m of this column is finite.

By Lemma 24.16, we can find elements $a, b, c, d, a', b', c', d' \in k$ such that (24.17.8) and (24.17.9) hold. Then take P_1, P_2, Q_1, Q_2 as described under (24.17.9). As in that proof, it is easy to see that $P_1 A = AQ_1$, $P_2 B = BQ_2$, and P_i and Q_i have the required forms. But, unlike that proof, we now have $\det Q_1[\infty, \infty] = \alpha$, because A (and hence Q_1) has only one infinite length label. □

THEOREM 24.20. *Assume Hypothesis 24.1, and suppose that* Λ *is strictly split. Then* $\text{eqstab}(M)$ *equals one of the following:*

(i) $k_1^\times \times 1$, *if* M *has rank* $(a, 0)$ *with* $a \neq 0$.
(ii) $1 \times k_2^\times$, *if* M *has rank* $(0, b)$ *with* $b \neq 0$.
(iii) $k_1^\times \times k_2^\times$, *if* M *has rank* (a, b) *with both* a *and* b *nonzero, and some direct summand* M_i *of* M *has rank* $(1, 0)$ *or* $(0, 1)$.
(iv) $k^\times$ *if every indecomposable direct summand* M_i *of* M *has rank (1,1). In particular,* $\text{eqstab}(\Lambda) = k^\times$.

PROOF. (iv) We need to prove that $\text{eqstab}(M)$ is the diagonal copy of k in $k_1 \times k_2$, that is, $\det Q_{11}[\infty, \infty] = \det Q_{12}[\infty, \infty]$, and every element of $k^\times$ occurs in this way.

First note that, by block triangularity and diagonal block equality for the strictly split case [Lemma 24.13] we have:

$$\det Q_{1i} = \big(\det Q_{1i}[\infty, \infty]\big) \cdot \det Q_{2i} \qquad (i = 1, 2) \tag{24.20.1}$$

We claim that each of the matrices A_1, A_2, B_1, and B_2 is square and invertible. If $M = M(A_1, A_2, B_1, B_2)$ is indecomposable of rank (1,1), then this claim follows from the canonical form described in [**KL2**, Definitions 7.2]. For general M, (A_1, A_2, B_1, B_2) is a direct sum of such 4-tuples, and so the claim still holds.

Therefore, taking determinants in (24.14.1), and using invertibility of all of the matrices, allows us to cancel $\det A_i$ and $\det B_i$ when they occur, so that $\det Q_{j1} = \det Q_{j2}$ $(j = 1, 2)$. Putting these facts together with (24.20.1) then yields the desired relation $\det Q_{11}[\infty, \infty] = \det Q_{12}[\infty, \infty]$.

To prove the converse, we may assume, as usual, that M is indecomposable, by Lemma 23.2. Note that relations (24.14.1) hold if we take each P_j and each Q_{ji} to be an identity matrix of the appropriate size. Then multiplying both sides of the equation by an arbitrary element $\alpha \in k^\times$ yields a matrix 4-tuple in which $\det Q_{11}[\infty, \infty] = \det Q_{12}[\infty, \infty] = \alpha$, because the canonical form for a deleted cycle 4-tuple with rank (1,1) has one infinite length label in each of A_1 and A_2. This completes the proof in situation (iv).

(i) As in the converse part of situation (iv), we may assume that M is indecomposable. It suffices to show that every element of the form $(\alpha, 1) \in k^\times \times k^\times$

occurs in eqstab(M). By the hypothesis of the present situation, A_1 has one infinite length label and A_2 has none. As before, relations (24.14.1) hold if we take each P_j and each Q_{ji} to be an identity matrix of the appropriate size. Then multiplying both sides of the equation by an arbitrary element $\alpha \in k^\times$ yields a matrix 4-tuple in which $\det Q_{11}[\infty, \infty] = \alpha$ and $\det Q_{12}[\infty, \infty] = 1$, because the canonical form for a deleted cycle 4-tuple with rank (1,0) has one infinite length label in A_1 and none in A_2. This completes the proof in situation (iv).

(ii) Similar to situation (i).

(iii) We may suppose that some direct summand M_i of M has rank (1,0), and eqstab(M_i) $= k^\times \times 1$, by (i). By hypothesis and the list of possible ranks of indecomposables at the beginning of subsection 24.9, M has another indecomposable direct summand M_j whose rank is either (1,1) or (0,1). In the former case, eqstab(M_m) $=$ diag($k^\times$) by (iv), while in the latter case, eqstab(M_j) $= 1 \times k^\times$ by (ii). Since eqstab(M) $\supseteq$ eqstab(M_i)$\cdot$ eqstab(M_j), [Lemma 23.2], the proof is complete. □

CHAPTER 6

Web of Class Groups

This chapter develops the category — called the "web of class groups" — whose objects are the genus class groups of modules in fingen(Λ), and whose morphisms are group homomorphisms between these class groups that we call "loss of structure ξ-maps".

Section 25 begins by reviewing the definition and (known) basic properties of genus class groups. Then we define the ξ maps and establish their basic properties. For the first appplication of ξ-maps, the section ends with a precise description of the relationship between genus class groups of Λ-modules and the Γ-genus class group of Γ, by means of "Mayer-Vietoris" exact sequences. The new feature of our Mayer-Vietoris sequence is that its proof does not require a conductor ideal for Λ and Γ (thereby enabling us to deal with the situation in which Γ is not module-finite over Λ). Instead, it makes use of residue inclusions, thus taking advantage of the existence of local conductors.

Section 26: Our next application of ξ-maps is to show that all failure of direct-sum cancellation results from kernels of ξ-maps being nonzero, and these kernels are always contained in the restricted genus class groups of the modules involved.

Sections 27 and 28: The picture of the web of class groups is completed by showing that the set of genus class groups of faithful Λ-modules has an inverse limit with respect to a partial order determined by the ξ-maps. We call this limit the "super genus class group" s$\mathcal{G}$. It maps onto every actual genus class group of Λ-modules, faithful or not. In the classical situation that ${}_\Lambda\Gamma$ is finitely generated, s$\mathcal{G}$ is an actual genus class group; that is, the genus class group $\mathcal{G}(P)$ of some $P \in$ fingen(Λ) maps in a natural way onto all genus class groups $\mathcal{G}(M)$, $M \in$ fingen(Λ).

Working in fingen$_\infty(\Lambda)$ instead of fingen(Λ) often results in considerable simplification of notation, and no loss of generality [Proposition 25.10]. We take advantage of this as soon as we begin connecting genus class groups and ξ-maps with our idèle action $M \to M^{\boldsymbol{u}}$ [which is defined only in fingen$_\infty(\Lambda)$]. However, most results stated for fingen$_\infty(\Lambda)$ that do not explicitly involve the idèle action in their statement can be easily extended to fingen(Λ) by means of Proposition 25.10.

25. Genus Class Group, ξ-map, Mayer-Vietoris Sequence

DEFINITION 25.1 (Genus class group). For $M \in$ fingen(Λ), the elements of the *genus class group* $\mathcal{G}(M)$ of M are the isomorphism classes $[M']$ of modules in genus(M). Addition in $\mathcal{G}(M)$ is defined by:

$$(25.1.1) \qquad [M'] + [M''] = [M'''] \iff M' \oplus M'' \cong M \oplus M'''$$

where $M', M'', M''' \in$ genus(M). It follows that the zero element of $\mathcal{G}(M)$ is $[M]$. That this actually defines an abelian group is shown in [**GL1**, Definition 1.4]. In

Lemma 25.2 below, we show that $\mathcal{G}(M)$ actually displays all direct-sum relations in genus(M).

We remark that, the fact that the sum $[M'''] = [M'] + [M'']$ is uniquely determined by $[M']$, $[M'']$, and $[M]$ in the direct-sum relation in (25.1.1) relies on the fact that *direct-sum cancellation holds in every genus of finitely generated Λ-modules.* This in turn follows because Λ is reduced and noetherian of dimension 1 [**GL2**, Corollary 5.10], as already noted in Lemma 9.2(ii). (For nonreduced, commutative, noetherian rings of dimension 1, the elements of $\mathcal{G}(M)$ must be defined to be "stable" isomorphism classes; see [**GL2**].)

We also remark that the isomorphism class $[M]$, as the zero element of $\mathcal{G}(M)$, is an arbitrarily selected base point in genus(M), in the sense that $\mathcal{G}(M) \cong \mathcal{G}(N)$ for all $N \in$ genus(M) [Corollary 25.8], even if $N \not\cong M$. (Another formula relating the addition in $\mathcal{G}(M)$ to the addition in $\mathcal{G}(N)$ is given in (25.3.2).)

When additional precision is required, we use the notation $\text{genus}_\Lambda(M)$ and $\mathcal{G}_\Lambda(M)$ respectively, for the genus and genus class group of $M \in$ fingen(Λ). For example, we sometimes refer to $\text{genus}_\Gamma(X)$ and $\mathcal{G}_\Gamma(X)$, the genus and genus class group respectively, of $X \in$ fingen(Γ).

The set of isomorphism classes $[M']$ of modules $M' \in$ rgen(M) is easily seen to form a subgroup of $\mathcal{G}(M)$. We denote this subgroup by r$\mathcal{G}(M)$, and call it the *restricted genus class group* of M.

LEMMA 25.2. *Let $M \in$ fingen(Λ) and $M'_i, M''_i \in$ genus(M) ($i = 1, 2, \ldots, n$). Then*

$$(25.2.1) \qquad \sum_{i=1}^{n} [M'_i] = \sum_{i=1}^{n} [M''_i] \quad (\textit{in } \mathcal{G}(M)) \iff \quad \oplus_{i=1}^{n} M'_i \cong \oplus_{i=1}^{n} M''_i$$

PROOF. Since direct-sum cancellation holds within every genus [Lemma 9.2(ii)], it suffices to prove the equivalence

$$(25.2.2) \qquad [S] = \sum_{i=1}^{n} [M'_i] \quad (\text{in } \mathcal{G}(N)) \iff \oplus_{i=1}^{n} M'_i \cong S \oplus N^{(n-1)}$$

whenever S and all terms M'_i are in genus(N).

We prove (25.2.2) by induction on n. When $n = 2$, this equivalence is simply definition (25.1.1). Suppose, inductively, that (25.2.2) holds for $n - 1$, and let $[T] = \sum_{i=2}^{n} [M'_i]$ and $[S] = [M'_1] + [T]$ in $\mathcal{G}(N)$. Then by definition, $M'_1 \oplus T \cong S \oplus N$, so that $M'_1 \oplus T \oplus N^{(n-2)} \cong S \oplus N^{(n-1)}$. By induction we have $\oplus_{i=2}^{n} M'_i \cong T \oplus N^{(n-2)}$, which substituted into the previous isomorphism yields the right-hand side of the equivalence in (25.2.2). Similarly, the right-hand side of this equivalence implies the left-hand side. □

REMARKS 25.3. (i) An interesting corollary of Lemma 25.2 is that, whenever N and all terms M'_i and M''_i are in genus(M), we have

$$(25.3.1) \quad \sum_{i=1}^{n} [M'_i] = \sum_{i=1}^{n} [M''_i] \quad (\text{in } \mathcal{G}(M)) \iff \sum_{i=1}^{n} [M'_i] = \sum_{i=1}^{n} [M''_i] \quad (\text{in } \mathcal{G}(N))$$

even though the value of the sum $\sum_{i=1}^{n} [M'_i]$ might be different in $\mathcal{G}(M)$ than it is in $\mathcal{G}(N)$, when $M \not\cong N$. (However, this equivalence can fail when both sums do not have the same number of terms. For example, if $M \not\cong N$, then adding $[M]$ to just one side of the equation preserves equality in $\mathcal{G}(M)$ but not in $\mathcal{G}(N)$.)

(ii) Let $M_1', M_2', N \in \text{genus}(M)$, and denote the sums in $\mathcal{G}(M)$ and $\mathcal{G}(N)$ by $+_M$ and $+_N$, respectively. We note that these two sums are related by the formula

$$[M_1'] +_M [M_2'] = [M_1'] +_N [M_2'] -_N [M] \tag{25.3.2}$$

To prove this, let $[S] = [M_1'] + [M_2']$ in $\mathcal{G}(M)$. Since $[M] = 0$ in $\mathcal{G}(M)$, we can rewrite this equation as $[S] + [M] = [M_1'] + [M_2']$. Then because both sides have the same number of terms, (25.3.1) shows that the latter equation also holds in $\mathcal{G}(N)$, which proves (25.3.2).

DEFINITION 25.4 (ξ-maps). We define three types of *loss-of-structure ξ-maps* here, and one more type in Proposition 27.8.

(i) Let $M, N \in \text{fingen}(\Lambda)$. We say that $\xi^{M,N}$ *is defined* if, for each module $M' \in \text{genus}(M)$, there exists a module $N' \in \text{genus}(N)$, *unique up to isomorphism*, such that $M' \oplus N \cong M \oplus N'$. When this is the case, we can view $\xi^{M,N}$ as a function $\mathcal{G}(M) \to \mathcal{G}(N)$, defined by the equivalence

$$\xi^{M,N}[M'] = [N'] \iff M \oplus N' \cong M' \oplus N \tag{25.4.1}$$

(ii) We use the notation $\xi_\Gamma^{X,Y}$ when refering to the version of $\xi^{X,Y}$ that applies to genus class groups of Γ-modules.

(iii) For $M \in \text{fingen}(\Lambda)$ and $X \in \text{fingen}(\Gamma)$, we define the function $\xi_{\Lambda,\Gamma}^{M,X}\colon \mathcal{G}(M) \to \mathcal{G}_\Gamma(X)$ by

$$\begin{gathered} \xi_{\Lambda,\Gamma}^{M,X}[M'] = [X'] \iff \\ (\Gamma \otimes_\Lambda M) \oplus X' \cong (\Gamma \otimes_\Lambda M') \oplus X \quad \text{as } \Gamma\text{-modules} \end{gathered} \tag{25.4.2}$$

provided that, for each $M' \in \text{genus}(M)$, there is a unique such $X' \in \text{genus}_\Gamma(X)$, up to Γ-isomorphism. (See Lemma 25.13 for more properties of this variation of the ξ-map.)

We note that, if ${}_\Lambda\Gamma$ is finitely generated and ${}_\Gamma X$ is projective, then $\xi_{\Lambda,\Gamma}^{M,X} = \xi^{M,X}$ [Proposition 32.3].

(iv) We define one more variation of this idea in Proposition 27.8, calling it $\xi^{s\mathcal{G},N}$ where $s\mathcal{G}$ denotes the "super genus class group," the inverse limit of the set of genus class groups of faithful modules in $\text{fingen}(\Lambda)$.

LEMMA 25.5. *Let $M, N \in \text{fingen}(\Lambda)$.*

(i) *If $\xi^{M,N}$ is defined, then it is a (group) homomorphism.*
(ii) $\ker(\xi^{M,N}) \subseteq r\mathcal{G}(M)$ *(the restricted genus class group of M).*
(iii) *If both $\xi^{M,N}$ and $\xi^{N,M}$ are defined, then they are mutually inverse isomorphisms.*

PROOF. (i) Write $\xi = \xi^{M,N}$. To see that ξ is a homomorphism, suppose that $\xi[M'] = [N']$ and $\xi[M''] = [N'']$. Let $[S] = [M'] + [M'']$ in $\mathcal{G}(M)$, and $[T] = [N'] + [N'']$ in $\mathcal{G}(N)$; we must show that $\xi[S] = [T]$. By the definition of ξ, we have $M' \oplus N \cong M \oplus N'$ and $M'' \oplus N \cong M \oplus N''$, and by the definition of S and T, we have $M' \oplus M'' \cong M \oplus S$ and $N' \oplus N'' \cong N \oplus T$. Taking the direct sum of the first pair of isomorphisms and substituting the second pair of isomorphisms yields

$$M \oplus S \oplus N^{(2)} \cong M^{(2)} \oplus N \oplus T \tag{25.5.1}$$

Since direct-sum cancellation holds in every genus of Λ-modules, in particular, in genus($M \oplus N$), we can cancel $M \oplus N$ from (25.5.1) to obtain $S \oplus N \cong M \oplus T$. That is, $\xi[S] = [T]$, as desired.

(ii) $[M'] \in \ker(\xi^{M,N})$ implies $M' \oplus N \cong M \oplus N$, and so local cancellation [Lemma 9.2(i)] implies that $M' \in$ genus(M). Tensoring the previous isomorphism with Γ and applying direct-sum cancellation of Γ-modules implies $\Gamma \otimes_\Lambda M' \cong \Gamma \otimes_\Lambda M$, completing the proof.

(iii) This follows immediately from (25.4.1). □

In every case in which we actually prove that the map $\xi^{M,N}$ is defined, we need the (explicit or implicit) hypothesis that $\mathcal{C}(M) \subseteq \mathcal{C}(N)$. This sufficient condition is not necessary. For example, if $\mathcal{G}(M) = 0$ for all M, then $\xi^{M,N}$ is defined for all M and N. But we know of no interesting situation where $\xi^{M,N}$ is defined but $\mathcal{C}(M) \not\subseteq \mathcal{C}(N)$.

LEMMA 25.6. *Let $M, N \in$ fingen(Λ) satisfy $\mathcal{C}(M) \subseteq \mathcal{C}(N)$.*

(i) *If, for every $N', N'' \in$ genus(N), $M \oplus N' \cong M \oplus N''$ implies $N' \cong N''$, then $\xi^{M,N}$ is defined.*

(ii) *If $\mathcal{C}(M) = \mathcal{C}(N)$ and $\xi^{M,N}$ is defined, then $\xi^{M,N}$ is surjective.*

PROOF. We claim that it suffices to show that, given $M' \in$ genus(M), there exists $N' \in$ genus(N) such that $M' \oplus N \cong M \oplus N'$. For then the "cancellation" hypothesis in (i) will imply the uniqueness of $[N']$, completing the proof of (i). Moreover, if $\mathcal{C}(N) \subseteq \mathcal{C}(M)$ also, then reversing the roles of M and N, given $N' \in$ genus(N), there must exist $M' \in$ genus(M) such that $M' \oplus N \cong M \oplus N'$, so that $\xi^{M,N}$ is surjective, proving (ii) as well.

Thus, we suppose that $M' \in$ genus(M) is given, and we show that there exists $N' \in$ genus(N) such that $M' \oplus N \cong M \oplus N'$. The idea is to apply Jacobinski's direct summand theorem [Lemma 9.3(ii)] to prove that M is isomorphic to a direct summand of $M' \oplus N$. Since $M' \in$ genus(M), certainly $M_\mathfrak{m}$ is isomorphic to $M'_\mathfrak{m}$, and hence to a direct summand of $M'_\mathfrak{m} \oplus N_\mathfrak{m}$, for every $\mathfrak{m} \in$ maxspec(Λ). By (4.3.1), the ring Γ_Q is the direct sum of the fields $(\Gamma_h)_Q$, and hence M_Q and N_Q have decompositions

$$M_Q = \oplus_{h \in \mathcal{C}(M)} V_h \quad \text{and} \quad N_Q \cong \oplus_{h \in \mathcal{C}(N)} W_h \tag{25.6.1}$$

where V_h and W_h are $(\Gamma_h)_Q$-vector spaces, with only the nonzero terms displayed in (25.6.1). Since $M' \in$ genus(M), we have that $M'_Q \cong M_Q$, so that $\mathcal{C}(M) \subseteq \mathcal{C}(N)$ and (25.6.1) imply that every indecomposable direct summand of M_Q appears strictly more often as a direct summand of $M'_Q \oplus N_Q$. Therefore, by Lemma 9.3(ii), we have that $M' \oplus N \cong M \oplus N'$ for some Λ-module N'. Moreover, since direct-sum cancellation holds locally [Lemma 9.2(i)], we see that $N' \in$ genus(N), proving the claim and hence completing the proof of the lemma. □

EXAMPLE 25.7. Even if $\mathcal{C}(M) \subseteq \mathcal{C}(N)$ and $\xi^{M,N}$ is defined, it need not be surjective. For example, if $\Lambda = \Gamma = \Gamma_1 \oplus \Gamma_2$, then $\xi^{\Gamma_1,\Gamma}$ is defined [Corollary 25.9] but obviously not surjective if $\mathcal{G}(\Gamma_2) \neq 0$.

We now establish the simplest situations in which $\xi^{M,N}$ is defined.

COROLLARY 25.8. *Let $M, N \in$ fingen(Λ). If genus(M) = genus(N), then $\mathcal{G}(M) \cong \mathcal{G}(N)$ via $\xi^{M,N}$.*

PROOF. Since $\operatorname{genus}(M) = \operatorname{genus}(N)$, we have $\mathcal{C}(M) = \mathcal{C}(N)$. Also, direct-sum cancellation holds in every genus of Λ-modules [Lemma 9.2]. So by Lemma 25.6, both $\xi^{M,N}$ and $\xi^{N,M}$ are defined. Then Lemma 25.5 shows that $\xi^{M,N}$ is an isomorphism of $\mathcal{G}(M)$ onto $\mathcal{G}(N)$. □

COROLLARY 25.9. *Let $A, B, M, N \in \operatorname{fingen}(\Lambda)$. Then:*

(i) *$\xi^{M,(M\oplus N)}$ is defined, and given by the explicit formula:*

$$\xi^{M,(M\oplus N)}[M'] = [M' \oplus N] \tag{25.9.1}$$

(ii) *If $M \in \operatorname{genus}(A \oplus B)$, then $\xi^{A,M}$ is defined.*

(iii) *If $M \in \operatorname{genus}(A^{(2)})$, then $\xi^{A,M}$ and $\xi^{M,A}$ are mutually inverse isomorphisms.*

PROOF. (i) First we show that $\xi^{M,(M\oplus N)}$ is defined. Since $\mathcal{C}(M) \subseteq \mathcal{C}(M \oplus N)$, it suffices to check the cancellation property in Lemma 25.6(i). So suppose that $M \oplus A' \cong M \oplus A''$, with $A', A'' \in \operatorname{genus}(M \oplus N)$. Recall that cancellation holds in every genus of Λ-modules [Lemma 9.2(ii)]. Therefore, after adding N to both sides of the previous isomorphism, we can cancel $M \oplus N$, concluding that $A' \cong A''$.

Now that we know that $\xi^{M,(M\oplus N)}$ is defined, it suffices to check that formula (25.9.1) satisfies the defining relation for $\xi^{M,(M\oplus N)}$, namely $M \oplus (M' \oplus N) \cong (M \oplus N) \oplus M'$, which is obvious.

(ii) As in the proof of (i), it suffices to show that $A \oplus M' \cong A \oplus M''$ implies $M' \cong M''$. After adding B to both sides of the given isomorphism we can cancel $A \oplus B$, as in the proof of (i).

(iii) In view of (ii), it suffices to prove that $\xi^{M,A}$ is defined [Lemma 25.5]. Since $\mathcal{C}(M) = \mathcal{C}(A)$ under the hypotheses of (iii), it suffices to prove, as in the proof of (i), that $M \oplus A' \cong M \oplus A''$ implies $A' \cong A''$. Adding A and (as before) canceling M, and then canceling A again [Lemma 9.2(ii)], yields the desired isomorphism. □

The statement that $\xi^{M,M\oplus N}$ is always defined [Corollary 25.9(i)] is a special case of the fact that $\xi^{M,X}$ is defined whenever $\mathcal{C}(M) \subseteq \mathcal{C}(X)$ and $\operatorname{istab}(M) \subseteq \operatorname{istab}(X)$ [Corollary 26.8]. In Proposition 26.10 we prove a basic transitivity property of ξ-maps.

The next easily proved proposition shows that there is no loss of generality if we restrict our study of genus class groups and ξ-maps to $\operatorname{fingen}_\infty$. Recall that every $M \in \operatorname{fingen}(\Lambda)$ has a decomposition $M = M_\infty \oplus M_0$ with $M_\infty \in \operatorname{fingen}_\infty(\Lambda)$ and $M_0 \in \operatorname{finlen}(\Lambda)$, and the terms in this decomposition are unique up to isomorphism [Lemma 7.1].

PROPOSITION 25.10. *In the notation of the previous paragraph, let $M = M_\infty \oplus M_0$ and $N = N_\infty \oplus N_0 \in \operatorname{fingen}(\Lambda)$. Then $\operatorname{genus}(M)$ coincides with the collection of isomorphism classes of modules of the form $M' = M'_\infty \oplus M_0$ where $M'_\infty \in \operatorname{genus}(M_\infty)$. Moreover:*

(i) *$\mathcal{G}(M) \cong \mathcal{G}(M_\infty)$ via ξ^{M,M_∞}; and*

(ii) *$\xi^{M,N}$ is defined if and only if ξ^{M_∞,N_∞} is defined. When the conditions hold we have:*

$$\xi^{M,N}[M'_\infty \oplus M_0] = N'_\infty \oplus N_0 \iff \xi^{M_\infty,N_\infty}[M'_\infty] = [N'_\infty]$$

PROOF. We have a decomposition $M' = M'_\infty \oplus M'_0$. Thus, to prove the non-trivial part of first assertion of the proposition it suffices to show that the genus of

M_0 equals its isomorphism class. This holds since every Λ-module of finite length is the direct sum of its localizations at maximal ideals of Λ [Corollary 6.4].

The remaining assertions of the proposition follow directly from the definitions involved, together with the fact that modules of finite length cancel from direct sums [**F**, 4.6 and 4.12]. $\square$

The basic connection between our idèle action and genus class groups is given by the following two results. Here, and for most of the rest of this memoir, we take advantage of the simplification of notation obtained by working in fingen_∞ [Proposition 25.10].

THEOREM 25.11. *For every $M \in \text{fingen}_\infty(\Lambda)$, an isomorphism $\mathcal{I}_{\mathcal{C}(M)}/\operatorname{istab}(M) \cong \mathrm{r}\mathcal{G}(M)$ is induced by $\boldsymbol{u} \mapsto [M^{\boldsymbol{u}}] \quad (\boldsymbol{u} \in \mathcal{I}_{\mathcal{C}(M)})$.*

PROOF. For $\boldsymbol{u} \in \mathcal{I}_{\mathcal{C}(M)}$, let $\mu(\boldsymbol{u}) = [M^{\boldsymbol{u}}]$. Then μ is a function from $\mathcal{I}_{\mathcal{C}(M)}$ onto $\mathrm{r}\mathcal{G}(M)$ by Theorem 22.6. Moreover, for $\boldsymbol{u}, \boldsymbol{v} \in \mathcal{I}_{\mathcal{C}(M)}$, we have $M^{\boldsymbol{u}} \oplus M^{\boldsymbol{v}} \cong M^{\boldsymbol{uv}} \oplus M$ because, by Theorem 22.7, both sides are isomorphic to $(M \oplus M)^{\boldsymbol{uv}}$. Thus, μ is a homomorphism. That $\ker(\mu) = \operatorname{istab}(M)$ is simply the definition $\operatorname{istab}(M)$ [Definition 23.1]. $\square$

COROLLARY 25.12. *Let $M, N \in \text{fingen}_\infty(\Lambda)$. If $\operatorname{genus}(M) = \operatorname{genus}(N)$, then $\mathrm{r}\mathcal{G}(M) \cong \mathrm{r}\mathcal{G}(N)$ via $[M^{\boldsymbol{u}}] \mapsto [N^{\boldsymbol{u}}]$, and this isomorphism is the restriction of $\xi^{M,N}$.*

PROOF. We already know that $\xi^{M,N}$ is an isomorphism in this situation [Corollary 25.8]. Therefore, it suffices to check that $\xi^{M,N}[M^{\boldsymbol{u}}] = [N^{\boldsymbol{u}}]$. By (25.4.1), this is equivalent to showing that $M^{\boldsymbol{u}} \oplus N \cong M \oplus N^{\boldsymbol{u}}$. But the latter holds because both sides are isomorphic to $(M \oplus N)^{\boldsymbol{u}}$ [Theorem 22.7]. $\square$

LEMMA 25.13. *Let $X \in \text{fingen}_\infty(\Gamma)$ and $M \in \text{fingen}_\infty(\Lambda)$. Then:*

(i) *$\mathcal{G}_\Gamma(\Gamma_{\mathcal{C}(X)})$ can be identified with a direct summand of $\mathcal{G}_\Gamma(\Gamma)$ via the injection $\xi_\Gamma^{\Gamma_{\mathcal{C}(X)},\Gamma}$. This injection takes $[W'] \to [W' \oplus (1-e)\Gamma]$, where e is the identity element of the ring $\Gamma_{\mathcal{C}(M)}$.*

(ii) *$\xi_{\Lambda,\Gamma}^{M,\Gamma_{\mathcal{C}(M)}}$ is defined, and $\operatorname{im}(\xi_{\Lambda,\Gamma}^{M,\Gamma_{\mathcal{C}(M)}}) = \mathcal{G}_\Gamma(\Gamma_{\mathcal{C}(M)})$.*

(iii) *$\xi_{\Lambda,\Gamma}^{M,\Gamma_{\mathcal{C}(M)}}$ is a group homomorphism with kernel $\mathrm{r}\mathcal{G}(M)$.*

PROOF. (i) Since $\Gamma = \Gamma_{\mathcal{C}(X)} \oplus (1-e)\Gamma$, the formula for $\mathcal{G}_\Gamma(\Gamma_{\mathcal{C}(X)})$ follows from Corollary 25.9, applied to Γ-modules. The rest of (i) holds because this decomposition of Γ is into the direct sum of fully invariant submodules.

(ii) and (iii) Let $\tau[M'] = [\Gamma \otimes_\Lambda M']$ where $M' \in \operatorname{genus}(M)$ and the isomorphism class on the right is the Γ-isomorphism class. It is easily checked that $\tau = \xi_{\Lambda,\Gamma}^{M,(\Gamma \otimes_\Lambda M)}$. We claim that

$$\xi_{\Lambda,\Gamma}^{M,\Gamma_{\mathcal{C}(M)}} = \xi_\Gamma^{(\Gamma \otimes_\Lambda M),\Gamma_{\mathcal{C}(M)}} \cdot \tau \tag{25.13.1}$$

and the factor $\xi_\Gamma^{(\Gamma \otimes_\Lambda M),\Gamma_{\mathcal{C}(M)}}$ is an isomorphism.

To prove the claim, note that the factor $\xi_\Gamma^{(\Gamma \otimes_\Lambda M),\Gamma_{\mathcal{C}(M)}}$ is defined since $\Gamma \otimes_\Lambda M$ and $\Gamma_{\mathcal{C}(M)}$ have the same torsionfree support set $\mathcal{C}(M)$ and direct-sum cancellatiom holds in fingen(Γ) [Lemma 25.6]. Similarly, $\xi_\Gamma^{\Gamma_{\mathcal{C}(M)},(\Gamma \otimes_\Lambda M)}$ is defined and hence is the inverse isomorphism to $\xi_\Gamma^{(\Gamma \otimes_\Lambda M),\Gamma_{\mathcal{C}(M)}}$ [Lemma 25.5]. The equality in (25.13.1) now follows immediately from the definitions involved and the fact that direct-sum cancellation holds in fingen(Γ).

In particular, $\xi_{\Lambda,\Gamma}^{M,\Gamma_{\mathcal{C}(M)}}$ is defined. Since τ is easily checked to be a group homomorphism $\mathcal{G}(M) \to \mathcal{G}_\Gamma(\Gamma \otimes_\Lambda M)$, (25.13.1) shows that $\xi_{\Lambda,\Gamma}^{M,\Gamma_{\mathcal{C}(M)}}$ is a group homomorphism also. Since $\xi_{\Gamma}^{(\Gamma\otimes_\Lambda M),\Gamma_{\mathcal{C}(M)}}$ is an isomorphism, (25.13.1) shows that $\ker(\xi_{\Lambda,\Gamma}^{M,\Gamma_{\mathcal{C}(M)}}) = \ker(\tau) = \mathrm{r}\mathcal{G}(M)$, completing the proof of (iii).

Obviously $\mathrm{im}(\xi_{\Lambda,\Gamma}^{M,\Gamma_{\mathcal{C}(M)}}) \subseteq \mathcal{G}_\Gamma(\Gamma_{\mathcal{C}(M)})$, so it remains only to prove the opposite inclusion, the most difficult part of the proof. Since $\xi_{\Gamma}^{(\Gamma\otimes_\Lambda M),\Gamma_{\mathcal{C}(M)}}$ is an isomorphism, it suffices to show that τ maps $\mathcal{G}(M)$ onto $\mathcal{G}_\Gamma(\Gamma \otimes_\Lambda M)$.

Thus, given any Γ-module $W \in \mathrm{genus}_\Gamma(\Gamma \otimes_\Lambda M)$, we must find a Λ-module $N \in \mathrm{genus}(M)$ such that $\Gamma \otimes_\Lambda N \cong W$. (Note that it is *not* true, in general, that every finitely generated Γ-module can be realized as $\Gamma \otimes_\Lambda H$ for some $H \in \mathrm{fingen}(\Lambda)$. See Lemma 17.5.) Since $W \in \mathrm{fingen}(\Gamma)$, we can write

$$W = Y_\infty \oplus t(W) \tag{25.13.2}$$

for some projective Γ-module Y_∞.

Choose a separated cover $S \twoheadrightarrow M$, with kernel K and associated Γ-module X, and fix a decomposition $X = X_\infty \oplus t(X)$. Then $X = \Gamma \otimes_\Lambda S = \Gamma S$ and $K \subseteq t(X)$ [Corollary 18.9], so by right exactness of the tensor product we have

$$\Gamma \otimes_\Lambda M \cong (\Gamma S)/(\Gamma K) = X/(\Gamma K) \cong X_\infty \oplus \frac{t(X)}{\Gamma K} \tag{25.13.3}$$

Since $\Gamma \otimes_\Lambda M$ is in the same Γ-genus as W, and since the isomorphism class of the torsion submodule is determined by its localizations [Corollary 6.4], comparing (25.13.3) with (25.13.2) shows that:

$$W \cong Y_\infty \oplus \frac{t(X)}{\Gamma K} \tag{25.13.4}$$

Therefore, local cancellation [Lemma 9.2(i)] yields $\mathrm{genus}_\Gamma(X_\infty) = \mathrm{genus}_\Gamma(Y_\infty)$. Since we are interested in W only up to isomorphism, we may assume that $Y_\infty \subseteq (X_\infty)_Q$, and hence $(Y_\infty)_Q = (X_\infty)_Q$ in $(X_\infty)_Q$.

Let $H = X_Q \oplus t(X)$. Since $X = X_\infty \oplus t(X)$ and $t(X)_Q = 0$, we have $H = (X_\infty)_Q \oplus t(X)$. We claim that, for every $\mathfrak{m} \in \mathrm{maxspec}(\Lambda)$, there exists a $\Gamma_\mathfrak{m}$-automorphism $\zeta(\mathfrak{m}) = \zeta_\infty(\mathfrak{m}) \oplus 1_t(\mathfrak{m})$ of $H_\mathfrak{m}$, where $1_t(\mathfrak{m})$ is the identity map on $t(X)_\mathfrak{m}$, such that both of the following conditions hold:

$$(X_\infty)_\mathfrak{m}\,\zeta_\infty(\mathfrak{m}) = (Y_\infty)_\mathfrak{m}\ (\forall \mathfrak{m}) \quad \text{and} \quad \zeta_\infty(\mathfrak{m}) = 1\ (\mathrm{a}\forall \mathfrak{m}) \tag{25.13.5}$$

Since $X_\infty, Y_\infty \in \mathrm{fingen}_\infty(\Gamma)$ we have that $X_\infty = \Gamma A$ and $Y_\infty = \Gamma B$ for some finitely generated Λ-submodules A and B of X_∞ and Y_∞, respectively. Moreover, $(X_\infty)_Q = A_Q$ and $(Y_\infty)_Q = B_Q$, because $\Lambda_Q = \Gamma_Q$. Therefore, $A_\mathfrak{m} = B_\mathfrak{m}$ in $(X_\infty)_Q$ ($\mathrm{a}\forall \mathfrak{m}$) [Lemma 8.2(iii)]. Multiplying by $\Gamma_\mathfrak{m}$, we see that $(X_\infty)_\mathfrak{m} = (Y_\infty)_\mathfrak{m}$ in $(X_\infty)_Q$ ($\mathrm{a}\forall \mathfrak{m}$). For each $\mathfrak{m}$ for which this equality holds, let $\zeta_\infty(\mathfrak{m}) = 1$. For each of the finitely many remaining maximal ideals $\mathfrak{m}$, let $\zeta'_\infty(\mathfrak{m})$ be any $\Gamma_\mathfrak{m}$-isomorphism of $(X_\infty)_\mathfrak{m}$ onto $(Y_\infty)_\mathfrak{m}$, which exists since $\mathrm{genus}_\Gamma(X_\infty) = \mathrm{genus}_\Gamma(Y_\infty)$. Each $\zeta'_\infty(\mathfrak{m})$ induces (by Q-localization) a Γ_Q-automorphism of $(X_\infty)_Q$; let $\zeta_\infty(\mathfrak{m})$ be this automorphism. Then for each maximal ideal $\mathfrak{m}$, setting $\zeta(\mathfrak{m})$ equal to the direct sum of the map $\zeta_\infty(\mathfrak{m})$ with the identity map $1_t(\mathfrak{m})$ completes the proof of the claim.

Consider the inclusion $S \subseteq X \subseteq X_Q \oplus t(X) = H$ of Λ-modules. Since $\Gamma S = X$ and $\Lambda_Q = \Gamma_Q$, we have $S_Q = X_Q = H_Q$ in H_Q. Therefore, by Theorem 8.5 and (25.13.5), there is a unique finitely generated Λ-submodule T of H such that $T_\mathfrak{m} =$

$(S_{\mathfrak{m}})\zeta(\mathfrak{m})$ in $H_{\mathfrak{m}}$ $(\forall\mathfrak{m})$; moreover, $T_Q = H_Q$. Since each $\zeta(\mathfrak{m})$ is an automorphism of H_Q we have $T \in \operatorname{genus}(S)$; and $\Gamma T = Y_\infty \oplus t(X)$ because this equation holds locally at all maximal ideals of Λ. Moreover, the natural map $\Gamma \otimes_\Lambda T \to \Gamma T$ is an isomorphism, which also can be checked locally at all maximal ideals of Λ. (Here we use the assumption that $\Gamma \otimes_\Lambda S \to \Gamma S = X$ is an isomorphism, together with the fact that each $\zeta(\mathfrak{m})$ is an automorphism of $H_{\mathfrak{m}}$.)

Let $N = T/K$, where K is the kernel of the separated cover $S \twoheadrightarrow M$. Clearly $N \in \operatorname{genus}(M)$, because each $\zeta(\mathfrak{m})$ is an isomorphism restricting to the identity on $t(X)_{\mathfrak{m}}$ (and hence on $K_{\mathfrak{m}}$). We claim that the natural map $T \twoheadrightarrow T/K$ is a separated cover of N, with associated Γ-module $Y_\infty \oplus t(X)$. Again it suffices to check this locally [Theorem 18.13], where it holds because each $\zeta(\mathfrak{m})$ is an automorphism and equals the identity on $t(X)_{\mathfrak{m}}$. Exactly as in the proof of (25.13.3), we now have that $\Gamma \otimes_\Lambda N \cong Y_\infty \oplus \big(t(X)/(\Gamma K)\big)$, and hence $\Gamma \otimes_\Lambda N \cong W$ by (25.13.4). □

We are now ready to put together the parts of this section's main result for genus class groups, namely, the existence of a Mayer-Vietoris sequence.

THEOREM 25.14 (Mayer-Vietoris sequence). *For every $M \in \operatorname{fingen}_\infty(\Lambda)$ there is a short exact sequence of natural homomorphisms*

$$0 \to \mathcal{I}_{\mathcal{C}(M)}/\operatorname{istab}(M) \xrightarrow{\mu^M} \mathcal{G}(M) \xrightarrow{\xi_{\Lambda,\Gamma}^{M,\Gamma_{\mathcal{C}(M)}}} \mathcal{G}_\Gamma(\Gamma_{\mathcal{C}(M)}) \to 0 \tag{25.14.1}$$

where the map μ^M is induced by $\boldsymbol{u} \mapsto [M^{\boldsymbol{u}}]$, and $\operatorname{im}(\mu^M) = \mathrm{r}\mathcal{G}(M)$.

PROOF. The fact that μ^M is a homomorphism from $\mathcal{I}_{\mathcal{C}(M)}/\operatorname{istab}(M)$ to $\mathcal{G}(M)$ with image $\mathrm{r}\mathcal{G}(M)$ was proved in Theorem 25.11. The needed properties of $\xi_{\Lambda,\Gamma}^{M,\Gamma_{\mathcal{C}(M)}}$ were proved in Lemma 25.13. □

We note that, if ${}_\Lambda\Gamma$ is finitely generated, (25.14.1) can be simplified by replacing $\xi_{\Lambda,\Gamma}^{M,\Gamma_{\mathcal{C}(M)}}$ with $\xi^{M,\Gamma_{\mathcal{C}(M)}}$ [Corollary 32.4]. We also note that the Mayer-Vietoris sequence (25.14.1) need not be split [Example 36.1].

26. Cancellation, Restricted Genus, and Idèles

LEMMA 26.1. *Suppose that $M \oplus N \cong M' \oplus N$ for $M, M', N \in \operatorname{fingen}(\Lambda)$. Then:*

(i) *$M' \in \operatorname{rgen}(M)$.*

(ii) *If N is in the genus of some direct summand of M, then $M \cong M'$.*

PROOF. (i) By local cancellation [Lemma 9.2(i)], $M' \in \operatorname{genus}(M)$. Tensoring the given isomorphism with Γ and applying cancellation of Γ-modules completes the proof of (i).

(ii) By hypothesis, there is a decomposition $M = A \oplus B$ in which $A \in \operatorname{genus}(N)$. Then $M \oplus (N \oplus B) \cong M' \oplus (N \oplus B)$, so since $N \oplus B \in \operatorname{genus}(M)$ and cancellation holds in each genus of Λ-modules [Lemma 9.2(ii)], it follows that $M \cong M'$. □

The key to understanding cancellation — and the extent to which it can fail — lies in understanding $\ker(\xi^{M,(M\oplus N)})$, which is always a subgroup of $\mathrm{r}\mathcal{G}(M)$. More precisely:

THEOREM 26.2. *Let $M, N \in \operatorname{fingen}(\Lambda)$ and let $M', M'' \in \operatorname{genus}(M)$ and $N' \in \operatorname{genus}(N)$. Then the following statements are equivalent:*

(i) *$M' \oplus N' \cong M'' \oplus N'$.*

(ii) $[M'] - [M''] \in \ker(\xi^{M,(M\oplus N)})$ *(difference computed in $\mathcal{G}(M)$).*
(iii) $M' \oplus N \cong M'' \oplus N$.

Moreover, $\ker(\xi^{M,(M\oplus N)}) \subseteq \mathrm{r}\mathcal{G}(M)$.

PROOF. For $A \in \mathrm{genus}(M)$, we note that (25.9.1) can be rewritten as the equivalence:

$$[A] \in \ker(\xi^{M,(M\oplus N)}) \iff A \oplus N \cong M \oplus N \tag{26.2.1}$$

For the rest of this proof, let $[A] = [M'] - [M'']$ in $\mathcal{G}(M)$. By (25.1.1), we have that $A \oplus M'' \cong M \oplus M'$. Adding $N \oplus N'$ to both sides yields the isomorphism:

$$(A \oplus N) \oplus (M'' \oplus N') \cong (M \oplus N) \oplus (M' \oplus N') \tag{26.2.2}$$

To prove (i) $\iff$ (ii), we use the fact that cancellation holds in every genus of Λ-modules — genus$(M \oplus N)$ in the present situation. If (i) holds, then cancel the second direct sum from each side of (26.2.2), yielding the right-hand side of the equivalence in (26.2.1), so that (ii) holds. Conversely, if (ii) holds, use (26.2.1) to cancel the first direct sum on each side of (26.2.2), obtaining (i).

Since we proved the equivalence (i) $\iff$ (ii) for arbitrary $N' \in \mathrm{genus}(N)$, the equivalence (ii) $\iff$ (iii) follows immediately.

The supplementary statement is part of Lemma 25.5. □

DEFINITION 26.3. We say that genus(N) *cancels from* genus(M) if $M' \oplus N' \cong M'' \oplus N'$ — with $M', M'' \in \mathrm{genus}(M)$ and $N' \in \mathrm{genus}(N)$ — implies that that $M' \cong M''$.

COROLLARY 26.4. *For $M, N \in \mathrm{fingen}(\Lambda)$, the following conditions are equivalent.*

(i) genus(N) *cancels from* genus(M).
(ii) $\ker(\xi^{M,(M\oplus N)}) = 0$.

If $M, N \in \mathrm{fingen}_\infty(\Lambda)$, then (i) *and* (ii) *are also equivalent to:*

(iii) *For all $\boldsymbol{u} \in \mathcal{I}_{C(M)}$, $M^{\boldsymbol{u}} \oplus N \cong M \oplus N$ implies $M^{\boldsymbol{u}} \cong M$.*

PROOF. (i) $\iff$ (ii) is an immediate consequence of Theorem 26.2, and (i) $\implies$ (iii) is obvious.

(iii) $\implies$ (ii). Take any $[A] \in \ker(\xi^{M,(M\oplus N)})$. Then $A \oplus N \cong M \oplus N$ by (26.2.1), and $[A] \in \mathrm{r}\mathcal{G}(M)$ by Theorem 26.2. Therefore, $A \cong M^{\boldsymbol{u}}$ for some $\boldsymbol{u} \in \mathcal{I}_{C(M)}$ [Theorem 22.6], and hence (iii) implies $A \cong M^{\boldsymbol{u}} \cong M$. That is, $[A] = 0$ in $\mathcal{G}(M)$, proving (ii).

(i) $\implies$ (iii) This holds since $M^{\boldsymbol{u}} \in \mathrm{rgen}(M)$ [Theorem 22.6]. □

We are now ready for the main result connecting direct sum cancellation with the idèle-action of §22.

THEOREM 26.5. *For $M, N \in \mathrm{fingen}_\infty(\Lambda)$ the following assertions are equivalent.*

(i) genus(N) *cancels from* genus(M).
(ii) *Whenever both $N^{\boldsymbol{u}}$ and $M^{\boldsymbol{u}}$ are defined and $N^{\boldsymbol{u}} \cong N$, then $M^{\boldsymbol{u}} \cong M$. In other words,* $\mathrm{istab}(N) \cap \mathcal{I}_{C(M)} \subseteq \mathrm{istab}(M)$.

PROOF. (i)⇒(ii). Suppose that $\boldsymbol{u} \in \operatorname{istab}(N) \cap \mathcal{I}_{\mathcal{C}(M)}$. Then using Theorem 22.7, we get that $M^{\boldsymbol{u}} \oplus N \cong (M \oplus N)^{\boldsymbol{u}} \cong M \oplus N^{\boldsymbol{u}} \cong M \oplus N$, so that by (i), $M^{\boldsymbol{u}} \cong M$, and hence $u \in \operatorname{istab}(M)$.

(ii)⇒(i). By Corollary 26.4, it suffices to show that $\boldsymbol{u} \in \mathcal{I}_{\mathcal{C}(M)}$, together with $M^{\boldsymbol{u}} \oplus N \cong M \oplus N$, imply that $M^{\boldsymbol{u}} \cong M$. Again using Theorem 22.7, we get that $M \oplus N \cong M^{\boldsymbol{u}} \oplus N \cong (M \oplus N)^{\boldsymbol{u}}$, so that $\boldsymbol{u} \in \operatorname{istab}(M \oplus N)$. By Corollary 24.7(ii), we can write $\boldsymbol{u} = \boldsymbol{v}\boldsymbol{w}$ for some $\boldsymbol{v} \in \operatorname{istab}(M)$ and $\boldsymbol{w} \in \operatorname{istab}(N)$. Then by (ii), $\boldsymbol{v}^{-1}\boldsymbol{u} = \boldsymbol{w} \in \operatorname{istab}(N) \cap \mathcal{I}_{\mathcal{C}(M)} \subseteq \operatorname{istab}(M)$, from which we conclude that $M \cong M^{\boldsymbol{v}^{-1}\boldsymbol{u}} \cong M^{\boldsymbol{u}}$, as desired. □

COROLLARY 26.6. *For* $M, N \in \operatorname{fingen}(\Lambda)$, *if* $\mathcal{C}(M) \cap \mathcal{C}(N) = \varnothing$, *then* $\operatorname{genus}(N)$ *cancels from* $\operatorname{genus}(M)$.

PROOF. We may assume that $M, N \in \operatorname{fingen}_{\infty}(\Lambda)$ [Lemma 7.1]. Because the group $\mathcal{I}_{\mathcal{C}(M)\cap\mathcal{C}(N)}$ consists of its identity element, the result follows immediately from Theorem 26.5. □

We remark that the conclusion of Corollary 26.6 will be considerably strengthened in Lemma 28.1.

COROLLARY 26.7. *For* $M, N \in \operatorname{fingen}_{\infty}(\Lambda)$ *such that* $\mathcal{C}(N) \subseteq \mathcal{C}(M)$, *the following assertions are equivalent.*

(i) $\operatorname{genus}(N)$ *cancels from* $\operatorname{genus}(M)$.
(ii) $\operatorname{istab}(N) \subseteq \operatorname{istab}(M)$.
(iii) $\xi^{N,M}$ *is defined.*

PROOF. (i)⟺(ii). Clearly it suffices to show that condition (ii) of the present corollary is equivalent to condition (ii) of Theorem 26.5. For the nontrivial half of this equivalence, suppose that $\operatorname{istab}(N) \cap \mathcal{I}_{\mathcal{C}(M)} \subseteq \operatorname{istab}(M)$. Now $\operatorname{istab}(N) = \operatorname{istab}(N) \cap \mathcal{I}_{\mathcal{C}(N)}$ (because $\operatorname{istab}(N) \subseteq \mathcal{I}_{\mathcal{C}(N)}$), and in turn $\mathcal{I}_{\mathcal{C}(N)} \subseteq \mathcal{I}_{\mathcal{C}(M)}$ (because, by assumption $\mathcal{C}(N) \subseteq \mathcal{C}(M)$), so that $\operatorname{istab}(N) \cap \mathcal{I}_{\mathcal{C}(N)} \subseteq \operatorname{istab}(N) \cap \mathcal{I}_{\mathcal{C}(M)}$. It follows that $\operatorname{istab}(N) \subseteq \operatorname{istab}(M)$, as required.

(iii)⇒(i) By Corollary 26.4, to prove that (i) holds, it suffices to show that $\ker(\xi^{M,(M\oplus N)}) = 0$. Now if $[M'] \in \ker(\xi^{M,(M\oplus N)})$, then by (26.2.1), $M' \oplus N \cong M \oplus N$. But then (iii) implies, by Definition 25.4(i), that $M' \cong M$; that is, $[M'] = [M]$ in $\mathcal{G}(M)$, and therefore $\ker(\xi^{M,(M\oplus N)}) = 0$, as required.

(i)⇒(iii) Since $\mathcal{C}(N) \subseteq \mathcal{C}(M)$, this is a restatement of Lemma 25.6(i), with the roles of M and N interchanged. □

COROLLARY 26.8. *Let* $M, N \in \operatorname{fingen}_{\infty}(\Lambda)$.

(i) *If we have* $\mathcal{C}(N) \subseteq \mathcal{C}(M)$ *and either* $\operatorname{istab}(N) \subseteq \operatorname{istab}(M)$ *or* $\operatorname{eqstab}(M) \subseteq \operatorname{eqstab}(N)$, *then* $\xi^{N,M}$ *is defined.*
(ii) *If we have* $\mathcal{C}(N) = \mathcal{C}(M)$ *and either* $\operatorname{istab}(N) = \operatorname{istab}(M)$ *or* $\operatorname{eqstab}(M) = \operatorname{eqstab}(N)$, *then* $\xi^{N,M}$ *is an isomorphism.*

PROOF. First note that $\operatorname{eqstab}(N) \subseteq \operatorname{eqstab}(M)$ implies $\operatorname{istab}(N) \subseteq \operatorname{istab}(M)$, and $\operatorname{eqstab}(N) = \operatorname{eqstab}(M)$ implies $\operatorname{istab}(N) = \operatorname{istab}(M)$ [Theorem 23.5]. Therefore we can ignore eqstab in the rest of this proof. Statement (i) then becomes part of Corollary 26.7. Two applications of statement (i) show that both $\xi^{N,M}$ and $\xi^{M,N}$ are defined. Therefore they are mutually inverse isomorphisms [Lemma 25.5]. □

Next we observe that repeated direct summands have no effect on cancellation.

LEMMA 26.9. *Let* $M, N \in \text{fingen}(\Lambda)$. *If* genus$(N)$ *cancels from* genus(M), *and if* L *is in the genus of some direct summand of* $M \oplus N$, *then* genus$(N \oplus L)$ *cancels from* genus(M).

PROOF. Suppose that $M \oplus A \cong M' \oplus A$, where $M' \in \text{genus}(M)$ and $A \in \text{genus}(N \oplus L)$. By Lemma 9.3(iii), there is a decomposition $A = N' \oplus L'$ with $N' \in \text{genus}(N)$ and $L' \in \text{genus}(L)$, and hence $M \oplus N' \oplus L' \cong M' \oplus N' \oplus L'$. We can cancel L' by Lemma 26.1, and we can cancel N' by hypothesis. □

We close this section with a basic property of ξ-maps. (The form in which we most often use it is stated in Theorem 27.1.)

PROPOSITION 26.10 (Transitivity). *Let* $A, B, C \in \text{fingen}_\infty(\Lambda)$. *Suppose:*

(i) $\mathcal{C}(A) \subseteq \mathcal{C}(B) \subseteq \mathcal{C}(C)$ *and* $\text{eqstab}(A) \subseteq \text{eqstab}(B) \subseteq \text{eqstab}(C)$; *or*

(ii) $\xi^{A,B}$, $\xi^{B,C}$, *and* $\xi^{A,C}$ *are defined,* $\mathcal{C}(B) \subseteq \mathcal{C}(A \oplus C)$ *and* $\text{istab}(B) \subseteq \text{istab}(A \oplus C)$.

Then $\xi^{B,C}\xi^{A,B} = \xi^{A,C}$.

PROOF. First we show that (i) $\Longrightarrow$ (ii). The given inclusions imply that all three ξ-maps in (ii) are defined [Corollary 26.8]. They obviously imply that $\mathcal{C}(B) \subseteq \mathcal{C}(A \oplus C)$. Also, the given $\mathcal{C}$-inclusions and eqstab-inclusions imply that $\text{istab}(B) \subseteq \text{istab}(A \oplus C)$, by the formulas in Corollary 24.7.

(ii) Let $\xi^{A,B}[A'] = [B']$ and $\xi^{B,C}[B'] = [C']$, so that $A \oplus B' \cong A' \oplus B$ and $B \oplus C' \cong B' \oplus C$. Adding yields

$$(A \oplus C') \oplus (B \oplus B') \cong (A' \oplus C) \oplus (B \oplus B') \tag{26.10.1}$$

We have $\text{istab}(B) \subseteq \text{istab}(A \oplus C)$, by the hypothesis of statement (ii). Therefore genus(B) cancels from genus$(A \oplus C)$ [Corollary 26.7], and hence genus$(B \oplus B)$ cancels from genus$(A \oplus C)$ [Lemma 26.9]. Therefore, we can cancel $B \oplus B'$ from (26.10.1), which implies that $\xi^{A,C}[A'] = [C']$, as desired. □

27. Super Mayer-Vietoris and faithful modules

In this section we show that the maps $\xi^{M,N}$ [Definition 25.4] organize the set of genus class groups of faithful modules modules in $\text{fingen}_\infty(\Lambda)$ into an inverse system, and we call the inverse limit of this system s$\mathcal{G}$, the "super genus class group". This is part of a "super Mayer-Vietoris sequence" that maps onto the (ordinary) Mayer-Vietoris sequence for every faithful Λ-module. In the next section, we extend these ideas to unfaithful modules.

When all residue fields of Λ are finite, we have a natural identification $\text{s}\mathcal{G} = \mathcal{G}(\Lambda)$ [Corollary 27.11]. When singspec(Λ) is finite, we can always take $\text{s}\mathcal{G} = \mathcal{G}(M)$ for some M of rank 2 [Theorem 27.10]. But if Λ has infinitely many singular maximal ideals $\mathfrak{m}$ at which the norm map $N_{k(\mathfrak{m})}^{F(\mathfrak{m})}$ is not surjective, we do not know whether an identification $\text{s}\mathcal{G} = \mathcal{G}(M)$ is possible. Unfortunately, we know too few examples of such rings.

Our starting point is to show that the relation $\text{istab}(M) \subseteq \text{istab}(N)$, when specialized to faithful modules, gives the family of genus class groups $\mathcal{G}(M)$ of faithful modules a partial order with respect to the family of maps $\xi^{M,N}$. This follows from the next theorem (which is a collection of results proved earlier in this memoir), because a module $M \in \text{fingen}(\Lambda)$ is faithful if and only if its torsionfree support set $\mathcal{C}(M)$ equals $\mathcal{C}(\Lambda)$ [see Notation 3.1]. For later use, we work in somewhat more generality than faithful modules.

THEOREM 27.1. *Let* $M, N \in \text{fingen}_\infty(\Lambda)$ *satisfy* $\mathcal{C}(M) = \mathcal{C}(N)$. *Then:*

(i) $\xi^{M,N}\colon \mathcal{G}(M) \to \mathcal{G}(N)$ *is defined if and only if* $\text{istab}(M) \subseteq \text{istab}(N)$.

(ii) *When* $\xi^{M,N}$ *is defined, it is surjective. If* $\xi^{N,M}$ *is also defined, then these maps are mutually inverse isomorphisms.*

(iii) *When* $\xi^{M,N}$ *is defined, it maps* $\mathrm{r}\mathcal{G}(M)$ *onto* $\mathrm{r}\mathcal{G}(N)$ *via* $\xi^{M,N}[M^{\boldsymbol{u}}] = [N^{\boldsymbol{u}}]$ $\big(\boldsymbol{u} \in \mathcal{I}_{\mathcal{C}(M)}\big)$.

(iv) *(Transitivity). Suppose, in addition, that* $\mathcal{C}(N) = \mathcal{C}(L)$ *and* $\text{istab}(M) \subseteq \text{istab}(N) \subseteq \text{istab}(L)$. *Then* $\xi^{N,L}\xi^{M,N} = \xi^{M,L}$.

PROOF. (i) This is a restatement of part of Corollary 26.7.

(ii) This is a restatement of Lemma 25.6(ii) and Lemma 25.5.

(iii) $\mathrm{r}\mathcal{G}(M)$ and $\mathrm{r}\mathcal{G}(N)$ are the sets of isomorphism classes $[M^{\boldsymbol{u}}]$ and $[N^{\boldsymbol{u}}]$, respectively, as $\boldsymbol{u}$ ranges over all residue unit idèles in $\mathcal{I}_{\mathcal{C}(M)} = \mathcal{I}_{\mathcal{C}(N)}$ [Theorem 22.6(ii)]. The fact that $\xi^{M,N}[M^{\boldsymbol{u}}] = [N^{\boldsymbol{u}}]$ follows from (25.4.1), because $M^{\boldsymbol{u}} \oplus N \cong (M \oplus N)^{\boldsymbol{u}} \cong M \oplus N^{\boldsymbol{u}}$ [Theorem 22.7].

(iv) Recall the formula $\text{istab}(A \oplus B) = \text{istab}(A)\,\text{istab}(B)$ [Corollary 24.7(ii)]. Therefore $\text{istab}(M) \subseteq \text{istab}(N) \subseteq \text{istab}(L)$ implies $\text{istab}(N) \subseteq \text{istab}(M \oplus L)$. The desired transitivity therefore follows from Proposition 26.10. □

Next, we observe that our partial-order structure on genus class groups (of faithful Λ-modules) extends naturally to a partial order of their Mayer-Vietoris sequences.

LEMMA 27.2 (Mayer-Vietoris consistency). *Let* $M, N \in \text{fingen}_\infty(\Lambda)$ *be such that* $\mathcal{C}(M) = \mathcal{C}(N)$ *and* $\text{istab}(M) \subseteq \text{istab}(N)$. *Then the following diagram commutes.*

(27.2.1)
$$\begin{array}{ccccc}
0 \to \mathcal{I}_{\mathcal{C}(M)}/\,\text{istab}(M) & \xrightarrow{\mu^M} & \mathcal{G}(M) & \xrightarrow{\xi_{\Lambda,\Gamma}^{M,\Gamma_{\mathcal{C}(M)}}} & \mathcal{G}_\Gamma(\Gamma_{\mathcal{C}(M)}) \to 0 \\
\Big\downarrow \text{nat} & & \Big\downarrow \xi^{M,N} & & \Big\downarrow 1 \\
0 \to \mathcal{I}_{\mathcal{C}(N)}/\,\text{istab}(N) & \xrightarrow{\mu^N} & \mathcal{G}(N) & \xrightarrow{\xi_{\Lambda,\Gamma}^{N,\Gamma_{\mathcal{C}(N)}}} & \mathcal{G}_\Gamma(\Gamma_{\mathcal{C}(N)}) \to 0
\end{array}$$

PROOF. The horizontal maps are from Theorem 25.14, and by Theorem 27.1, $\xi^{M,N}$ is defined. Commutativity of the left-hand square follows from Theorem 27.1(iii), because the map μ^M is induced by $\boldsymbol{u} \mapsto [M^{\boldsymbol{u}}]$ [Theorem 25.14].

To prove commutativity of the right-hand square, first note that $\xi_\Gamma^{X,Y}$ is always defined for Γ-modules X, Y such that $\mathcal{C}(X) = \mathcal{C}(Y)$, by Theorem 27.1 applied to Γ-modules, because the istab condition becomes trivial (since $\text{singspec}(\Gamma)$ is empty). Therefore, $\xi_{\Lambda,\Gamma}^{M,\Gamma_{\mathcal{C}(M)}}$ and $\xi_{\Lambda,\Gamma}^{N,\Gamma_{\mathcal{C}(N)}}$ are defined. Let $\xi^{M,N}[M'] = [N']$ and $\xi_{\Lambda,\Gamma}^{N,\Gamma_{\mathcal{C}(N)}}[N'] = [Y']$. Then the following two isomorphisms hold.

$$(\Gamma \otimes_\Lambda M) \oplus (\Gamma \otimes_\Lambda N') \cong (\Gamma \otimes_\Lambda M') \oplus (\Gamma \otimes_\Lambda N)$$
$$(\Gamma \otimes_\Lambda N) \oplus Y' \cong (\Gamma \otimes_\Lambda N') \oplus \Gamma_{\mathcal{C}(N)}$$

Taking the direct sum of these relations and applying direct-sum cancellation of Γ-modules yields $(\Gamma \otimes_\Lambda M) \oplus Y' \cong (\Gamma \otimes_\Lambda M') \oplus \Gamma_{\mathcal{C}(N)}$. In other words, $\xi_{\Lambda,\Gamma}^{M,\Gamma_{\mathcal{C}(M)}}[M'] = [Y']$, as desired. □

The next step is to show that our partially ordered system is an inverse system. Again we work in somewhat more generality than faithful modules, for later use.

LEMMA 27.3 (Inverse-directedness). *Let $M, N \in \mathrm{fingen}_\infty(\Lambda)$ satisfy $\mathcal{C}(M) = \mathcal{C}(N)$. Then there exists $L \in \mathrm{fingen}(\Lambda)$ such that $\mathcal{C}(L) = \mathcal{C}(M)$ and both $\xi^{L,M}$ and $\xi^{L,N}$ are defined.*

PROOF. By Theorem 27.1, it suffices to find a module $L \in \mathrm{fingen}_\infty(\Lambda)$ such that $\mathcal{C}(L) = \mathcal{C}(M)$ and $\mathrm{istab}(L) \subseteq \mathrm{istab}(M) \cap \mathrm{istab}(N)$. This last relation will hold if we have, for all $\mathfrak{m} \in \mathrm{singspec}(\Lambda)$,

$$\mathrm{eqstab}(L_\mathfrak{m}) \subseteq \mathrm{eqstab}(M_\mathfrak{m}) \cap \mathrm{eqstab}(N_\mathfrak{m}) \tag{27.3.1}$$

because eqstab is determined locally [Lemma 23.3], and because of the relation $\mathrm{istab}(L) = \mathrm{eqstab}(L)\,\mathrm{im}\big((\Gamma_{\mathcal{C}(L)})^\times\big)$ [Theorem 23.5] (and the corresponding relations for M and N).

We construct L in two approximations, making use of the decomposition $\Gamma = \oplus_{h \in \mathcal{H}} \Gamma_h$. Let $V = \oplus_h V_h$, where each V_h is a $(\Gamma_h)_Q$-vector space of dimension 2 if $\Gamma_h \in \mathcal{C}(M)$, and $V_h = 0$ if $\Gamma_h \notin \mathcal{C}(M)$. For the first approximation, let L be the Λ-submodule of V generated by the numerators of some finite Γ_Q-generating set of V, so that $L_Q = V$. Then for each $\mathfrak{m} \in \mathrm{singspec}(\Lambda)$, we have $(L_\mathfrak{m})_Q = V_\mathfrak{m}$. If $\mathfrak{m}$ is in the support of a unique Γ_h, with $\Gamma_h \in \mathcal{C}(M)$, we therefore have $(L_\mathfrak{m})_Q = V_h$. On the other hand, if $\mathfrak{m}$ is in the support of two coordinate rings Γ_{h_1} and Γ_{h_2}, then $(L_\mathfrak{m})_Q$ equals 0, or just one of V_{h_i}, or $V_{h_1} \oplus V_{h_2}$, respectively, according as none, one, or both of Γ_{h_1} and Γ_{h_2} belongs to $\mathcal{C}(M)$.

$L_\mathfrak{m}$, $M_\mathfrak{m}$, and $N_\mathfrak{m}$ are free (possibly zero) for almost all $\mathfrak{m} \in \mathrm{singspec}(\Lambda)$ [Lemma 8.2], and if any of them equals zero, then all three are zero. For the rest of this proof, we need only consider those $\mathfrak{m}$ such that all three modules are nonzero. For those $\mathfrak{m}$ such that all three modules are free and nonzero, their equality stabilizers all equal $k(\mathfrak{m})^\times$ [Theorem 24.5], and therefore (27.3.1) holds.

Let $\mathcal{E}$ be the (finite) set of "exceptional" $\mathfrak{m}$, at which at least one of $L_\mathfrak{m}$, $M_\mathfrak{m}$, and $N_\mathfrak{m}$ is nonfree (and hence all three are nonzero). We build the second approximation by changing the localization $L_\mathfrak{m}$ at the finitely many $\mathfrak{m} \in \mathcal{E}$, in such a way that (27.3.1) holds for these $\mathfrak{m}$ as well. We can do this arbitrarily and still obtain a Λ-module L with the specified localizations, provided that we never change $(L_\mathfrak{m})_Q$ [Lemma 8.4]. We consider three cases.

Case 1: $\mathfrak{m}$ is unsplit. Since $(L_\mathfrak{m})_Q$ has dimension 2 as a $(\Gamma_\mathfrak{m})_Q$-vector space, we can change $L_\mathfrak{m}$ to an indecomposable $\Lambda_\mathfrak{m}$-module of torsionfree rank 2, such that $\mathrm{eqstab}(L_\mathfrak{m}) = \mathrm{im}(N_{k_\mathfrak{m}}^{F(\mathfrak{m})})$ [Theorem 24.5(i)(a)]. On the other hand, the only possibilities for the other equality stabilizers are $\mathrm{im}(N_{k_\mathfrak{m}}^{F(\mathfrak{m})})$, $k(\mathfrak{m})^\times$, and $F(\mathfrak{m})^\times$ [Theorem 24.5(i)]. Therefore, (27.3.1) holds.

Case 2: $\mathfrak{m}$ is split, and both of the coordinate rings Γ_h supported by $\mathfrak{m}$ belong to $\mathcal{C}(M)$. Here, all of $L_\mathfrak{m}$, $M_\mathfrak{m}$, $N_\mathfrak{m}$ are faithful. Change $L_\mathfrak{m}$ to a free $\Lambda_\mathfrak{m}$-module of rank 2, so that $\mathrm{eqstab}(L_\mathfrak{m}) = k(\mathfrak{m})^\times$ [Theorem 24.5(ii)(a)]. Since the equality stabilizer of each faithful $\Lambda_\mathfrak{m}$-module equals either $k(\mathfrak{m})^\times$ or $k(\mathfrak{m})_1^\times \times k(\mathfrak{m})_2^\times$ [Theorem 24.5(ii)], we again have (27.3.1).

Case 3: $\mathfrak{m}$ is split, but only one of the coordinate rings Γ_{h_1} and Γ_{h_2} supported by $\mathfrak{m}$ belongs to $\mathcal{C}(M)$, say $\Gamma_{h_1} \in \mathcal{C}(M)$. Then all three equality stabilizers equal $k(\mathfrak{m})_1^\times \times 1$ [Theorem 24.5(ii)(b)], so no change is necessary. □

PROPOSITION 27.4. *Let $\mathcal{F}_0$ be any finite set of modules in* $\text{fingen}_\infty(\Lambda)$. *Then there is a faithful* $M \in \text{fingen}_\infty(\Lambda)$ *such that diagram* (27.2.1) *commutes for every* $N \in \mathcal{F}_0$.

PROOF. Because of transitivity of ξ [Theorem 27.1(iv)] we can apply Lemma 27.3 a finite number of times, obtaining a faithful module M such that $\xi^{M,N}$ maps $\mathcal{G}(M)$ onto $\mathcal{G}(N)$ for every $N \in \mathcal{F}_0$. Mayer-Vietoris consistency [Lemma 27.2] then completes the proof. □

We now have the tools needed to define the "super" genus class group and place it in its own Mayer-Vietoris sequence.

DEFINITION 27.5 (s$\mathcal{G}$, rs$\mathcal{G}$). By Theorem 27.1 and Lemma 27.3, the collection all genus class groups of faithful modules in $\text{fingen}_\infty(\Lambda)$ forms an inverse system with respect to the set of ξ-maps. We define the *super genus class group* s$\mathcal{G}$ of the Dedekind-like ring Λ to be the inverse limit of this system. Similarly, the collection of all restricted genus class groups of faithful modules in $\text{fingen}_\infty(\Lambda)$ forms an inverse system, whose inverse limit rs$\mathcal{G}$ we call the *restricted super genus class group* of Λ.

In more detail, let $\mathcal{F}$ be a set of representatives of the isomorphism classes of faithful modules in $\text{fingen}_\infty(\Lambda)$. [See Definition 27.7 for alternative choices $\mathcal{F}'$ for $\mathcal{F}$.] Form the (strong) direct product $\mathcal{P} = \prod_{M\in\mathcal{F}} \mathcal{G}(M)$ of all genus class groups $\mathcal{G}(M)$, as M ranges over $\mathcal{F}$. We write elements of $\mathcal{P}$ as indexed families $[\boldsymbol{A}] = \{[A_M]\}_{M\in\mathcal{F}}$ where each $A_M \in \text{genus}(M)$. Then s$\mathcal{G}$ is the subgroup of $\mathcal{P}$ consisting of those elements $[\boldsymbol{A}]$ such that $\xi^{M,N}[A_M] = [A_N]$ whenever $\xi^{M,N}$ is defined, and rs$\mathcal{G}$ is the subgroup of s$\mathcal{G}$ consisting of those elements satisfying the further restriction that $[A_M] \in \text{r}\mathcal{G}(M)$ for each index $M \in \mathcal{F}$.

If there exists a faithful $M \in \text{fingen}(\Lambda)$ such that $\xi^{M,N}$ is defined for all faithful $N \in \text{fingen}(\Lambda)$, we write

$$\text{s}\mathcal{G} = \mathcal{G}(M) \qquad \text{(canonically)} \tag{27.5.1}$$

because s$\mathcal{G}$ can then be identified (via coordinate projection) with $\mathcal{G}(M)$.

THEOREM 27.6 (Super Mayer-Vietoris sequence). *The collection of all Mayer-Vietoris sequences for faithful modules in* $\text{fingen}_\infty(\Lambda)$ *induces a short exact sequence of inverse limits that we call the "super Mayer-Vietoris sequence":*

$$0 \to \text{rs}\mathcal{G} \xrightarrow{\mu^{\text{s}\mathcal{G}}} \text{s}\mathcal{G} \xrightarrow{\xi_{\Lambda,\Gamma}^{\text{s}\mathcal{G},\Gamma}} \mathcal{G}_\Gamma(\Gamma) \to 0 \tag{27.6.1}$$

where the map $\xi_{\Lambda,\Gamma}^{\text{s}\mathcal{G},\Gamma}$ *is defined as follows. Given* $[\boldsymbol{A}] = \{[A_M]\}_{M\in\mathcal{F}} \in \text{s}\mathcal{G}$, *choose any index* $M \in \mathcal{F}$, *and set* $\xi_{\Lambda,\Gamma}^{\text{s}\mathcal{G},\Gamma}[\boldsymbol{A}] = \xi_\Gamma^{M,\Gamma}[A_M]$. *(The subscript* Γ *indicates that we are dealing with the version of* ξ *that applies to* Γ*-modules.)*

PROOF. In (27.6.1) we view $\mathcal{G}_\Gamma(\Gamma_{C(M)})$ as the inverse limit of a constant inverse system. Then Mayer-Vietoris consistency [Lemma 27.2] states that the collection of all Mayer-Vietoris sequences forms a map of inverse systems, and hence induces maps of their inverse limits, as shown in (27.6.1). The fact that the sequence in (27.6.1) is short exact follows from the fact that all three inverse systems are surjective systems [**AM**, Proposition 10.2]; that is, the maps ξ used in forming the limits s$\mathcal{G}$ and rs$\mathcal{G}$ are surjective [Theorem 27.1 (ii) and (iii)]. In particular, since $\mathcal{G}_\Gamma(\Gamma)$ is the inverse limit of a constant system, the map $\xi_{\Lambda,\Gamma}^{\text{s}\mathcal{G},\Gamma}$ is evaluated as given. □

DEFINITION 27.7 (Cofinal subets). Let $\mathcal{F}'$ be a subset of $\mathcal{F}$, the index set from which s$\mathcal{G}$ is computed [Definition 27.5]. We call $\mathcal{F}'$ *cofinal in* $\mathcal{F}$ if, for every faithful $N \in \text{fingen}_\infty(\Lambda)$, there exists $M \in \mathcal{F}'$ such that $\text{istab}(M) \subseteq \text{istab}(N)$ (and hence $\xi^{M,N}$ is defined) [Corollary 26.7]. The following observation is an immediate consequence of the definitions involved.

Let $\mathcal{F}'$ be cofinal in $\mathcal{F}$. Then $\{\mathcal{G}(M) \mid M \in \mathcal{F}'\}$ is an inverse system, and its inverse limit can be canonically identified with s$\mathcal{G}$ *via coordinate projection and injection.*

For example, if $\mathcal{F}'$ is a subset of $\mathcal{F}$ consisting of one module from each genus of faithful modules in fingen(Λ), then $\mathcal{F}'$ is cofinal in $\mathcal{F}$, because $\xi^{M,N}$ is an isomorphism whenever M and N are in the same genus [Corollary 25.8]. (Recall that, in the Introduction to this paper, we described the "web of class groups" by means of a set $\mathcal{E}$ consisting of one module from each genus in fingen(Λ). The subset of $\mathcal{E}$ consisting of the faithful modules in $\mathcal{E}$ is an example of such a cofinal subset $\mathcal{F}'$.)

As a second example, suppose that some $M \in \mathcal{F}$ has the property that $\xi^{M,N}$ is defined for all faithful $N \in \text{fingen}(\Lambda)$. Then the one-element set $\{M\}$ is cofinal in $\mathcal{F}$. This is the situation encountered when we write $\text{s}\mathcal{G} = \mathcal{G}(M)$, as in (27.5.1).

We shall see another example in Theorem 27.9.

The next simple proposition extends consistency to the super Mayer-Vietoris sequence as well.

PROPOSITION 27.8 (Super Mayer-Vietoris consistency). *Let* s$\mathcal{G}$ *be computed from a cofinal subset $\mathcal{F}'$ of $\mathcal{F}$. Then the following diagram commutes, for every faithful $N \in \text{fingen}_\infty(\Lambda)$:*

$$\begin{array}{ccccccc} 0 \to & \text{rs}\mathcal{G} & \xrightarrow{\mu^{\text{s}\mathcal{G}}} & \text{s}\mathcal{G} & \xrightarrow{\xi_{\Lambda,\Gamma}^{\text{s}\mathcal{G},\Gamma}} & \mathcal{G}_\Gamma(\Gamma) \to 0 \\ & \downarrow \text{nat} & & \downarrow \xi^{\text{s}\mathcal{G},N} & & \downarrow 1 \\ 0 \to \mathcal{I}/\,\text{istab}(N) & & \xrightarrow{\mu^N} & \mathcal{G}(N) & \xrightarrow{\xi_{\Lambda,\Gamma}^{N,\Gamma}} & \mathcal{G}_\Gamma(\Gamma) \to 0 \end{array} \tag{27.8.1}$$

where the maps "nat" and $\xi^{\text{s}\mathcal{G},N}$ are defined as follows. By cofinality of $\mathcal{F}'$, choose any $M' \in \mathcal{F}'$ such that $\text{istab}(M') \subseteq \text{istab}(N)$, let $\xi^{\text{s}G,N}$ be the composition of projection to coordinate M' with $\xi^{M',N}$, and let "nat" be the restriction of $\xi^{\text{s}G,N}$ to rs$\mathcal{G}$ *(using the identification of $\mathcal{I}/\,\text{istab}(N)$ with $\text{r}\mathcal{G}(N)$).*

Moreover, if $\text{s}\mathcal{G} = \mathcal{G}(M)$ (natural identification), then $\xi^{\text{s}\mathcal{G},N} = \xi^{M,N}$ and $\text{rs}\mathcal{G} = \text{r}\mathcal{G}(M)$.

PROOF. In this proof we identify $\mathcal{I}/\,\text{istab}(N)$ with $\text{r}\mathcal{G}(N)$, as described in Theorem 25.14. If $\mathcal{F}' = \mathcal{F}$, then $\xi^{\text{s}G,N}$ and "nat" are the maps induced by coordinate projection. Now let $\mathcal{F}'$ be any cofinal subset of $\mathcal{F}$, and let "nat" and $\xi^{\text{s}\mathcal{G},N}$ be defined as in the statement of the proposition. We prove that this definition is independent of our choice of M' and that the diagram commutes.

Choose $[\boldsymbol{A}] = \{[A_M]\}_{M\in\mathcal{F}'} \in \text{s}\mathcal{G}$, and choose $M', M'' \in \mathcal{F}'$ such that $\text{istab}(M') \subseteq \text{istab}(N)$ and $\text{istab}(M'') \subseteq \text{istab}(N)$. By inverse-directedness [Lemma 27.3] there exists a faithful $X \in \text{fingen}(\Lambda)$ such that $\text{istab}(X) \subseteq \text{istab}(M') \cap \text{istab}(M'')$, and by cofinality of $\mathcal{F}'$ there exists $Y \in \mathcal{F}'$ such that $\text{istab}(Y) \subseteq \text{istab}(X)$. Then the definition of $[\boldsymbol{A}]$ implies that $\xi^{Y,M'}[A_Y] = [A_{M'}]$ and therefore transitivity of ξ shows that $\xi^{Y,N}[A_Y] = \xi^{M',N}\xi^{Y,M'}[A_Y] = \xi^{M',N}[A_{M'}]$. Similarly $\xi^{Y,N}[A_Y] =$

$\xi^{M'',N}[A_{M''}]$, showing that the definition of $\xi^{sG,N}$ is independent of the choice of M'.

Commutativity of the diagram now follows immediately from ordinary Mayer-Vietoris consistency, that is, from commutativity of diagram (27.2.1), with M' in place of M.

The supplementary statement is the special case that $\mathcal{F}'$ consists of a single element, and motivates our notation $\xi^{s\mathcal{G},N}$. □

The remainder of this section is devoted to describing the ways in which $s\mathcal{G}$ and the super Mayer-Vietoris sequence can be simplified in various special situations.

THEOREM 27.9 (Countable case). *Suppose that the number of unsplit maximal ideals* $\mathfrak{m}$, *such that* $N_{k(\mathfrak{m})}^{F(\mathfrak{m})}$ *is not surjective, is countable (finite or infinite). Then:*

(i) *The index set* $\mathcal{F}$ *[Definition 27.5] contains a cofinal sequence* $M_0, M_1, \ldots$, *where all ranks of each* M_i *equal 2 and* ξ^{M_i,M_j} *is defined whenever* $i \geq j$. *(Thus,* $s\mathcal{G}$ *is the inverse limit of the sequence* $\mathcal{G}(M_0), \mathcal{G}(M_1), \mathcal{G}(M_2), \ldots$.*)*

(ii) *The natural map* $\mathcal{I} \to \mathrm{rs}\mathcal{G}$, *sending* $\boldsymbol{u} \mapsto \{[M^{\boldsymbol{u}}]\}_{M\in\mathcal{F}} \in \mathrm{rs}\mathcal{G}$, *is surjective. (Thus,* $\mathrm{rs}\mathcal{G}$ *is a quotient of the group* $\mathcal{I}$ *of residue unit idèles.)*

PROOF. We begin by reviewing some facts that we use repeatedly in the proof, usually without explicit mention.

(27.9.1) Let M, N be faithful modules in $\mathrm{fingen}_\infty(\Lambda)$. Then:

(i) $\mathrm{eqstab}(M)(\mathfrak{m}) = \mathrm{eqstab}(\hat{M}_\mathfrak{m})$.

(ii) $\mathrm{eqstab}(M) \subseteq \mathrm{eqstab}(N) \iff \mathrm{eqstab}(\hat{M}_\mathfrak{m}) \subseteq \mathrm{eqstab}(\hat{N}_\mathfrak{m})$ for all $\mathfrak{m} \in \mathrm{singspec}(\Lambda)$.

(iii) $\mathrm{eqstab}(M)(\mathfrak{m}) \subseteq \mathrm{eqstab}(N)(\mathfrak{m})$ $\big(\forall \mathfrak{m} \in \mathrm{singspec}(\Lambda)\big)$ implies that $\xi^{M,N}$ is defined.

Fact (i) holds by Lemma 24.3. (ii) holds by Lemma 24.3. The hypothesis of (iii) implies, in view of (i) and (ii), that $\mathrm{eqstab}(M) \subseteq \mathrm{eqstab}(N)$. This implies that $\mathrm{istab}(M) \subseteq \mathrm{istab}(N)$ [Theorem 23.5], and hence Theorem 27.1(i) shows that $\xi^{M,N}$ is defined.

Let $M \in \mathrm{fingen}_\infty(\Lambda)$ be faithful. Then for $\mathfrak{m} \in \mathrm{singspec}(\Lambda)$ we claim:

$$\mathrm{eqstab}(\hat{M}_\mathfrak{m}) = \begin{cases} k(\mathfrak{m})^\times \text{ or } k(\mathfrak{m})_1 \times k(\mathfrak{m})_2 & \text{if } \mathfrak{m} \text{ is split} \\ \mathrm{im}(N_{k(\mathfrak{m})}^{F(\mathfrak{m})}) \text{ or } k(\mathfrak{m})^\times \text{ or } F(\mathfrak{m})^\times & \text{if } \mathfrak{m} \text{ is unsplit} \end{cases} \tag{27.9.2}$$

Since M is faithful, every $\hat{M}_\mathfrak{m}$ is $\hat{\Lambda}_\mathfrak{m}$-faithful, and therefore the claim is part of Theorem 24.5. Recall that, in the split case, the notation $k(\mathfrak{m})$ denotes the diagonal copy of $k(\mathfrak{m})$ in $k(\mathfrak{m})_1 \times k(\mathfrak{m})_2$.

Let $\mathfrak{m}_1, \mathfrak{m}_2, \ldots$ be the countable (finite or infinite) set of unsplit maximal ideals of Λ for which the indicated norm map is not surjective. Recall that the index set $\mathcal{F}$ used in building the inverse limit $s\mathcal{G}$ contains one module from each isomorphism class of faithful finitely generated Λ-modules [Definition 27.5].

Claim 1: There exists a countable (finite or infinite) set of modules $M_0, M_1, \ldots$, each an element of $\mathcal{F}$, such that, for each index n,

$$\mathrm{eqstab}\big((M_n)\hat{}_\mathfrak{m}\big) = \begin{cases} \mathrm{im}(N_{k(\mathfrak{m})}^{F(\mathfrak{m})}) & \text{if } \mathfrak{m} \in \{\mathfrak{m}_1, \mathfrak{m}_2, \ldots, \mathfrak{m}_n\} \\ k(\mathfrak{m})^\times & \text{otherwise.} \end{cases} \tag{27.9.3}$$

Let M_0 be the element of $\mathcal{F}$ such that $M_0 \cong \Lambda^{(2)}$. By (27.9.1)(i) and Theorem 24.5 we have $\mathrm{eqstab}\big((M_0)\hat{_\mathfrak{m}}\big) = k(\mathfrak{m})^\times$ for every $\mathfrak{m} \in \mathrm{singspec}(\Lambda)$.

Choose $n > 0$. We construct M_n in two approximations. For the first approximation, take $M_n = M_0$. For the second approximation, we change the isomorphism class of finitely many completions $(M_0)\hat{_\mathfrak{m}}$. The Package Deal Theorem for Completions 15.6 allows us to make such changes provided that we never change the rank of any completion that we change, that is, provided that each new $(M_0)\hat{_\mathfrak{m}}$ still has all ranks 2.

For each $\mathfrak{m} \in \{\mathfrak{m}_1, \mathfrak{m}_2, \ldots, \mathfrak{m}_n\}$, replace $(\hat{M}_0)_\mathfrak{m}$ by an indecomposable module in $\mathrm{fingen}(\hat{\Lambda}_\mathfrak{m})$ of rank 2. Such indecomposable modules exist and have the desired equality stabilizer, by Theorem 24.5 (and are never torsionfree!). The resulting new M_n is isomorphic to an element of $\mathcal{F}$, which we may suppose equals M_n, proving Claim 1.

Claim 2: Given M_i and M_j with $i \geq j$, the map ξ^{M_i,M_j} is defined. Moreover, for every faithful $M \in \mathrm{fingen}(\Lambda)$, there exists n such that $\xi^{M_n,M}$ is defined.

By (27.9.3), $\mathrm{eqstab}(M_i) \subseteq \mathrm{eqstab}(M_j)$ whenever $i \geq j$, and therefore (27.9.1) shows that ξ^{M_i,M_j} is defined.

For the second assertion of the claim, let M be given. It suffices, as in the previous paragraph, to find n such that $\mathrm{eqstab}(M_n) \subseteq \mathrm{eqstab}(M)$.

First we compare $\mathrm{eqstab}(M_0)$ with $\mathrm{eqstab}(M)$. $(M_0)\hat{_\mathfrak{m}}$ is $\hat{\Lambda}_\mathfrak{m}$-free for all $\mathfrak{m} \in$ singspec, and therefore $\mathrm{eqstab}\big((M_0)\hat{_\mathfrak{m}}\big) = k(\mathfrak{m})^\times$ for all such $\mathfrak{m}$. Therefore (27.9.2) shows that $\mathrm{eqstab}\big((M_0)\hat{_\mathfrak{m}}\big) \subseteq \mathrm{eqstab}(\hat{M}_\mathfrak{m})$ for all split $\mathfrak{m}$ and for all unsplit $\mathfrak{m}$ such that $N^{F(\mathfrak{m})}_{k(\mathfrak{m})}$ is surjective. It remains to consider the unsplit maximal ideals $\mathfrak{m}_i$ whose norm map is not surjective.

$\hat{M}_\mathfrak{m}$ is $\hat{\Lambda}_\mathfrak{m}$-free $\big(\mathrm{a}\forall \mathfrak{m} \in \mathrm{singspec}(\Lambda)\big)$, by the Package Deal Theorem for Completions, and therefore $\mathrm{eqstab}(\hat{M}_\mathfrak{m}) = k(\mathfrak{m})^\times = \mathrm{eqstab}\big((M_0)\hat{_\mathfrak{m}}\big)$ for almost all of the maximal ideals $\mathfrak{m} = \mathfrak{m}_i$. Therefore there exists n such that every exception to the inclusion $\mathrm{eqstab}\big((M_0)\hat{_\mathfrak{m}}\big) \subseteq \mathrm{eqstab}(\hat{M}_\mathfrak{m})$ lies in the set $\mathfrak{m}_1, \ldots, \mathfrak{m}_n$. Furthermore, by (27.9.2) applied to M_0, these exceptions can be removed if we replace $\mathrm{eqstab}\big((M_0)\hat{_\mathfrak{m}}\big)$ by $\mathrm{im}(N^{F(\mathfrak{m})}_{k(\mathfrak{m})})$. Since this is exactly what we did in the definition of M_n — see (27.9.3) — M_n satisfies $\mathrm{eqstab}(M_n) \subseteq \mathrm{eqstab}(M)$, and the proof of Claim 2 is now complete.

From Claim 2, it now follows immediately that $\mathrm{s}\mathcal{G}$ is the inverse limit of the sequence $\mathcal{G}(M_0), \mathcal{G}(M_1), \mathcal{G}(M_2), \ldots$, proving part (i) of the theorem.

For part (ii) of the theorem, take any element $[\boldsymbol{A}] \in \mathrm{rs}\mathcal{G}$. By Theorem 22.6(ii), we can write $[\boldsymbol{A}] = \{[M^{\boldsymbol{u}(M)}]\}_{M \in \mathcal{F}}$, for suitable residue unit idèles $\boldsymbol{u}(M)$ for each $M \in \mathcal{F}$. To prove part (ii) of the theorem, we must find a single residue unit idèle $\boldsymbol{w} \in \mathcal{I}$ such that

$$M^{\boldsymbol{w}} \cong M^{\boldsymbol{u}(M)} \qquad (\forall M \in \mathcal{F}) \tag{27.9.4}$$

Note that, for $F, G \in \mathcal{F}$:

$$\xi^{F,G} \text{ is defined} \implies [G^{\boldsymbol{u}(F)}] = \xi^{F,G}[F^{\boldsymbol{u}(F)}] = [G^{\boldsymbol{u}(G)}] \tag{27.9.5}$$

(The first equality holds by Theorem 27.1(iii); the second equality holds because the map $\xi^{F,G}$ takes the F-coordinate of $[\boldsymbol{A}]$ to the G-coordinate of $[\boldsymbol{A}]$, since $[\boldsymbol{A}]$ is an element of the inverse limit defining $\mathrm{s}\mathcal{G}$.)

We first construct a residue unit idèle $\boldsymbol{w}$ for which (27.9.4) holds whenever M equals some M_n.

Claim 3: Given M_n, we can find a residue unit idèle $\boldsymbol{s}_n$ such that

$$M_n^{\boldsymbol{u}(M_n)} \cong M_n^{\boldsymbol{u}(M_{n-1})\boldsymbol{s}_n} \tag{27.9.6}$$

where the residue unit idèle $\boldsymbol{s}_n$ also satisfies:

$$\begin{cases} \boldsymbol{s}_n \in \operatorname{eqstab}(M_{n-1}) & \text{and} \\ \boldsymbol{s}_n(\mathfrak{m}) = 1 & \text{if } \mathfrak{m} \neq \mathfrak{m}_n \end{cases} \tag{27.9.7}$$

Since $\xi^{M_n,M_{n-1}}$ is defined [Claim 2], (27.9.5) implies that $M_{n-1}^{\boldsymbol{u}(M_n)} \cong M_{n-1}^{\boldsymbol{u}(M_{n-1})}$. Therefore, we have an expression $\boldsymbol{u}(M_n) = \boldsymbol{u}(M_{n-1})\boldsymbol{st}$, where $\boldsymbol{s} \in \operatorname{eqstab}(M_{n-1})$ and $\boldsymbol{t} \in \operatorname{im}(\Gamma^\times)$ [Theorem 23.5]. Let the residue unit idèle $\boldsymbol{s}_n$ be defined by $s_n(\mathfrak{m}_n) = s(\mathfrak{m}_n)$, and $s_n(\mathfrak{m}) = 1$ for all $\mathfrak{m} \neq \mathfrak{m}_n$, so that s_n satisfies (27.9.7). Then because $\operatorname{im}(\Gamma^\times) \subseteq \operatorname{istab}(M_n)$ [Theorem 23.5], we get $M_n^{\boldsymbol{u}(M_n)} = M_n^{\boldsymbol{u}(M_{n-1})st} \cong M_n^{\boldsymbol{u}(M_{n-1})s}$. But by construction, $\operatorname{eqstab}(M_n)(\mathfrak{m}) = \operatorname{eqstab}(M_{n-1})(\mathfrak{m})$ for every $\mathfrak{m} \neq \mathfrak{m}_n$, so using (27.9.7) and the definition of $\boldsymbol{s}_n$, we also obtain $M_n^{\boldsymbol{u}(M_{n-1})s} \cong M_n^{\boldsymbol{u}(M_{n-1})s_n}$, which establishes (27.9.6) and completes the proof of Claim 3.

Now by Claim 3, for each M_n beginning with $n = 1$, we can find s_n satisfying (27.9.6) and (27.9.7), and so we replace $\boldsymbol{u}(M_n)$ by $\boldsymbol{u}(M_{n-1})s_n$. Then for each M_n we obtain the formula:

$$\boldsymbol{u}(M_n) = \boldsymbol{u}(M_0)\boldsymbol{s}_1\boldsymbol{s}_2\dots\boldsymbol{s}_n \tag{27.9.8}$$

Since each $\boldsymbol{s}_n$ has only one nontrivial coordinate, namely $s_n(\mathfrak{m}_n)$, we can let $\boldsymbol{w}$ be the product of $\boldsymbol{u}(M_0)$ with the infinitely many s_n. That is, for each n, the $\mathfrak{m}_n$-coordinate of $\boldsymbol{w}(\mathfrak{m}_n) = \boldsymbol{u}(M_0)(\mathfrak{m}_n) \cdot s_n(\mathfrak{m}_n)$, while $\boldsymbol{w}(\mathfrak{m}) = \boldsymbol{u}(M_0)(\mathfrak{m})$ if $\mathfrak{m} \notin \{\mathfrak{m}_1, \mathfrak{m}_2, \dots\}$.

Claim 4: $\boldsymbol{w}$ is a residue unit idèle in $\mathcal{I}$. To prove this claim, we need only show that $w(\mathfrak{m}) \in k(\mathfrak{m})^\times$ (a$\forall\mathfrak{m}$). This in turn holds because $\boldsymbol{u}(M_0)$ is itself a residue unit idèle, and each coordinate of $\boldsymbol{u}(M_0)$ altered to form a coordinate of $\boldsymbol{w}$ is in fact multiplied by an element of $k(\mathfrak{m})^\times$.

Claim 5: (27.9.4) holds for each module M_n. If we set $\boldsymbol{s}' = \boldsymbol{w}\boldsymbol{u}(M_n)^{-1}$, then to prove this claim, it suffices to show that $\boldsymbol{s}' \in \operatorname{eqstab}(M_n)$, which we can check one coordinate at a time. But for $i > n$, we have $s'(\mathfrak{m}_i) \in k(\mathfrak{m}_i)^\times = \operatorname{eqstab}(M_n)(\mathfrak{m}_i)$, while for all other $\mathfrak{m}$ we have $s'(\mathfrak{m}) = 1$, where the required verification is trivial.

Finally, we can complete the proof of the second part of the theorem by proving (27.9.4) for every $M \in \mathcal{F}$. Given $M \in \mathcal{F}$, $\xi^{M_n,M}$ is defined for some n, by Claim 2. By Claim 5, $[M_n^{\boldsymbol{w}}] = [M_n^{\boldsymbol{u}(M_n)}]$, and applying $\xi^{M_n,M}$ to this relation, and using (27.9.5), yields (27.9.4). □

For brevity, we say that a module has *rank 2* if all of its ranks equal 2.

THEOREM 27.10 (Finite case). *We have the following canonical identifications (via ξ-maps, as in* (27.5.1)*).*

(i) *If there are only finitely many unsplit maximal ideals* $\mathfrak{m} \in \operatorname{singspec}(\Lambda)$ *such that* $N_{k(\mathfrak{m})}^{F(\mathfrak{m})}$ *is not surjective, then* $\mathrm{s}\mathcal{G} = \mathcal{G}(M)$ *for some [indeed, any] faithful* $M \in \operatorname{fingen}_\infty(\Lambda)$ *of rank 2 such that* $\operatorname{eqstab}(M)(\mathfrak{m}) = \operatorname{im}(N_{k(\mathfrak{m})}^{F(\mathfrak{m})})$ *for every unsplit* $\mathfrak{m}$ *and* $\operatorname{eqstab}(M) = k(\mathfrak{m})^\times$ *for all split* $\mathfrak{m}$.

(ii) *If* $N_{k(\mathfrak{m})}^{F(\mathfrak{m})}$ *is surjective for every unsplit* $\mathfrak{m}$, *then* $\mathrm{s}\mathcal{G} = \mathcal{G}(\Lambda)$.

PROOF. According to the definition of these canonical identifications, it suffices to show that the required M has the property that $\xi^{M,N}$ is defined for all faithful $N \in \text{fingen}_\infty(\Lambda)$; and for this it suffices to verify that $\text{eqstab}(M) \subseteq \text{eqstab}(N)$ for all faithful $N \in \text{fingen}_\infty(\Lambda)$ [Corollary 26.8]. This last condition can be complete-locally verified [Lemma 24.3].

In situation (i), we can let let M be the last module M_n in the sequence $M_0, M_1, \ldots$ in Theorem 27.9. [See (27.9.3).]

In situation (ii), the last M_n in the sequence $M_0, M_1, \ldots$ in Theorem 27.9 is M_0 [see (27.9.3)], which is isomorphic to $\Lambda^{(2)}$. Therefore, as in situation (i), we have $\text{s}\mathcal{G} = \mathcal{G}(\Lambda^{(2)})$. To complete the proof, recall that $\mathcal{G}(\Lambda) \cong \mathcal{G}(\Lambda^{(2)})$ via the mutually inverse isomorphisms $\xi^{\Lambda,\Lambda^{(2)}}$ and $\xi^{\Lambda^{(2)},\Lambda}$ [Corollary 25.9]. □

For an example where $\text{s}\mathcal{G} = \mathcal{G}(\Lambda)$ and is nontrivial, see Example 31.2 (and invoke Corollary 27.11). For an example where $\text{s}\mathcal{G} = \mathcal{G}(M)$ for some M, but $\mathcal{G}(M) \not\cong \mathcal{G}(\Lambda)$, see Example 34.1.

The following Corollary was proved (in slightly different terminology) in [**L2**, Remarks 9.9A], in the special case that $\text{singspec}(\Lambda)$ is finite and all singular maximal ideals are split.

COROLLARY 27.11. *Under either of the following hypotheses we have* $\text{s}\mathcal{G} = \mathcal{G}(\Lambda)$ *(canonical identification).*

(i) *All residue fields of* Λ *are finite.*
(ii) Λ *is a finitely generated algebra over an algebraically closed field.*

PROOF. (i) Here the norm map is always a surjection [**LN**, Theorem 2.28(ii)].

(ii) Here all residue fields are algebraically closed, so that all $\mathfrak{m} \in \text{singspec}(\Lambda)$ are split. □

We call a module M over a Dedekind-like ring *torsionfree* if it has no nonzero submodules of finite length. Recall that we say that M has *rank 2* if all ranks of M equal 2.

THEOREM 27.12. *Suppose that* ${}_\Lambda\Gamma$ *is finitely generated and* $\text{s}\mathcal{G} \neq \mathcal{G}(\Lambda)$. *Then* $\text{s}\mathcal{G} = \mathcal{G}(M)$ *where* M *has (torsionfree) rank 2 and cannot be torsionfree. [See Example 34.1.]*

PROOF. Since ${}_\Lambda\Gamma$ is finitely generated, $\text{singspec}(\Lambda)$ is finite [Proposition 10.11] and $\text{s}\mathcal{G} = \mathcal{G}(M)$ for some faithful $M \in \text{fingen}_\infty(\Lambda)$ of torsionfree rank 2 such that $\text{eqstab}(M)(\mathfrak{m}) = \text{im}(N^{F(\mathfrak{m})}_{k(\mathfrak{m})})$ for every unsplit $\mathfrak{m}$ [Theorem 27.10(i)]. Moreover, since $\text{s}\mathcal{G} \neq \mathcal{G}(\Lambda)$, there must be at least one unsplit $\mathfrak{m}$ such that $N^{F(\mathfrak{m})}_{k(\mathfrak{m})}$ is not surjective [Theorem 27.10(ii)], say $\mathfrak{m} = \mathfrak{n}$.

To prove that M must have a nonzero submodule of finite length, it suffices to show that $\hat{M}_\mathfrak{n}$ has a nonzero $\hat{\Lambda}_\mathfrak{n}$-submodule of finite length [Proposition 7.4 and the fact that all maximal ideals of Dedekind-like rings have height 1].

We have $\hat{M}_\mathfrak{n} \in \text{fingen}_\infty(\hat{\Lambda}_\mathfrak{n})$ [Proposition 7.3], $\text{eqstab}(M)(\mathfrak{n}) = \text{eqstab}(\hat{M}_\mathfrak{n})$ [Lemma 24.3], and since ranks are complete-locally determined [Lemma 15.2] the rank of $\hat{M}_\mathfrak{n}$ must be 2. Moreover, $\text{eqstab}(\hat{M}_\mathfrak{n}) = \text{im}(N^{F(\mathfrak{n})}_{k(\mathfrak{n})}) \subset k(\mathfrak{n})^\times$. Write $\hat{M}_\mathfrak{n}$ as a direct sum of indecomposables. Since the eqstab of a direct sum is the product of the eqstabs of the summands [Theorem 24.5], the list of indecomposable modules and their ranks in the complete local case given in Theorem 24.5 shows that the only possibility is that $\hat{M}_{\mathfrak{n}}$ be indecomposable of rank 2. In this situation, Lemma 16.2(i) asserts that $\hat{M}_{\mathfrak{n}}$ has a nonzero submodule of finite length. Therefore the same is true of M itself [Proposition 7.4]. □

28. Super Mayer-Vietoris and unfaithful modules

This section completes the proof that the super genus class group s$\mathcal{G}$ maps onto every actual genus class group in $\text{fingen}_\infty(\Lambda)$ in a natural way, *but not via ξ-maps when the module is unfaithful* [Lemma and Definition 28.2], and that the super Mayer-Vietoris sequence has the analogous property [Theorem 28.3]. Recall that $\mathcal{C}(M)$ denotes the torsionfree support set of the module M, and $\Gamma_{\mathcal{C}(M)} = \oplus_{h\in\mathcal{C}(M)}\Gamma_h$ [Notation 3.1 and Definitions 22.1]. Our main tool is the following Lemma. We do not make our usual restriction of working in $\text{fingen}_\infty(\Lambda)$.

LEMMA 28.1. *Let $M, N \in \text{fingen}(\Lambda)$ be such that $\mathcal{C}(M)\cap\mathcal{C}(N) = \varnothing$. Then every module $L \in \text{genus}(M \oplus N)$ has a decomposition $L = M' \oplus N'$ with $M' \in \text{genus}(M)$ and $N' \in \text{genus}(N)$. Moreover, M' and N' are determined up to isomorphism by L.*

PROOF. The existence of M' and N' (over any commutative noetherian ring of Krull dimension 1) is given in Lemma 9.3(iii), so we need only prove prove the uniqueness. Thus, we can suppose that

$$M \oplus N \cong M' \oplus N' \tag{28.1.1}$$

for some $M' \in \text{genus}(M)$ and $N' \in \text{genus}(N)$, where $\mathcal{C}(M) \cap \mathcal{C}(N) = \varnothing$. We want to prove that $M \cong M'$ and $N \cong N'$. First we reduce to the case that none of the four terms has nonzero direct summands of finite length.

Recall that indecomposable modules of finite length have local endomorphism rings, and therefore cancel from direct sums [**F**, 4.5]. Moreover, M and M' remain in the same genus after the cancellation [Lemma 9.2(i)]. Therefore, if we can show that M and M' have nonzero indecomposable direct summands $X \cong X'$, then we can cancel X and X' from the direct sums in (28.1.1) and repeat the procedure until M and M' have no nonzero direct summands of finite length. Giving the same treatment to N and N' then completes the reduction.

Let X be any indecomposable finite-length direct summand of M. Then X is the direct sum of its localizations at maximal ideals [Corollary 6.4(i)], and hence indecomposability implies that $X = X_{\mathfrak{m}}$ for some maximal ideal $\mathfrak{m}$. Localize (28.1.1) at $\mathfrak{m}$. Then $\text{genus}(M) = \text{genus}(M')$ implies that $M_{\mathfrak{m}} \cong M'_{\mathfrak{m}}$. Since $X = X_{\mathfrak{m}}$, X remains of direct summand of $M_{\mathfrak{m}}$. Therefore $M'_{\mathfrak{m}}$ has a direct summand $X' \cong X$. Since every finite-length direct summand of $M'_{\mathfrak{m}}$ is also a finite-length direct summand of M' [Lemma 6.2(iii)], our reduction is now complete. Thus we assume, from now on, that *none of M, M', N, N' has nonzero direct summands of finite length.*

First, we claim that $\Gamma \otimes_\Lambda M \cong \Gamma \otimes_\Lambda M'$ and $\Gamma \otimes_\Lambda N \cong \Gamma \otimes_\Lambda N'$ as Γ-modules. Since Γ is a direct sum of Dedekind-domains, it follows that they have decompositions $\Gamma \otimes_\Lambda M = A \oplus T$, $\Gamma \otimes_\Lambda N = B \oplus U$, $\Gamma \otimes_\Lambda M' = A' \oplus T'$,and $\Gamma \otimes_\Lambda N' = B' \oplus U'$ with A, B, A', B' projective and T, U, T', U' of finite length. Therefore, tensoring (28.1.1) with Γ yields the Γ-isomorphism

$$(A \oplus T) \oplus (B \oplus U) \cong (A' \oplus T') \oplus (B' \oplus U')$$

Then equating the projective summands gives the Γ-isomorphism

$$A \oplus B \cong A' \oplus B' \tag{28.1.2}$$

Since $\Lambda_Q = \Gamma_Q$ and $T_Q = 0$, localizing the relation $\Gamma \otimes_\Lambda M = A \oplus T$ at Q yields $M_Q \cong A_Q$. By definition, $\mathcal{C}(M)$ is the set of all coordinate rings Γ_h such that $\Gamma_h M_Q \neq 0$. Since $M_Q \cong A_Q$, we conclude that A is a projective $\Gamma_{\mathcal{C}(M)}$-module. Similarly, A' is a projective module over $\Gamma_{\mathcal{C}(M')} = \Gamma_{\mathcal{C}(M)}$, and B and B' are projective modules over $\Gamma_{\mathcal{C}(N)} = \Gamma_{\mathcal{C}(N')}$. Since $\mathcal{C}(M) \cap \mathcal{C}(N) = \varnothing$, it follows from (28.1.2) that $A \cong A'$ and $B \cong B'$.

Thus, we have shown that the projective part of $\Gamma \otimes_\Lambda M$ is isomorphic to the projective part of $\Gamma \otimes_\Lambda M'$. Since Γ is a direct sum of Dedekind domains, to complete the proof that $\Gamma \otimes_\Lambda M \cong \Gamma \otimes_\Lambda M'$, it now suffices to show that both sides are in the same Γ-genus. But this follows immediately from the fact that M and M' are in the same Λ-genus. Similarly, the isomorphism $\Gamma \otimes_\Lambda N \cong \Gamma \otimes_\Lambda N'$ follows, completing the proof of the claim.

Given that $M' \in \operatorname{genus}(M)$ and $N' \in \operatorname{genus}(N)$, the claim implies that in fact $M' \in \operatorname{rgen}(M)$ and $N' \in \operatorname{rgen}(N)$, so that we can write $M' \cong M^{\boldsymbol{u}}$ and $N' \cong N^{\boldsymbol{v}}$ for some residue unit idèles $\boldsymbol{u} \in \mathcal{I}_{\mathcal{C}(M)}$ and $\boldsymbol{v} \in \mathcal{I}_{\mathcal{C}(N)}$ [Theorem 22.6(ii)]. Therefore, $M \oplus N \cong M^{\boldsymbol{u}} \oplus N^{\boldsymbol{v}} \cong (M \oplus N)^{\boldsymbol{uv}}$ [Theorem 22.7], and hence $\boldsymbol{uv} \in \operatorname{istab}(M \oplus N) = \operatorname{istab}(M) \operatorname{istab}(N)$ [Corollary 24.7(ii)]. Thus, we have a relation $\boldsymbol{uv} = \boldsymbol{xy}$, where $\boldsymbol{x} \in \operatorname{istab}(M)$ and $\boldsymbol{y} \in \operatorname{istab}(N)$. It now suffices to show that we can take $\boldsymbol{x} = \boldsymbol{u}$ and $\boldsymbol{y} = \boldsymbol{v}$, for then we have $M' \cong M^{\boldsymbol{u}} \cong M^{\boldsymbol{x}} \cong M$ and $N' \cong N^{\boldsymbol{v}} \cong N^{\boldsymbol{y}} \cong N$.

In fact, it suffices to show that, for each $\mathfrak{m} \in \operatorname{singspec}(\Lambda)$, we can take $x(\mathfrak{m}) = u(\mathfrak{m})$ *or* $y(\mathfrak{m}) = v(\mathfrak{m})$, for the relation $u(\mathfrak{m})v(\mathfrak{m}) = x(\mathfrak{m})y(\mathfrak{m})$ shows that, if one equation holds, then both must hold. Therefore, choose $\mathfrak{m}$, and recall that $\mathfrak{m}$ is in the support of either one or two coordinate rings Γ_h of Γ [Lemma 11.6(i)]. We consider four cases.

Case 1. The one or two coordinate rings supported by $\mathfrak{m}$ belong to $\mathcal{C}(M)$. Since $\mathcal{C}(M) \cap \mathcal{C}(N) = \varnothing$, it follows that $\mathfrak{m}$ is in the support of no coordinate ring in $\mathcal{C}(N)$, and therefore $v(\mathfrak{m}) = y(\mathfrak{m}) = 1$ [Definitions 22.1], as desired.

Case 2. The one or two coordinate rings supported by $\mathfrak{m}$ belong to $\mathcal{C}(N)$. Then $u(\mathfrak{m}) = x(\mathfrak{m}) = 1$, by the same argument as in Case 1.

Case 3. No coordinate ring supported by $\mathfrak{m}$ belongs to $\mathcal{C}(M)$ or to $\mathcal{C}(N)$. Then $u(\mathfrak{m}) = x(\mathfrak{m}) = v(\mathfrak{m}) = y(\mathfrak{m}) = 1$, as desired.

Case 4. $\mathfrak{m}$ is in the support of two coordinate rings Γ_{h_1} and Γ_{h_2}, with $\Gamma_{h_1} \in \mathcal{C}(M)$ and $\Gamma_{h_2} \in \mathcal{C}(N)$. Then $\mathcal{C}(M) \cap \mathcal{C}(N) = \varnothing$ implies that $\Gamma_{h_2} \notin \mathcal{C}(M)$. Choose a separated cover of M, with associated Γ-module X. Then some direct summand of X is an ideal of Γ_{h_1}, but no ideal of Γ_{h_2} occurs as a summand of X. Therefore, $\mathfrak{m}$ is strictly split, and $\operatorname{eqstab}(M)(\mathfrak{m}) = k(\mathfrak{m})_1^\times \times 1$ [Theorem 24.5(ii)(b)]. Moreover, since $\boldsymbol{u} \in \mathcal{I}_{\mathcal{C}(M)}$ and $\Gamma_{h_2} \notin \mathcal{C}(M)$, it must be the case that $u(\mathfrak{m}) = (\alpha, 1)$ for some $\alpha \in k(\mathfrak{m})_1^\times$ [Definitions 22.1], and hence we can take $x(\mathfrak{m}) = u(\mathfrak{m})$, as desired. □

We now have the tools necessary to show that the super genus class group $\mathrm{s}\mathcal{G}$ maps onto each class group $\mathcal{G}(N)$ in a natural way.

LEMMA AND DEFINITION 28.2 ($\psi^{\mathrm{s}\mathcal{G},N}$). *Let* $\mathrm{s}\mathcal{G}$ *be the super genus class group and* $N \in \operatorname{fingen}_\infty(\Lambda)$. *A surjective group homomorphism* $\psi^{\mathrm{s}\mathcal{G},N}\colon \mathrm{s}\mathcal{G} \twoheadrightarrow \mathcal{G}(N)$ *can be defined as follows. Let* $A \in \operatorname{fingen}_\infty(\Lambda)$ *be such that* $N \oplus A$ *is faithful and* $\mathcal{C}(N) \cap \mathcal{C}(A) = \varnothing$, *and let* $\psi^{\mathrm{s}\mathcal{G},N}$ *be the composition:*

$$(28.2.1) \qquad \psi^{\mathrm{s}\mathcal{G},N}\colon \quad \mathrm{s}\mathcal{G} \xrightarrow{\ \xi^{\mathrm{s}\mathcal{G},(N\oplus A)}\ } \mathcal{G}(N \oplus A) \xrightarrow{\mathrm{proj}} \mathcal{G}(N)$$

where proj *is the map defined by* $[N' \oplus A'] \mapsto [N']$ *(for* $N' \in \operatorname{genus}(N)$ *and* $A' \in \operatorname{genus}(A)$*), and* ξ *is the map defined in Proposition 27.8. Then the map* $\psi^{s\mathcal{G},N}$ *is independent of the choice of* A*. Moreover, if* N *is faithful, then* $\psi^{s\mathcal{G},N} = \xi^{s\mathcal{G},N}$*.*

PROOF. Note that one possibility for the choice of A is simply the projection of Λ in $\oplus_{h \notin \mathcal{C}(N)} \Gamma_h$ (which we take to be 0 if N is faithful).

Given A, the map proj is well-defined, by Lemma 28.1, and is obviously surjective.

To prove independence of A, suppose that B is also a finitely generated Λ-module such that $N \oplus B$ is faithful, and $\mathcal{C}(N) \cap \mathcal{C}(B) = \varnothing$. We refer to the version of $\psi = \psi^{s\mathcal{G},N}$ defined using B instead of A as the "B-version of ψ." By inverse-directedness [Lemma 27.3] there is a module L such that $\mathcal{C}(L) = \mathcal{C}(A) = \mathcal{C}(B)$, and such that $\xi^{L,A}$ and $\xi^{L,B}$ are defined. If we can show that the L-version of ψ equals the A-version, then the L-version of ψ (by similar reasoning) also equals the B-version, and we will be done. Therefore, without loss of generality, we can assume that $\xi^{B,A}$ is defined.

Now since $\mathcal{C}(B) = \mathcal{C}(A)$, and $\xi^{B,A}$ is defined, it follows that $\operatorname{istab}(B) \subseteq \operatorname{istab}(A)$ [Theorem 27.1(i)]. Then $\operatorname{istab}(N \oplus B) = \operatorname{istab}(N)\operatorname{istab}(B) \subseteq \operatorname{istab}(N)\operatorname{istab}(A) = \operatorname{istab}(N \oplus A)$ [Corollary 24.7(ii)]. Since $\mathcal{C}(N \oplus B) = \mathcal{C}(N \oplus A)$, it follows again from Theorem 27.1(i) that $\xi^{(N\oplus B),(N\oplus A)}$ is also defined.

It now suffices to show that the two vertical paths from $s\mathcal{G}$ to $\mathcal{G}(N)$ in the following diagram describe the same homomorphism. For this, it suffices to show that the upper and lower squares commute.

$$\begin{array}{ccc}
s\mathcal{G} & \xrightarrow{\quad 1 \quad} & s\mathcal{G} \\
\big\downarrow \xi^{s\mathcal{G},(N\oplus B)} & & \big\downarrow \xi^{s\mathcal{G},(N\oplus A)} \\
\mathcal{G}(N \oplus B) & \xrightarrow{\xi^{(N\oplus B),(N\oplus A)}} & \mathcal{G}(N \oplus A) \\
\big\downarrow \operatorname{proj} & & \big\downarrow \operatorname{proj} \\
\mathcal{G}(N) & \xrightarrow{\quad 1 \quad} & \mathcal{G}(N)
\end{array}$$

Note that the maps $\xi^{s\mathcal{G},(N\oplus B)}$ and $\xi^{s\mathcal{G},(N\oplus A)}$ are just projection maps from the inverse limit that defines $s\mathcal{G}$ [Proposition 27.8]. Thus, given $[\boldsymbol{A}] = \{[A_M]\}_{M \in \mathcal{F}} \in s\mathcal{G}$, we have $\xi^{s\mathcal{G},(N\oplus B)}[\boldsymbol{A}] = [A_{N\oplus B}]$ and $\xi^{s\mathcal{G},(N\oplus A)}[\boldsymbol{A}] = [A_{N\oplus A}]$. But since $[\boldsymbol{A}] \in s\mathcal{G}$ and $\xi^{(N\oplus B),(N\oplus A)}$ is defined, it follows that $\xi^{(N\oplus B),(N\oplus A)}[A_{N\oplus B}] = [A_{N\oplus A}]$ [Definition 27.5], so the top square commutes.

For the bottom square, suppose that $\xi^{(N\oplus B),(N\oplus A)}[N' \oplus B'] = [N'' \oplus A']$ for some $N', N'' \in \operatorname{genus}(N)$, $A' \in \operatorname{genus}(A)$, and $B' \in \operatorname{genus}(B)$. Then, by Definition 25.4 of the ξ-maps, $(N \oplus B) \oplus (N'' \oplus A') \cong (N \oplus A) \oplus (N' \oplus B')$. Therefore by Lemma 28.1, we get that $N \oplus N'' \cong N \oplus N'$. But then by cancellation within $\operatorname{genus}(N)$ [Lemma 9.2(ii)], it follows that $N'' \cong N'$, which shows that the bottom square commutes as well.

Since we can take $A = 0$ if N is faithful, the supplementary statement follows as well. □

The final result in this section completes our proof that the super Mayer-Vietoris sequence maps onto every Mayer-Vietoris sequence in a natural way.

THEOREM 28.3. *The following diagram commutes for every* $N \in \mathrm{fingen}_\infty(\Lambda)$ *(faithful or not).*

$$\text{(28.3.1)} \qquad \begin{array}{ccccccc} 0 \to & \mathrm{rs}\mathcal{G} & \xrightarrow{\mu^{\mathrm{s}\mathcal{G}}} & \mathrm{s}\mathcal{G} & \xrightarrow{\xi_{\Lambda,\Gamma}^{\mathrm{s}\mathcal{G},\Gamma}} & \mathcal{G}_\Gamma(\Gamma) & \to 0 \\ & \Big\downarrow{\scriptstyle \psi^{\mathrm{s}\mathcal{G},N}} & & \Big\downarrow{\scriptstyle \psi^{\mathrm{s}\mathcal{G},N}} & & \Big\downarrow{\scriptstyle \mathrm{proj}} & \\ 0 \to & \mathrm{r}\mathcal{G}(N) & \xrightarrow{\mu^{N}} & \mathcal{G}(N) & \xrightarrow{\xi_{\Lambda,\Gamma}^{N,\Gamma_{\mathcal{C}(N)}}} & \mathcal{G}_\Gamma(\Gamma_{\mathcal{C}(N)}) & \to 0 \end{array}$$

Here, the map $\psi^{\mathrm{s}\mathcal{G},N}$ *is as defined in Lemma and Definition 28.2, and* proj *is the natural projection map. All three vertical maps are surjective, and the vertical maps* ψ *equal* ξ *if* N *is faithful.*

PROOF. Let $A \in \mathrm{fingen}_\infty(\Lambda)$ be such that $N \oplus A$ is faithful and $\mathcal{C}(N) \cap \mathcal{C}(A) = \varnothing$, and consider the following diagram, whose top and bottom rows are the same as in (28.3.1). Each map "res" is the restriction of the map immediately to its right; and each map "proj" is the natural projection map, which exists by Lemma 28.1.

$$\text{(28.3.2)} \qquad \begin{array}{ccccccc} 0 \to & \mathrm{rs}\mathcal{G} & \xrightarrow{\mu^{\mathrm{s}\mathcal{G}}} & \mathrm{s}\mathcal{G} & \xrightarrow{\xi_{\Lambda,\Gamma}^{\mathrm{s}\mathcal{G},\Gamma}} & \mathcal{G}_\Gamma(\Gamma) & \to 0 \\ & \Big\downarrow{\scriptstyle \mathrm{res}} & & \Big\downarrow{\scriptstyle \xi^{\mathrm{s}\mathcal{G},(N\oplus A)}} & & \Big\downarrow{\scriptstyle 1} & \\ 0 \to & \mathrm{r}\mathcal{G}(N \oplus A) & \xrightarrow{\mu^{N\oplus A}} & \mathcal{G}(N \oplus A) & \xrightarrow{\xi_{\Lambda,\Gamma}^{(N\oplus A),\Gamma}} & \mathcal{G}_\Gamma(\Gamma) & \to 0 \\ & \Big\downarrow{\scriptstyle \mathrm{res}} & & \Big\downarrow{\scriptstyle \mathrm{proj}} & & \Big\downarrow{\scriptstyle \mathrm{proj}} & \\ 0 \to & \mathrm{r}\mathcal{G}(N) & \xrightarrow{\mu^{N}} & \mathcal{G}(N) & \xrightarrow{\xi_{\Lambda,\Gamma}^{N,\Gamma_{\mathcal{C}(N)}}} & \mathcal{G}_\Gamma(\Gamma_{\mathcal{C}(N)}) & \to 0 \end{array}$$

The two vertical arrows under $\mathrm{s}\mathcal{G}$ form a copy of diagram (28.2.1), and their composition is therefore $\psi^{\mathrm{s}\mathcal{G},N}$, which is a surjection. The top two squares commute, by Super Mayer-Vietoris consistency [Proposition 27.8]. Hence, we will be done if we show that the two lower squares commute. Commutativity of the lower left square is obvious; proof of commutativity of the lower right square is a minor variation of commutativity of the right-hand square in (27.2.1) (Mayer-Vietoris consistency), so we omit the details. □

CHAPTER 7

Direct Sums

This chapter completes our proof of the fact that the non-local part of the description of direct-sum behavior of Λ-modules is governed by local and group-theoretic considerations. This was already done for direct-sum cancellation [Section 26], which we had to do in an earlier section because it was needed in the development of the super genus class group. We formalize this notion in Section 29.

Let $M \in \text{fingen}_\infty(\Lambda)$. Section 30 examines the questions of how $\mathcal{G}(M)$ is related to the genus class groups of direct summands X of M, and when a module locally isomorphic to a direct summand of M is itself isomorphic to a direct summand of M. Let $X \oplus Y$. We prove that $\text{im}(\xi^{X,M})$ and $\text{im}(\xi^{Y,M})$ are always direct summands of $\mathcal{G}(M)$. But these summands can overlap. The extreme situation is the case that X, Y, and M are faithful. Here $\xi^{X,M}$ and $\xi^{Y,M}$ are both surjections [Lemma 25.6 and Corollary 25.9], and hence $\text{im}(\xi^{X,M}) = \text{im}(\xi^{Y,M})$. In the general situation we prove that $\text{im}(\xi^{X,M}) \cap \text{im}(\xi^{Y,M})$ is a direct summand of $\mathcal{G}(M)$ and can be identified with $\mathcal{G}(B)$ for some B such that $\mathcal{C}(B) = \mathcal{C}(X) \cap \mathcal{C}(Y)$.

Section 31 gives the simplest nontrivial examples of what can happen.

29. General Results

Viewing direct-sum structure via genus class groups and ξ-maps involves making a choice of base point in each genus class group, to serve as its zero element. This leads to two possible points of view. (i) Arbitrarily choose a base point in each genus, yielding one genus class group $\mathcal{G}(M)$ for each genus. This gives a good overall view, but probably cannot actually be carried out. (ii) When studying a specific direct-sum problem, one only needs to choose base points for the genera that actually occur, and natural choices sometimes exist. (See, for example, Corollary 29.2 and Example 31.3.)

Our starting point is the following simple result. Informally, it expresses the isomorphism class of a direct sum $\oplus_i A'_i$, where the terms come from possibly distinct genera, say $[A'_i] \in \mathcal{G}(A_i)$ as the sum of the "natural images" of the summands $[A'_i]$ in the direct-sum genus class group, plus a (unpleasant) correction term that is due to the arbitrary way in which base points are chosen. The 2-summand case of this is illustrated in diagram (2.2.4). Here, and in a few similar places, there is no advantage to working in $\text{fingen}_\infty(\Lambda)$.

THEOREM 29.1. *Let $A_1, \ldots, A_m \in \text{fingen}(\Lambda)$ and $M \in \text{genus}(\oplus_i A_i)$. Then for $A'_i \in \text{genus}(A_i)$ $(i = 1, \ldots, m)$ we have:*

$$[\oplus_i A'_i] = \sum_i \xi^{A_i,M}[A'_i] \oplus [M_0] \qquad \text{in } \mathcal{G}(M). \tag{29.1.1}$$

where the "correction term" $[M_0]$ depends only on the base points $A_1, \ldots, A_m, M$, and not on $A'_1, \ldots, A'_m$.

PROOF. Let $\xi^{A_i,M}[A_i'] = [M_i']$, and hence $M_i' \in \text{genus}(M)$. Then $A_i \oplus M_i' \cong A_i' \oplus M$ for each i. Direct-summing over all i yields

$$(\oplus_i A_i) \oplus (\oplus_i M_i') \cong (\oplus_i A_i') \oplus M^{(m)}$$

Let $M_0 = \oplus_i A_i$. Then since each of M_0, M_i', and $\oplus_i A_i'$ is an element of genus(M), and $[M] = 0 \in \mathcal{G}(M)$, formula (29.1.1) now follows from (25.2.1). □

The presence of the correction term seems unavoidable. For example, if we choose the base point M to be $\oplus_i A_i$, then the correction term becomes zero, giving the more pleasant formula on the left-hand side of implication (29.2.1). On the other hand, if we choose any other base point for genus$(\oplus_i A_i)$, we get $[M_0] \neq 0$, by (29.2.1) again. However, it is conceivable that one can choose the whole set of base points to be closed under direct sums. [See Problem 6 at the end of this memoir.]

When studying a specific direct-sum problem, one can often choose the base point for the direct-sum genus so that the formula on the left-hand side of (29.2.1) holds (as we do several times in what follows).

COROLLARY 29.2. *Keep the notation of Theorem 29.1. Then* $[M_0] = [\oplus_i A_i]$. *In particular,*

(29.2.1) $$[\oplus_i A_i'] = \sum_i \xi^{A_i,M}[A_i'] \quad \textit{in } \mathcal{G}(M) \iff \oplus_i A_i \cong M$$

PROOF. The value of $[M_i]$ in the first assertion is what we used in the proof of Theorem 29.1. But it also follows from (29.1.1) if we take every $A_i' = A_i$. For then $[A_i'] = 0$ in $\mathcal{G}(A_i)$, and hence $\xi^{A_i,M}[A_i'] = 0$ in $\mathcal{G}(M)$. □

REMARK 29.3. Theorem 29.1 shows that *all direct-sum behavior is determined by local and group-theoretical considerations.*

To make this connection more explicit, consider two direct sums $M' = \oplus_i A_i'$ and $N' = \oplus_j B_j'$ in fingen(Λ). How can we tell whether $M' \cong N'$? The first step is to determine whether genus(M') = genus(N'), a local question. If in fact genus(M') = genus(N'), then we can select base points M, A_i, B_i in the genera of M', A_i', B_i', respectively, after which two applications of formula (29.1.1) determine whether or not $[M'] = [N']$ in $\mathcal{G}(M)$.

Even if $M \not\cong \oplus_i A_i$ the following "comparison" version of Theorem 29.1 holds.

COROLLARY 29.4. *Let* $M \in \text{genus}(\oplus_i A_i)$ *and* $A_i', A_i'' \in \text{genus}(A_i)$. *Then:*

(29.4.1) $$\oplus_i A_i' \cong \oplus_i A_i'' \iff \sum_i \xi^{A_i,M}[A_i'] = \sum_i \xi^{A_i,M}[A_i''] \quad \textit{in } \mathcal{G}(M).$$

PROOF. This holds since two applications of Theorem 29.1 yields an equation with the same correction term on both sides; and hence it drops out. □

REMARKS 29.5 (All decompositions of M). Let $M \in \text{fingen}(\Lambda)$ be given. We observe that the present memoir contains machinery sufficient to find all decompositions $M = \oplus_i M_i$ (with each $M_i \neq 0$), although the procedure can be quite difficult to carry out in specific cases.

Call two decompositions $M = \oplus_{i=1}^m A_i$ and $M = \oplus_{j=1}^n B_j$ *genus-equivalent* if $m = n$ and, after a suitable permutation of the summands, every $B_i \in \text{genus}(A_i)$. The module M has only finitely many genus-inequivalent decompositions, because it has only finitely many genera of direct summands (even when singspec(Λ) is infinite), by Theorem 16.11. Moreover, finding a full set of genus-inequivalent decompositions of M is a finite combinatorial problem using Package Deal Theorem 15.6 for completions, because the Krull-Schmidt theorem holds when Λ is complete. To actually carry out this procedure, one must know the decomposition, into indecomposables, of the finitely many nonfree $\hat{M}_{\mathfrak{m}}$.

Fixing a particular decomposition $M = \oplus_{i=1}^{m} A_i$, the question remains as to which direct sums $N = \oplus_i A_i'$, with each $A_i' \in \text{genus}(A_i)$, satisfy $N \cong M$. Theorem 29.1 and its corollary give a criterion, in terms of genus class groups and ξ-maps, for $N \cong M$, namely $\sum_{i=1}^{n} \xi^{A,M}[A_i'] = 0$ in $\mathcal{G}(M)$.

30. Decomposition of $\mathcal{G}(M)$; relation to direct summands of M

In this section we study decompositions of $\mathcal{G}(M) = \oplus_i \mathcal{G}(B_i)$ where the modules B_i are simpler than M in the sense that their torsionfree support sets $\mathcal{C}(B_i)$ are smaller than $\mathcal{C}(M)$ and mutually disjoint. We show that, for every direct summand X of M, $\text{im}(\xi^{X,M})$ is a direct summand of $\mathcal{G}(M)$, giving one realization of this in Theorem 30.9 and another in Theorem 30.11. At the end of the section we consider when a module X that is locally a direct summand of a torsionfree module M is an actual direct summand, showing that an obviously necessary condition is also sufficient.

To keep notation from getting out of hand, we make more "natural identifications" (always in terms of our natural ξ-maps) than in the rest of this memoir. Our first lemma gives the simplest type of disjoint-torsionfree-support decomposition.

LEMMA 30.1. *Let $B = B_1 \oplus \ldots \oplus B_n \in \text{fingen}_\infty(\Lambda)$, and $\mathcal{C}(B_i) \cap \mathcal{C}(B_j) = \varnothing$ when $i \neq j$. Then we have a canonical identification (group isomorphism):*

$$\oplus_i \mathcal{G}(B_i) = \mathcal{G}(\oplus_i B_i) \quad \textit{via}$$

$$([B_1'], [B_2'], \ldots, [B_n']) \to [B_1' \oplus B_2' \oplus \ldots \oplus B_n'] \tag{30.1.1}$$

This isomorphism equals $\sum_i \xi^{B_i,B}$.

PROOF. We have $\xi^{B_1,B}[B_1'] = [B_1' \oplus B_2 \oplus \ldots \oplus B_n]$ by Corollary 25.9(i). A similar formula holds for each other $\xi^{B_i,B}$. The fact that $\sum_i \xi^{B_i,B}[B_i'] = [B_1' \oplus \ldots \oplus B_n']$ then follows from Lemma 25.2.

Because of the disjoint torsionfree support, every $B' \in \text{genus}(\oplus_i B_i)$ has a unique (up to isomorphism) decomposition of the form $B = \oplus_i B_i'$ with each $B_i' \in \text{genus}(B_i)$ [Lemmas 9.3 and 28.1]; and this shows that $\sum_i \xi^{B_i,B}$ is a bijection. □

DEFINITION 30.2 (eqstab-equivalent). We call $M, B \in \text{fingen}_\infty(\Lambda)$ *eqstab-equivalent* if $\text{eqstab}(M) = \text{eqstab}(B)$ and $\mathcal{C}(M) = \mathcal{C}(B)$. This condition is useful because it is complete-locally verifiable [Lemmas 15.2 and 24.3], it is a sufficient condition for $\xi^{B,M}$ and $\xi^{M,B}$ to be mutually inverse isomorphisms [Corollary 26.8 and Lemma 25.5], and because suitable choices of B yield more useful decompositions of $\mathcal{G}(M)$ than decompositions of M itself.

When $\mathcal{G}(B) \cong \mathcal{G}(M)$ via $\xi^{B,M}$ we often write this isomorphism as equality.

NOTATION 30.3 ($e(\mathcal{D})$). For a subset $\mathcal{D} \subseteq \mathcal{C}(\Gamma)$ we define $e(\mathcal{D}) = \sum\{1_h \in \mathcal{D}\}$ where 1_h denotes the identity element of Γ_h. Thus $e(\mathcal{D})$ is an idempotent element of Γ. If M is torsionfree, then $M \subseteq M_Q$, and hence we can regard the product $e(\mathcal{D}) \cdot M$ as a Λ-submodule of M_Q. It is not necessarily contained in M.

All of our decompositions of $\mathcal{G}(M)$ will arise from disjoint-torsionfree-support decompositions of some B that is eqstab-equivalent to M, as in the next lemma.

LEMMA 30.4. *Let M and $B_1 \oplus \ldots B_n = B$ be* eqstab-*equivalent modules in* $\text{fingen}_\infty(\Lambda)$ *such that the torsionfree support sets $\mathcal{C}(B_i)$ are disjoint. Then there is an identification (group isomorphism)*

$$\mathcal{G}(B_1) \oplus \ldots \oplus \mathcal{G}(B_n) = \mathcal{G}(M) \qquad \textit{via} \quad \sum_{i=1}^{n} \xi^{B_i,M} = \xi^{B,M} \tag{30.4.1}$$

Replacing any B_i within its eqstab-*equivalence class leaves leaves the subgroup $\mathcal{G}(B_i)$ of M in* (30.4.1) *(i.e.* $\text{im}(\xi^{B_i,M})$*) unchanged, and also leaves the* eqstab-*equivalence class of B unchanged.*

Suppose M is torsionfree, and let $e_i = e\big(\mathcal{C}(B_i)\big)$. Then we can replace each B_i by $B_i = e_i M$, and the isomorphism $\xi^{M,B}$ is then given by:

$$\xi^{M,B}[M'] = [e_1 M' \oplus \ldots \oplus e_n M'] \tag{30.4.2}$$

PROOF. Since M and B are eqstab-equivalent, $\xi^{B,M}$ and $\xi^{M,B}$ are mutually inverse isomorphisms. We have the inclusions $\text{eqstab}(B_i) \subseteq \text{eqstab}(B) = \text{eqstab}(M)$ [the first of these by Corollary 24.7] and $\mathcal{C}(B_i) \subseteq \mathcal{C}(B) = \mathcal{C}(M)$. Therefore we have the transitivity formulas $\xi^{B,M}\xi^{B_i,B} = \xi^{B_i,M}$ [Proposition 26.10]. Summing on i and using the identification in Lemma 30.1 yields (30.4.1).

Replacement of B_i within its eqstab-*equivalence class.* Let A_i be eqstab-equivalent to B_i, so that ξ^{A_i,B_i} is an isomorphism. As before, we have the transitivity formula $\xi^{B_i,M}\xi^{A_i,B_i} = \xi^{A_i,M}$; and this shows that $\text{im}(\xi^{A_i,M}) = \text{im}(\xi^{B_i,M})$ as desired. The fact that replacement of B_i by A_i does not change the eqstab-equivalence class of B is part of Corollary 24.8.

Torsionfree case. Note that each $\mathcal{C}(B_i) = \mathcal{C}(e_i M)$, and recall that B is eqstab-equivalent to M. Therefore the fact that both decompositions $B = \oplus_i B_i$ and $M = \oplus_i e_i M$ are disjoint-torsionfree-support decompositions implies that each B_i is eqstab-equivalent to $e_i M$ [Corollary 24.8].

To get formula (30.4.2), choose $M' \in \text{genus}(M)$. Then $\xi^{M,B}[M'] = [\oplus_i B'_i]$ for suitable modules $B'_i \in \text{genus}(B_i)$ [Lemma 9.3]. This implies the first isomorphism in (30.4.3).

$$M \oplus (\oplus_i B'_i) \cong M' \oplus B \qquad e_i M \oplus B'_i \cong e_i M' \oplus B_i \tag{30.4.3}$$

Since we are dealing with torsionfree modules, we can extend this Λ-isomorphism to a Λ_Q-isomorphism of the Q-localization of both sides. Multiplying by e_i then gives the second isomorphism in (30.4.3). Since $B_i = e_i M$ and direct-sum cancellation holds in every genus of Λ-modules [Lemma 9.2], the second isomorphism in (30.4.3) implies $B'_i \cong e_i M'$, as desired. □

DEFINITION 30.5 (connect, totally connect). Let $M \in \text{fingen}_\infty(\Lambda)$, $\mathfrak{m}$ a strictly split maximal ideal of Λ, and Γ_h and Γ_k the two coordinate rings of Γ supported by $\mathfrak{m}$. Recall that we say that $\mathfrak{m}$ *connects* Γ_h to Γ_k (in M) if $\hat{M}_\mathfrak{m}$ has an indecomposable direct summand of rank (1,1) with respect to the two coordinate rings $(\Gamma_h)\hat{}_\mathfrak{m}$ and $(\Gamma_k)\hat{}_\mathfrak{m}$ of $\hat{\Gamma}_\mathfrak{m}$ [Definition 16.5].

We say that $\mathfrak{m}$ *totally connects* Γ_h to Γ_k (in M), or *the connection is total* if $\mathfrak{m}$ connects Γ_h to Γ_k (in M) and $\hat{M}_\mathfrak{m}$ is a direct sum of indecomposable modules of rank (1,1).

LEMMA 30.6. *Let $M \in \operatorname{fingen}_\infty(\Lambda)$ and let $\mathfrak{m}$ be one of the finitely many strictly split maximal ideal of Λ that connect but do not totally connect coordinate rings Γ_h and Γ_k of Γ (in M). Then there exists $B \in \operatorname{fingen}_\infty(\Lambda)$ such that:*

(i) *$\mathfrak{m}$ does not connect Γ_h to Γ_k (in B);*

(ii) *The Γ_h and Γ_k-ranks of B equal the corresponding ranks of M; and*

(iii) *$\hat{B}_\mathfrak{n} \cong \hat{M}_\mathfrak{n}$ for all maximal ideals $\mathfrak{n} \neq \mathfrak{m}$ of Λ.*

Every such B is eqstab-*equivalent to M.*

PROOF. (First recall that only finitely many maximal ideals of Λ are strictly split [Proposition 11.4].) Let the Γ_h- and Γ_k-ranks of M be a and b respectively. Then the rank of $\hat{M}_\mathfrak{m}$ at its two coordinate rings $\hat{\Gamma}_h$ and $\hat{\Gamma}_k$ is (a, b). By Package Deal Theorem 15.6 for completions, there exists $B \in \operatorname{fingen}(\Lambda)$ such that $\hat{B}_\mathfrak{m} \cong (\Gamma_h)\hat{}_\mathfrak{m}{}^{(a)} \oplus (\Gamma_k)\hat{}_\mathfrak{m}{}^{(b)}$ and $\hat{B}_\mathfrak{n} \cong \hat{M}_\mathfrak{n}$ whever $\mathfrak{n} \neq \mathfrak{m}$, because this changes the isomorphism class of only one completion of M and leaves all ranks of M unchanged. Since a finitely generated Λ-module is in $\operatorname{fingen}_\infty(\Lambda)$ if and only if this is true of all of its completions at maximal ideals [Proposition 7.3], B is again in $\operatorname{fingen}_\infty(\Lambda)$. Thus (i)–(iii) hold.

Since passing from M to B does not change any ranks of Γ, we have $\mathcal{C}(B) = \mathcal{C}(M)$. To complete the proof that $\operatorname{eqstab}(B) = \operatorname{eqstab}(M)$, it suffices, in view of (iii), to check that $\operatorname{eqstab}(\hat{B}_\mathfrak{m}) = \operatorname{eqstab}(\hat{M}_\mathfrak{m})$. Recall the following facts [Theorem 24.5]:

(30.6.1) (Since $\mathfrak{m}$ is strictly split:) Each indecomposable direct summand of $\hat{M}_\mathfrak{m}$ has rank (1,1), (1,0) or (0,1). The eqstab in each case is respectively $k(\mathfrak{m})^\times$ [the diagonal copy of $k(\mathfrak{m})^\times$ in $k(\mathfrak{m})_1 \times k(\mathfrak{m})_2$], or $k(\mathfrak{m})_1^\times \times 1$, or $1 \times k(\mathfrak{m})_2^\times$. Moreover, the eqstab of a direct sum is the product of the eqstabs of the summands.

Since Γ_h is connected, but not totally connected, to Γ_k by $\mathfrak{m}$ (in M), any decomposition of $\hat{M}_\mathfrak{m}$ into a direct sum of indecomposables must contain at least one summand of rank (1,1), and at least one summand of one of the other two types. We conclude from (30.6.1) that a and b are both nonzero and $\operatorname{eqstab}(\hat{M}_\mathfrak{m}) = k(\mathfrak{m})_1^\times \times k(\mathfrak{m})_2^\times$.

Recall that $\hat{B}_\mathfrak{m} \cong (\Gamma_h)\hat{}_\mathfrak{m}{}^{(a)} \oplus (\Gamma_k)\hat{}_\mathfrak{m}{}^{(b)}$. Therefore, again by (30.6.1), we have $\operatorname{eqstab}(\hat{B}_\mathfrak{m}) = k(\mathfrak{m})_1^\times \times k(\mathfrak{m})_2^\times$, and so the lemma is proved. □

The next lemma yields the disjoint-torsionfree-support decompositions needed in all of the main results of this section.

LEMMA 30.7. *Let $\mathcal{C}_1, \ldots, \mathcal{C}_n$ be a partition of $\mathcal{C}(M)$ for some $M \in \operatorname{fingen}_\infty(\Lambda)$. Suppose that, for each two distinct sets $\mathcal{C}_i \neq \mathcal{C}_j$ and each maximal ideal $\mathfrak{m}$ that connects some $\Gamma_h \in \mathcal{C}_i$ to some $\Gamma_k \in \mathcal{C}_j$ (in M), this connection is not total. Then there is a decomposition*

$$\mathcal{G}(B_1) \oplus \ldots \oplus \mathcal{G}(B_n) = \mathcal{G}(M) \qquad \Big(\textit{isomorphism via } \sum_i \xi^{B_i, M}\Big) \tag{30.7.1}$$

where $\oplus_i B_i = B$ is eqstab-*equivalent to M and each $\mathcal{C}(B_i) = \mathcal{C}_i$.*

PROOF. By repeated use of Lemma 30.6, there exists B, eqstab-equivalent to M, such that if $\Gamma_h \in \mathcal{C}_i$ and $\Gamma_k \in \mathcal{C}_j$ with $i \neq j$, then Γ_h is not connected to Γ_k (in M). We claim that this implies that there is a decomposition $B = B_1 \oplus \ldots \oplus B_n$ such that each $\mathcal{C}(B_i) = \mathcal{C}_i$. Such a decomposition can be found by repeated use of (16.6.1): First get a decomposition $B = B_1 \oplus D$ such that $\mathcal{C}(B_1) = \mathcal{C}_1$ and $\mathcal{C}(D) = \cup_{i\geq 2}\mathcal{C}_i$. Then get a decomposition $D = B_2 \oplus D'$ such that $\mathcal{C}(B_2) = \mathcal{C}_2$ and $\mathcal{C}(D') = \cup_{i\geq 3}\mathcal{C}_i$, and so on until we get the desired decomposition of B. □

THEOREM 30.8 (Equal-ranks decomposition). *Let* $M \in \text{fingen}_\infty(\Lambda)$. *Then there is a decomposition*

$$\mathcal{G}(B_{r(1)}) \oplus \ldots \oplus \mathcal{G}(B_{r(n)}) = \mathcal{G}(M) \qquad \Big(\textit{isomorphism via } \sum_i \xi^{B_{r(i)},M}\Big) \tag{30.8.1}$$

such that:

(i) $r(1) < \cdots < r(n)$;

(ii) *Each* $\mathcal{C}(B_{r(i)})$ *is the set of* Γ_h *at which* M *has* Γ_h*-rank* $r(i)$, *and the sets* $\mathcal{C}(B_{r(i)})$ *form a partition of* $\mathcal{C}(M)$

(iii) *The module* $\oplus_i B_{r(i)} = B$ *is* eqstab*-equivalent to* M.

PROOF. For each $r \geq 1$ let $\mathcal{C}_r$ be the set of all coordinate rings Γ_h of Γ such that the Γ_h-rank of M is r. Only finitely many of these sets can be nonempty, since M is finitely generated, and the collection of those $\mathcal{C}_r$ that are nonempty clearly form a partition of $\mathcal{C}(M)$. The remaining conclusions of the theorem follow immediately from Lemma 30.7, as soon as we prove the following fact: If the Γ_h- and Γ_k-ranks of M are $r \neq s$ respectively, and if $\mathfrak{m}$ is a strictly split maximal ideal connecting Γ_h to Γ_k (in M), then the connection is not total.

Since ranks are complete-locally determined [Lemma 15.2], the rank of $\hat{M}_\mathfrak{m}$ with respect to its two coordinate rings is (r, s). But if the connection were total, $\hat{M}_\mathfrak{m}$ would be a direct sum of modules of rank (1,1), and we would have the contradiction $r = s$. □

THEOREM 30.9 (Overlapping-support decomposition). *Let* $M = X \oplus Y \in \text{fingen}_\infty(\Lambda)$. *Then there is a decomposition*

$$\mathcal{G}(B_{10}) \oplus \mathcal{G}(B_{11}) \oplus \mathcal{G}(B_{01}) = \mathcal{G}(M) \quad \Big(\textit{isomorphism via } \sum_{i,j} \xi^{B_{ij},M}\Big) \tag{30.9.1}$$

where $B_{10} \oplus B_{11} \oplus B_{01} = B$ *is* eqstab*-equivalent to* M, $\mathcal{C}(B_{10}) = \mathcal{C}(X) - \mathcal{C}(Y)$, $\mathcal{C}(B_{11}) = \mathcal{C}(X) \cap \mathcal{C}(Y)$, *and* $\mathcal{C}(B_{01}) = \mathcal{C}(Y) - \mathcal{C}(X)$.

For any such decompostion of $\mathcal{G}(M)$ *we have* $\text{im}(\xi^{X,M}) = \mathcal{G}(B_{10}) \oplus \mathcal{G}(B_{11})$. *(But* $\xi^{X,M}$ *is not necessarily one-to-one.)*

PROOF. *Existence of the decomposition.* Note that that the sets $\mathcal{C}_{10} = \mathcal{C}(X) - \mathcal{C}(Y)$, $\mathcal{C}_{11} = \mathcal{C}(X) \cap \mathcal{C}(Y)$, and $\mathcal{C}_{01} = \mathcal{C}(Y) - \mathcal{C}(X)$ form a partition of $\mathcal{C}(M)$. Therefore it suffices to show that if $\mathfrak{m}$ is a strictly split maximal ideal connecting coordinate rings Γ_h to Γ_k (in M) and these coordinate rings belong to distinct sets $\mathcal{C}_{ij}$, then the connection is not total [Lemma 30.7].

If this is a total connection, then $\hat{M}_\mathfrak{m}$ is a direct sum of indecomposable modules, each of rank (1,1). Since the Krull-Schmidt theorem holds over complete local rings, this shows that $\hat{M}_\mathfrak{m}$ has no direct summands of rank $(a, 0)$ or $(0, b)$, for nonzero a, b.

Since Γ_h and Γ_k belong to distinct sets $\mathcal{C}_{ij}$, we may assume that $\Gamma_h \in \mathcal{C}_{10}$ (otherwise interchange the roles of X and Y). Then the Γ_h-rank of X is nonzero; and since ranks are complete-locally determined [Lemma 24.3], the $(\Gamma_h)\hat{}_\mathfrak{m}$-rank of $\hat{X}_\mathfrak{m}$ is nonzero. On the other hand, since $\Gamma_h \in \mathcal{C}_{10}$, the Γ_k-rank of X is zero, and hence the $(\Gamma_k)\hat{}_\mathfrak{m}$-rank of $\hat{X}_\mathfrak{m}$ is zero. Therefore the direct summand $\hat{X}_\mathfrak{m}$ of $\hat{M}_\mathfrak{m}$, has an indecomposable direct summand of rank $(1,0)$. This shows that $\mathfrak{m}$ does not totally connect Γ_h to Γ_k (in M), as desired.

Image of $\xi^{X,M}$. This ξ-map is defined because X is a direct summand of M [Corollary 25.9]. Let $B_1 = B_{10} \oplus B_{11}$. Then we can make the identification $\mathcal{G}(B_{01} \oplus B_{11}) = \mathcal{G}(B_1)$ in $\mathcal{G}(M)$ [Lemma 30.1]. Because of our identification of $\mathcal{G}(B_1)$ with a subgroup of $\mathcal{G}(M)$, it suffices to prove that ξ^{M,B_1} is defined and is a surjection, because of the transitivity formula $\xi^{X,M} = \xi^{B_1,M}\xi^{X,B_1}$ [Proposition 26.10].

Since $\mathcal{C}(X) = \mathcal{C}(B_1)$ it suffices to show that $\text{eqstab}(X) \subseteq \text{eqstab}(B_1)$ [Corollary 26.8 and Lemma 25.6]. This is easily checked at $\mathfrak{m}$-adic completions at singular maximal ideals, because $\text{eqstab}(X) \subseteq \text{eqstab}(M) = \text{eqstab}(B)$ and $B = B_1 \oplus B_{01}$ (decomposition with disjoint torsionfree support), both of which persist after $\mathfrak{m}$-adic completion [see (30.6.1)]. □

THEOREM 30.10 (Universal decomposition). *Let $M \in \text{fingen}_\infty(\Lambda)$. Then there is a decomposition*

$$(30.10.1) \qquad \mathcal{G}(B_1) \oplus \ldots \oplus \mathcal{G}(B_n) = \mathcal{G}(M) \qquad \Big(\textit{isomorphism via } \sum_i \xi^{B_i,M}\Big)$$

where $\oplus_i B_i = B$ is eqstab*-equivalent to M, the sets $\mathcal{C}(B_i)$ are disjoint, and whenever a strictly split maximal ideal $\mathfrak{m}$ of Λ connects coordinate rings Γ_h to Γ_k (in B), the connection is total (and Γ_h and Γ_k necessarily belong to the same $\mathcal{C}(B_i)$). We call any such decomposition of $\mathcal{G}(M)$ a "universal decomposition" of $\mathcal{G}(M)$.*

Any other universal decomposition of $\mathcal{G}(M)$ consists of exactly the same subgroups of $\mathcal{G}(M)$. [See (30.10.3) *for more precision here.]*

PROOF. Define the *total connections graph* $\mathcal{K}^{\text{tc}}(M)$ as follows. The vertices of $\mathcal{K}^{\text{tc}}(M)$ consist of one point, labeled Γ_h, for every Γ_h such that the Γ_h-rank of M is nonzero. An edge connects Γ_h to Γ_k if and only if some strictly split maximal ideal $\mathfrak{m}$ of Λ totally connects Γ_h to Γ_k (in M). We prove:

(30.10.2) If B is eqstab-equivalent to M, then $\mathcal{K}^{\text{tc}}(B) = \mathcal{K}^{\text{tc}}(M)$.

Part of the definition of eqstab-equivalence is that $\mathcal{C}(B) = \mathcal{C}(M)$. Thus we only need to prove that, for each pair of coordinate rings $\Gamma_h \neq \Gamma_k$ and every strictly split maximal ideal $\mathfrak{m}$ connecting Γ_h to Γ_k (in Γ), the connection is total in $\mathcal{K}^{\text{tc}}(B)$ if and only if it is total in $\mathcal{K}^{\text{tc}}(M)$. By hypothesis, $\text{eqstab}(B) = \text{eqstab}(M)$, and therefore $\text{eqstab}(\hat{B}_\mathfrak{m}) = \text{eqstab}(\hat{M}_\mathfrak{m})$. It therefore suffices to prove: $\hat{M}_\mathfrak{m}$ is a direct sum of indecomposable modules of rank (1,1) if and only if $\text{eqstab}(\hat{M}_\mathfrak{m}) = k(\mathfrak{m})^\times$. Thus the proof of (30.10.2) is completed by (30.6.1).

Existence of decomposition (30.10.1). Repeated use of Lemma 30.6 yields a module B, eqstab-equivalent to M, such that whenever a maximal ideal $\mathfrak{m}$ of Λ connects coordinate rings of Γ (in B), the connection is total. Let $\mathcal{C}_1, \ldots, \mathcal{C}_n$ denote the connected components of $\mathcal{K}^{\text{tc}}(M)$. Then Lemma 30.7 yields decomposition (30.10.1), with each $\mathcal{C}(B_i) = \mathcal{C}_i$.

Uniqueness statement. Let $\oplus_{i=1}^m A_i = A$ afford a universal decomposition of $\mathcal{G}(M)$ analogous to (30.10.1). We prove that $m = n$ and, after suitable renumbering, the following statements hold for all i.

(30.10.3) (a) A_i is eqstab-equivalent to B_i (and hence $\mathcal{G}(A_i) \cong \mathcal{G}(B_i)$ via ξ^{A_i,B_i}).
(b) $\text{im}(\xi^{A_i,M}) = \text{im}(\xi^{B_i,M})$.

By definition of "universal decomposition" we have $\text{eqstab}(A) = \text{eqstab}(M) = \text{eqstab}(B)$. Therefore $\mathcal{K}^{\text{tc}}(A) = \mathcal{K}^{\text{tc}}(M)$ by (30.10.2), and hence the connected components of these graphs are the same. In particular, $m = n$ and we can renumber the A_i so that each $\mathcal{C}(A_i) = \mathcal{C}(B_i)$.

We also have $\text{eqstab}(A) = \text{eqstab}(M) = \text{eqstab}(B)$ by the definition of "universal decomposition" of $\mathcal{G}(M)$. Since the terms in the decomposition of B have disjoint torsionfree support, and (by hypothesis) the same is true of the terms in the decomposition of A, we have that each $\text{eqstab}(A_i) = \text{eqstab}(B_i)$ [Corollary 24.8]. In other words, A_i is eqstab-equivalent to B_i, as claimed in (30.10.3)(a).

Since, in addition, we have $\mathcal{C}(B_i) \subseteq \mathcal{C}(M)$ and $\text{eqstab}(B_i) \subseteq \text{eqstab}(B) = \text{eqstab}(M)$ [obvious] we have the transitivity formula $\xi^{B_i,M}\xi^{A_i,B_i} = \xi^{A_i,M}$ [Proposition 26.10], and hence $\text{im}(\xi^{A_i,M}) = \text{im}(\xi^{B_i,M})$, as claimed in (30.10.3)(b). □

THEOREM 30.11. *Let $M \in \text{fingen}_\infty(\Lambda)$ and consider any universal decomposition (30.10.1) of $\mathcal{G}(M)$. If X any nonzero direct summand of M, then $\text{im}(\xi^{X,M})$ is the direct sum of all $\mathcal{G}(B_i)$ such that $\mathcal{C}(B_i) \subseteq \mathcal{C}(X)$.*

PROOF. First we claim that $\mathcal{C}(X)$ is the union (necessarily disjoint) of those $\mathcal{C}(B_i)$ that are contained in $\mathcal{C}(X)$.

Let $M = X \oplus Y$. In the notation of the overlapping-support decomposition of $\mathcal{G}(M)$ [Theorem 30.9], let $X' = B_{10} \oplus B_{11}$, so that $X' \oplus B_{01} = B$, and note that, in each of the two foregoing direct sums, the terms have disjoint torsionfree support. Take universal decompositions of $\mathcal{G}(X')$ and $\mathcal{G}(B_{01})$. The direct sum of all of the terms in these decompositions is a universal decomposition of $\mathcal{G}(B) = \mathcal{G}(M)$, since direct sums never introduce new "connections" between summands. Call the terms in the union of the terms in these last two direct sums B_i $(i = 1, 2, \ldots, n)$. Thus we have expressed $\oplus_i B_i = B$ as the direct sum of terms B_i, some with $\mathcal{C}(B_i) \subseteq \mathcal{C}(X)$, and some with $\mathcal{C}(B_i) \cap \mathcal{C}(X) = \varnothing$. Since $\mathcal{C}(B) = \mathcal{C}(X)$, the claim is proved. (This proof implicitly makes use of the uniqueness assertion about universal decompositions, in Theorem 30.9.)

In Theorem 30.9 we proved that $\text{im}(\xi^{X,M}) = \text{im}(\xi^{X',M})$; and so the present proof is complete. □

THEOREM 30.12 (Local vs. global summands). *Let X', M be torsionfree modules in $\text{fingen}_\infty(\Lambda)$, and let $e = \sum_h \{1_h \mid \Gamma_h\text{-rank}(X) = \Gamma_h\text{-rank}(M)\}$.*

Suppose that $X'_\mathfrak{m}$ is isomorphic to a direct summand of $M_\mathfrak{m}$ $(\forall \mathfrak{m} \in \text{maxspec}(\Lambda))$. Then X' is isomorphic to a direct summand of M if (and only if) $eX' \cong eM$ as Λ-modules.

PROOF. For the easy part of the proof suppose that $X' \oplus Y \cong M$ for some Y. This isomorphism induces, by localization, a Λ_Q-isomorphism $X'_Q \oplus Y_Q \cong M_Q$ that takes eX' onto eM, as desired.

Conversely, suppose that $eX' \cong eM$. Since X' is locally isomorphic to a direct summand of M, some $X \in \text{genus}(X')$ is an actual direct summand of M [Lemma 9.3]. Say $M = X \oplus Y$. We use the overlapping-support decomposition of $\mathcal{G}(M)$ furnished by $M = X \oplus Y$ in Theorem 30.9. Thus, in the notation of that theorem, we have $\mathcal{G}(M) = \mathcal{G}(B_{10}) \oplus \mathcal{G}(B_{11}) \oplus \mathcal{G}(B_{01})$ and $B = B_{10} \oplus B_{11} \oplus B_{01}$ is eqstab-equivalent to M and $\mathcal{C}(B_{10}) = \mathcal{C}(X) - \mathcal{C}(Y)$, $\mathcal{C}(B_{11}) = \mathcal{C}(X) \cap \mathcal{C}(Y)$, and $\mathcal{C}(B_{01}) = \mathcal{C}(Y) - \mathcal{C}(X)$.

Let $e_{ij} = e(\mathcal{C}(B_{ij}))$. Since M is torsionfree,we can take each $B_{ij} = e_{ij}M$, and the isomorphism $\xi^{M,B}$ is then given by [Lemma 30.4]:

$$\xi^{M,B}[M'] = [e_{10}M' \oplus e_{11}M' \oplus e_{01}M'] \tag{30.12.1}$$

We construct the desired Y'' such that $X' \oplus Y'' \cong M$ in three parts, corresponding to the three terms on the right-hand side of (30.12.1)

Since $e_{11}M \cong e_{11}X \oplus e_{11}Y$, the maps $\xi^{e_{11}X,e_{11}M}$ and $\xi^{e_{11}Y,e_{11}M}$ are both defined [Corollary 25.9], and since $\mathcal{C}(e_{11}M) = \mathcal{C}(e_{11}X) = \mathcal{C}(e_{11}Y)$, these maps are surjections [Lemma 25.6]. Therefore there exists $Y_{11} \in \text{genus}(e_{11}Y)$ such that

$$\xi^{e_{11}X,e_{11}M}[e_{11}X'] + \xi^{e_{11}Y,e_{11}M}[Y_{11}] = 0 \quad \text{in } \mathcal{G}(e_{11}M)$$

Since $e_{11}M \cong e_{11}X \oplus e_{11}Y$, this implies [Corollary 29.2]:

$$e_{11}X' \oplus Y_{11} \cong e_{11}M \tag{30.12.2}$$

As with $\mathcal{G}(M)$ we have $\mathcal{G}(Y) = \mathcal{G}(e_{11}Y \oplus e_{01}Y)$ via $[Y'] \to [e_{11}Y' \oplus e_{01}Y']$. Therefore we can choose $Y'' \in \text{genus}(Y)$ such that $e_{11}Y'' \cong Y_{11}$ and $e_{01}Y'' \cong e_{01}Y$.

It now suffices to prove that $X' \oplus Y'' \cong M$. Setting $M' = M$ in (30.12.1), we see that it suffices to check the three parts of this separately.

Since $e_{11}Y'' \cong Y_{11}$, the (1,1)-part of what we are checking is given by (30.12.2).

(1,0)-part. This is given by the hypothesis that $eX' \cong eM$, together with the fact that $e = e_{10}$.

(0,1)-part. This is given by our choice $e_{01}Y'' \cong e_{01}Y$ and the fact that $e_{01}Y \cong e_{01}M$ (because $e_{01}X = 0$). □

Problem. There is doubtless a version of this theorem that does not require M to be torsionfree. But, in order to have a simple enough path through the maze of ξ-maps to make the more general version useful in examples, one needs a simple substitute for formula (30.12.1), which no longer holds in the non-torsionfree case. [Problem 9 at the end of this memoir]

31. Examples: Simplest $\mathcal{G}(\Lambda)$; Direct-Sum Decompositions

In this section we give one of the two simplest examples we know of a nontrivial genus class group, and use it to illustrate the simplest nontrivial examples of overlapping-support decompositions and local versus global decompositions. (The other simplest example is Example 33.3.) Recall that singspec(Λ) is a finite set when $\Lambda\Gamma$ is finitely generated [Proposition 10.11].

LEMMA 31.1. *Suppose that* ${}_\Lambda\Gamma$ *is finitely generated, and let* $\mathfrak{m}_1, \ldots, \mathfrak{m}_n$ *be the finitely many singular maximal ideals of* Λ. *Then:*

(i) $\mathcal{I} = \oplus_{i=1}^n \bar{\Gamma}(\mathfrak{m}_i)^\times$ *(group of residue unit idèles),*
(ii) $\text{istab}({}_\Lambda\Lambda) = [\oplus_{i=1}^n k(\mathfrak{m}_i)^\times]\cdot \text{im}(\Gamma^\times)$ (im = *image in* $\mathcal{I}$), *and*
(iii) $\mathcal{I}/\,\text{istab}({}_\Lambda\Lambda) \cong \text{r}\mathcal{G}(\Lambda)$ *(as described in Theorem 25.14).*

(Recall that, when $\mathfrak{m}_i$ *is split, the notation* $k(\mathfrak{m}_i)$ *in (ii) denotes the diagonal copy of* $k(\mathfrak{m}_i)^\times$ *in* $\bar{\Gamma}(\mathfrak{m}_i) = k(\mathfrak{m}_i) \times k(\mathfrak{m}_i)$.*)*

PROOF. Statememt (i) is the definition of $\mathcal{I}$ in this simple situation [see (22.1.2)], and statement (iii) contains its own proof. Therefore, it suffices to prove statement (ii).

First recall the formula $\text{istab}(\Lambda) = \text{eqstab}(\Lambda)\cdot \text{im}(\Gamma^\times)$ [Theorem 23.5]. Thus, it suffices to show that $\text{eqstab}(\Lambda) = \oplus_{i=1}^n k(\mathfrak{m}_i)^\times$. But $k(\mathfrak{m})^\times = \text{eqstab}(\hat{\Lambda}_\mathfrak{m}) =$

eqstab($\Lambda_{\mathfrak{m}}$) [Theorem 24.5 and Lemma 24.3], and this is the $\mathfrak{m}_i$-component of eqstab(Λ) [Lemma 23.3]. □

The following is one of the simplest Dedekind-like rings with nontrivial $\mathcal{G}(\Lambda)$. For other Dedekind-like rings with explicitly computable $\mathcal{G}(\Lambda) = \mathrm{r}\mathcal{G}(\Lambda) \neq 0$, and which are algebras over fields, see Examples 33.3 and 34.1. In the latter example Λ has a unique singular maximal ideal $\mathfrak{m}$, and $\mathfrak{m}$ is unsplit. For an explicitly computable example where $\mathcal{G}(\Lambda) \neq \mathrm{r}\mathcal{G}(\Lambda) \neq 0$ see Example 36.2.

EXAMPLE 31.2 (Simplest nontrivial $\mathcal{G}(\Lambda)$). Let $\Gamma = \oplus_{h=1}^{n} \Gamma_h$ for some positive integer n, where each $\Gamma_h = \mathbb{Z}$ (the ring of integers). Let $p_1, \ldots, p_{n-1}$ be odd prime numbers such that each $p_h \neq p_{h+1}$, let $k_h = \mathbb{Z}/p_h\mathbb{Z}$ (residue field) for each index h, and let

$$\Lambda = \{\boldsymbol{z} = (z_1, \ldots, z_n) \in \Gamma \mid z_h \equiv z_{h+1} \pmod{p_h} \quad (\forall h < n)\}$$

Then Λ is a Dedekind-like ring with finite normalization Γ, Λ has exactly $n-1$ singular maximal ideals, namely the kernels $\mathfrak{m}_h$ of the natural maps $\boldsymbol{z} \to z_h + p\mathbb{Z} \in k_h$, and every $\mathfrak{m}_h$ is strictly split [Proposition 33.2].

Note that $\mathcal{G}(\Gamma)$ is trivial, so that $\mathcal{G}(\Lambda) = \mathrm{r}\mathcal{G}(\Lambda)$ [Theorem 25.14]. We claim that:

$$(31.2.1) \qquad \mathcal{G}(\Lambda) = \mathrm{r}\mathcal{G}(\Lambda) \cong \prod_{h=1}^{n-1} \frac{k_h^\times}{\{\pm 1\}} \quad \text{and} \quad |\mathcal{G}(\Lambda)| = \Big(\prod_{h=1}^{n-1} (p_h - 1)\Big) \Big/ 2^{n-1}$$

The proof of this claim is a straightforward computation involving Lemma 31.1 and the fact that the only units of $\mathbb{Z}$ are ± 1. The only step requiring care is the proof that $\mathrm{im}(\Gamma^\times)$ reduces each individual factor $k_h^\times$ modulo $\{\pm 1\}$.

NOTATION 31.3. Let Λ be the ring in Example 31.2, constructed with $n = 5$ and arbitrary odd primes such that each $p_h \neq p_{h+1}$. The graphs in (31.3.1) illustrate Λ and several torsionfree Λ-modules that we study in the examples below. We proceed to explain these graphs.

Λ. Each Γ_h displayed in this diagram is one of the coordinate rings of Γ, the "1" above each Γ_h displays the fact that each Γ_h-rank of Λ equals 1. Each edge connecting Γ_h to Γ_{h+1} displays the fact that these coordinate modules are "connected" by one of the congruence conditions defining Λ.

A. This is the projection of Λ in the direct sum of the first four coordinate modules of Λ.

D. This is the projection of Λ in the direct sum of the last three coordinate modules of Λ.

$M = A \oplus D$. The 2 above coordinates Γ_3 and Γ_4 signify that these ranks of M equal 2. The 2 below the edge connecting Γ_3 to Γ_4 indicates that two connections are involved here: one from A and one from D.

$B = \Lambda_{12} \oplus \Lambda_{34} \oplus \Gamma_5$. Here Λ_{12} and Λ_{34} denote the projections of Λ in $\Gamma_1 \oplus \Gamma_2$ and $\Gamma_3 \oplus \Gamma_4$ respectively.

(31.3.1)

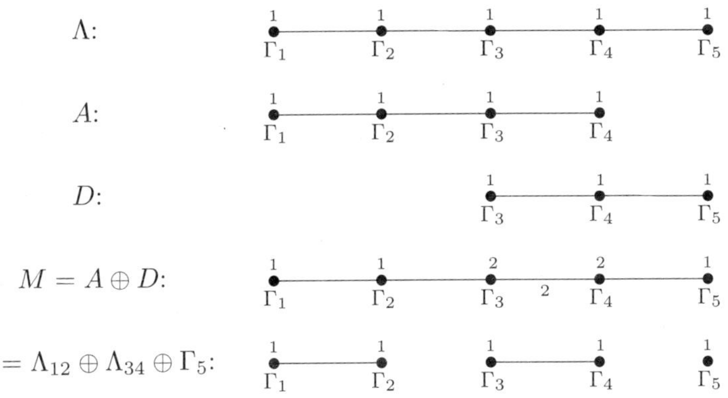

Although we shall not dwell on it here, the graph displayed for each module N is its connections graph $\mathcal{K}(N)$ [Definition 16.5]. For example, let $\mathfrak{m}_3 = \ker(\Lambda \to k_3)$, k_3 the residue field in Example 31.2. Then the 2 displayed in $\mathcal{K}(A \oplus D)$, below the edge connecting Γ_3 to Γ_4, indicates that the $\mathfrak{m}_3$-adic completion of M has two indecomposable direct summands of rank (1,1) with respect to the $\mathfrak{m}_3$-adic completions of Γ_3 and Γ_4, the two coordinate rings of $\hat{\Gamma}_{\mathfrak{m}_3}$.

Before proceeding, we note that *the* Λ*-modules* Λ, A, D, Λ_{12}, *and* Λ_{34} *in* (31.3.1) *are indecomposable,* and we display the orders of their genus class groups.

$$
\begin{aligned}
&|\mathcal{G}(\Lambda)| = (p_1 - 1)(p_2 - 1)(p_3 - 1)(p_4 - 1)/16 \\
&|\mathcal{G}(A)| = (p_1 - 1)(p_2 - 1)(p_3 - 1)/8 \\
&|\mathcal{G}(D)| = (p_3 - 1)(p_4 - 1)/4 \\
&|\mathcal{G}(\Lambda_{12})| = (p_1 - 1)/2 \qquad |\mathcal{G}(\Lambda_{34})| = (p_3 - 1)/2
\end{aligned} \tag{31.3.2}
$$

The order of $\mathcal{G}(\Lambda)$ is given by (31.2.1). Indecomposability is a special case of Theorem 16.8 (or directly verify the lack of nontrivial idempotents in Λ). The other listed modules can be handled similarly, as soon as one realizes that the relevant module can be viewed as a ring as well as a module.

EXAMPLE 31.4 (overlapping-support decomposition). *The order of* $\mathcal{G}(A \oplus D)$ *can be greater than, equal to, or less than that of* $\mathcal{G}(D)$, *depending upon the values of* p_1 *and* p_4. *On the other hand,* $|\mathcal{G}(A \oplus D)| \leq |\mathcal{G}(A)|$, *with equality if and only if* $p_2 = 3$.

We prove, below, that:

$$|\mathcal{G}(A \oplus D)| = |\mathcal{G}(B)| = (p_1 - 1)(p_3 - 1)/4 \tag{31.4.1}$$

The claims in the example then follow by comparing (31.4.1) with (31.3.2).

The easy part of (31.4.1) is the second equality because of the obvious isomorphism $\mathcal{G}(B) \cong \mathcal{G}(\Lambda_{12}) \oplus \mathcal{G}(\Lambda_{34})$ [which is also a special case of Lemma 30.1]; and the orders of the groups on the right-hand side are given in (31.3.2).

Consider $\mathcal{G}(A \oplus D)$. In the notation of the overlapping-support decomposition of $M = A \oplus D$ [Theorem 30.9], we have $\mathcal{G}(A \oplus D) \cong \mathcal{G}(B_{10}) \oplus \mathcal{G}(B_{11}) \oplus \mathcal{G}(B_{01})$ where $\mathcal{C}(B_{10}) = \mathcal{C}(A) - \mathcal{C}(D) = \{\Gamma_1, \Gamma_2\}$, $\mathcal{C}(B_{11}) = \mathcal{C}(A) \cap \mathcal{C}(D) = \{\Gamma_3, \Gamma_4\}$, and $\mathcal{C}(B_{01}) = \mathcal{C}(D) - \mathcal{C}(A) = \{\Gamma_5\}$.

Let $e_{10} = 1_1 + 1_2$ (the sum of the identity elements of Γ_1 and Γ_2), and similarly $e_{11} = 1_3 + 1_4$, and $e_{01} = 1_5$. Then, since we are dealing with torsionfree modules, Lemma 30.4 allows us to take $B_{10} = e_{10}M = \Lambda_{12}$, $B_{11} = e_{11}M = \Lambda_{34}^{(2)}$, and $B_{01} = e_{01}M = \Gamma_5$. Moreover the genus class group of any module is isomorphic to the genus class group of the direct sum of two copies of that module [Corollary 25.9]. Therefore we can take $B_{11} = \Lambda_{34}$. Finally, $\mathcal{G}(\Gamma_5) = \{0\}$ since Γ_5 is a principal ideal domain.

We conclude that $\mathcal{G}(M) \cong \mathcal{G}(\Lambda_{12}) \oplus \mathcal{G}(\Lambda_{34}) \oplus \{0\}$; and the order of this is $(p_1 - 1)(p_3 - 1)/4$, by (31.3.2). □

We now give two examples of local versus global summands of modules, in situations not covered by the previously-known results in Lemma 9.3.

EXAMPLE 31.5. *Every isomorphism class* $[D']$ *in* genus(D) *can occur as a direct summand of* $M = A \oplus D$, *even if* $|\mathcal{G}(D)|$ *is larger than* $|\mathcal{G}(A \oplus D)|$.

Note that D' is locally isomorphic to a direct summand of M because $D' \in$ genus(D). Let $e = 1_5$, the identity element of Γ_5. Then $eD' \cong eM \cong \Gamma_5$ because Γ_5 is a principal ideal domain. The desired conclusion follows from Theorem 30.12. □

Note that the conclusion would be different if the Dedekind domain Γ_5 were not a PID. The necessary and sufficient condition for D' to be isomorphic to a direct summand of M would then be that $eD' \cong \Gamma_5$.

Note also that the way that Γ_4 is connected to Γ_5 in D' (see the graph of D) is of no importance in this direct-summand question.

The next example is a more subtle version of the previous one.

EXAMPLE 31.6. Let $e = 1_1 + 1_2$ (the projection of the identity element of Λ in Λ_{12}). Then *a module* $A' \in$ genus(A) *is isomorphic to a direct summand of* $M = A \oplus D$ *if and only if* $eA' \cong \Lambda_{12}$.

The proof is the same as in Example 31.5. But here $\mathcal{G}(\Lambda_{12})$ contains $(p_1 - 1)/2$ isomorphism classes [see (31.3.2)], even though every Γ_h is a principal ideal domain.

Note that the way that Γ_2 is connected to Γ_3 in A' (see the graph of A) is of no importance in this direct-summand question.

The only decompositions of $M = A \oplus D$ that were considered in Examples 31.5 and 31.6 were those of the form $M = A' \oplus D'$, where $A' \in$ genus(A) and $D' \in$ genus(D), because we were examining the subtleties of such decompositions. But modules over Dedekind-like rings can have dramatically different decompositions, as the next example shows.

EXAMPLE 31.7. If m is an integer greater than or equal to 2, then there is a Dedekind-like ring Λ and a torsionfree module $M \in$ fingen(Λ) such that M is the direct sum of s indecomposable modules for every s in the interval $2 \leq s \leq m$.

Such an example is given in [**L1**, Example 3.2]. When this example is constructed from the ring of integers $\mathbb{Z}$, the resulting ring is Dedekind-like with finite normalization isomorphic to a direct sum of copies of $\mathbb{Z}$ [Proposition 33.2].

Examples also exist where $\Lambda = \mathbb{Z}G_n$, the integral group ring of a cyclic group of suitable squarefree order n [**L3**].

CHAPTER 8

Finite Normalization

In this chapter we restrict our attention to the classical situation in which our Dedekind-like ring Λ has *finite normalization,* that is, ${}_\Lambda\Gamma$ is finitely generated. In this situation singspec(Λ) is a finite set [Proposition 10.11], and there is an explicit construction of an arbitrary Dedekind-like ring Λ from a direct sum Γ of Dedekind domains, in such a way that Γ is the normalization of Λ [Section 33]. Moreover, we show that it is possible to construct any finite configuration of singspec(Λ) — i.e. the number and location of unsplit, strictly split, and nonstrictly split maximal ideals — inside of the product of (finitely many) copies of any quadratic number field with itself [Example 33.4].

Results on direct-sum behavior include determining when power isomorphism holds within a genus [Theorem 35.1]; using ${}_\Lambda\Gamma$ as a test module for cancellation [Theorem 35.2]; determining when power cancellation holds [Theorem 35.3]; relating the super genus class group to cancellation [Corollary 35.4]; and showing that ${}_\Lambda\Lambda$ always cancels from isomorphisms of torsionfree modules [Theorem 35.5].

Recall that, when ${}_\Lambda\Gamma$ is finitely generated, there is always an identification $s\mathcal{G} = \mathcal{G}(M)$ of the super genus class group with an actual genus class group, and M can often be taken to be Λ [Theorem 27.10, Corollary 27.11]. We construct an example showing that M cannot always be taken to be Λ [Example 34.1].

The final section specializes further, to quadratic orders. We construct an example showing that Mayer-Vietoris sequences do not always split [Example 36.1]. A by-product is an example of an explicitly computable $\mathcal{G}(\Lambda)$ in the situation where Γ is not a principal ideal domain and $r\mathcal{G}(\Lambda) \neq 0$ [Example 36.2] We also determine when direct-sum cancellation holds in fingen($\mathbb{Z}[\sqrt{n}\,]$), for arbitrary squarefree integers n [Example 36.3]. (These orders are always Dedekind-like.)

32. ${}_\Lambda\Gamma$ versus ${}_\Gamma\Gamma$

LEMMA 32.1. *Suppose that ${}_\Lambda\Gamma$ is finitely generated, $X \in \text{fingen}_\infty(\Gamma)$, and $X' \in \text{genus}_\Lambda(X)$. Then X' has a natural structure as a module in* $\text{genus}_\Gamma(X)$ *and the following Γ-isomorphisms hold for some $Y, Y' \in \text{finlen}(\Gamma)$.*

$$\text{(32.1.1)} \qquad \begin{gathered} \Gamma \otimes_\Lambda X \cong_\Gamma X \oplus Y \quad \textit{and} \quad \Gamma \otimes_\Lambda X' \cong_\Gamma X' \oplus Y' \,, \\ \textit{where } Y \cong_\Gamma t(\Gamma \otimes_\Lambda X) \cong_\Gamma Y' \end{gathered}$$

PROOF. Since Γ is a direct sum of Dedekind domains, the Γ-module $X \in \text{fingen}_\infty(\Gamma)$ is torsionfree, and hence is torsionfree as a Λ-module, since $\Lambda_Q = \Gamma_Q$. Since $X' \in \text{genus}_\Lambda(X)$, and torsionfreeness can be checked locally, at maximal ideals of Λ, we see that the Λ-module X' is torsionfree. Therefore the natural map $X' \to X'_Q$ is an injection, which we write as inclusion, $X' \subseteq X'_Q$. The product $\Gamma X'$ now makes sense, and satisfies $\Gamma X' = X'$, because this property holds with X in place of X' and the property can be checked locally, at maximal ideals of Λ. Thus X' is a torsionfree module in $\text{fingen}_\infty(\Gamma)$.

Since Γ is a direct sum of Dedekind domains, its finitely generated torsionfree modules X and X' are projective. Now we prove properties (32.1.1)

The natural map $\tau\colon \Gamma \otimes_\Lambda X \to \Gamma X = X$ is a surjection. Since X is projective, we have a decomposition $\Gamma \otimes_\Lambda X \cong_\Gamma X \oplus Y$, where $Y = \ker(\tau)$. To see that $Y = t(\Gamma \otimes_\Lambda X)$, it suffices to show (since X is torsionfree) that there is a regular element $d \in \Gamma$ such that $dY = 0$.

By finite normalization there exists d regular in Λ such that $d\Gamma \subseteq \Lambda$ [Proposition 10.11]. This d is also regular in Γ since $\Lambda_Q = \Gamma_Q$. Choose an arbitrary element $y = \sum_i \gamma_i \otimes x_i \in \ker(\tau)$. Then we have $\sum_i \gamma_i x_i = 0$. Since $d\Gamma \subseteq \Lambda$, we have $dy = 1 \otimes (\sum_i d\gamma_i x_i) = 0$. Thus $dY = 0$ and we have $Y = t(\Gamma \otimes_\Lambda X)$.

Similarly, $\Gamma \otimes_\Lambda X' \cong_\Gamma X' \oplus Y'$, where $Y' = t(\Gamma \otimes_\Lambda X')$. Therefore, to complete the proof of (32.1.1), it now suffices to prove that $t(\Gamma \otimes_\Lambda X) \cong_\Gamma t(\Gamma \otimes_\Lambda X')$.

By hypothesis, $X_\mathfrak{m} \cong_{\Lambda_\mathfrak{m}} X'_\mathfrak{m}$ for every $\mathfrak{m} \in \text{maxspec}(\Lambda)$, so that $(\Gamma \otimes_\Lambda X)_\mathfrak{m} \cong_{\Gamma_\mathfrak{m}} (\Gamma \otimes_\Lambda X')_\mathfrak{m}$ for every $\mathfrak{m} \in \text{maxspec}(\Lambda)$. But for every Γ-module H and every maximal ideal $\mathfrak{n}$ of Γ, $H_\mathfrak{n}$ can be viewed as a localization of $H_\mathfrak{m}$, where $\mathfrak{m} = \mathfrak{n} \cap \Lambda$ (i.e. the denominator set used in $H_\mathfrak{n}$ is as least as large as the denominator set used in $H_\mathfrak{m}$). Therefore $\Gamma \otimes_\Lambda X'$ is in the Γ-genus of $\Gamma \otimes_\Lambda X$, and hence the same is therefore true of their torsion submodules. Finally, since finitely generated torsion Γ-modules have finite length, their genus equals their isomorphism class [Corollary 6.4], completing the proof of (32.1.1). □

PROPOSITION 32.2. *Suppose that* ${}_\Lambda\Gamma$ *is finitely generated, and let* $X \in \text{fingen}(\Gamma)$ (*hence also* $X \in \text{fingen}(\Lambda)$). *Then* $\text{r}\mathcal{G}_\Lambda(X)$ *consists of the single isomorphism class* $[{}_\Lambda X]$.

PROOF. Let $X' \in \text{rgen}({}_\Lambda X)$. We want to prove that $X' \cong_\Lambda X$, where $\cong_\Lambda$ denotes isomorphism of Λ-modules. Since Γ is a direct sum of Dedekind domains we have a decomposition ${}_\Gamma X = {}_\Gamma P \oplus {}_\Gamma T$ where P is projective and $T = t(X)$, the torsion submodule of ${}_\Gamma X$. Since $P, T \in \text{fingen}(\Lambda)$, we have a decomposition ${}_\Lambda X' = {}_\Lambda P' \oplus {}_\Lambda T'$, where $P' \in \text{genus}({}_\Lambda P)$ and $T' \in \text{genus}({}_\Lambda T)$ [Lemma 9.3]. Since ${}_\Gamma T$ has finite length, ${}_\Lambda T$ also has finite length [Lemma 10.10]. Therefore, the genus of ${}_\Lambda T$ equals its isomorphism class [Corollary 6.4]; that is, $T' \cong_\Lambda T$. Thus, it now suffices to show that $P' \cong_\Lambda P$.

We have proved that $X \cong_\Gamma P \oplus T$ and $X' \cong_\Lambda P' \oplus T$. In particular, the first isomorphism is also a Λ-isomorphism. Therefore, localizing at an arbitrary maximal ideal $\mathfrak{m}$ of Λ, and remembering that $X' \in \text{genus}_\Lambda(X)$ — and using local cancellation [Lemma 9.2] — shows that $P' \in \text{genus}_\Lambda P$. Moreover, we have $P' \in \text{rgen}({}_\Lambda P)$, as is easily seen by tensoring X and X' by Γ (over Λ), and using direct-sum cancellation of Γ-modules. That is, we have reduced to proof to the case that $X, X' \in \text{fingen}_\infty(\Gamma)$ which we assume from now on.

Since X' is in the restricted genus of X we have $\Gamma \otimes_\Lambda X' \cong_\Gamma \Gamma \otimes_\Lambda X$. This, together with (32.1.1) and direct-sum cancellation of finitely generated Γ-modules shows that $X' \cong_\Gamma X$, and hence $X' \cong_\Lambda X$, as desired. □

Next, we examine the loss-of-structure ξ-maps in the in the case where Λ has finite normalization Γ.

PROPOSITION 32.3. *Suppose that* ${}_\Lambda\Gamma$ *is finitely generated. If* $A \in \text{fingen}_\infty(\Lambda)$, $X \in \text{fingen}_\infty(\Gamma)$, *and* $\mathcal{C}(A) \subseteq \mathcal{C}(X)$, *then* $\xi^{A,X}$ *is defined and equals* $\xi^{A,X}_{\Lambda,\Gamma}$.

PROOF. Let $X' \in \text{genus}({}_\Lambda X)$. We claim that, if $A \oplus X \cong_\Lambda A \oplus X'$ (Λ-module isomorphism), then $X \cong_\Lambda X'$.

By local cancellation we have $X' \in \text{genus}_\Lambda(X)$ [Lemma 9.2]. Next, tensor both sides of the given isomorphism with Γ (over Λ). Then apply direct-sum cancellation of finitely generated Γ-modules to get that $\Gamma \otimes_\Lambda X \cong_\Gamma \Gamma \otimes_\Lambda X'$. Thus, $X' \in \text{rgen}({}_\Lambda X)$. Since $X \in \text{fingen}_\infty(\Lambda)$ this implies $X' \cong_\Lambda X$ [Proposition 32.2]. Since X' and X are torsionfree, any such isomorphism can be extended, by Q-localization, to a $\Lambda_Q = \Gamma_Q$-isomorphism of X'_Q onto X_Q; and this extended map takes $X' = \Gamma X'$ onto $X = \Gamma X$, proving the claim.

The claim, together with the hypothesis $\mathcal{C}(A) \subseteq \mathcal{C}(X)$ shows that $\xi^{A,X}$ is defined [Lemma 25.6].

To see that $\xi^{A,X}_{\Lambda,\Gamma}$ is defined, we need to show that, given $A' \in \text{genus}_\Lambda(A)$, there exists $X' \in \text{genus}_\Gamma(X)$ — unique up to isomorphism — such that

$$(\Gamma \otimes_\Lambda A) \oplus X' \cong_\Gamma (\Gamma \otimes_\Lambda A') \oplus X \tag{32.3.1}$$

Since direct-sum cancellation holds for Γ-modules, uniqueness of X' is obvious. We prove existence at the same time that we show that the two ξ-maps are the same.

Let $\xi^{A,X}[A'] = [X']$. Then $A \oplus X' \cong A' \oplus X$. Tensoring with Γ, we get:

$$(\Gamma \otimes_\Lambda A) \oplus (\Gamma \otimes_\Lambda X') \cong_\Gamma (\Gamma \otimes_\Lambda A') \oplus (\Gamma \otimes_\Lambda X) \tag{32.3.2}$$

By (32.1.1) we have $\Gamma \otimes_\Lambda X' \cong_\Gamma X' \oplus Y$ and $\Gamma \otimes_\Lambda X \cong_\Gamma X \oplus Y$ for an appropriate $Y \in \text{finlen}(\Gamma)$. Substituting these into (32.3.2) and canceling Y yields (32.3.1), completing the proof. □

As an immediate consequence of this proposition, we can refine the description of the surjective map in the Mayer-Vietoris sequence (25.14.1) of a finitely generated Λ-module, when Λ has finite normalization.

COROLLARY 32.4. *Suppose that ${}_\Lambda\Gamma$ is finitely generated, and let $M \in \text{fingen}_\infty(\Lambda)$. Then the map $\xi^{M,\Gamma_{\mathcal{C}(M)}}_{\Lambda,\Gamma} : \mathcal{G}(M) \twoheadrightarrow \mathcal{G}_\Gamma(\Gamma_{\mathcal{C}(M)})$, in the Mayer-Vietoris sequence for M, equals $\xi^{M,\Gamma_{\mathcal{C}(M)}}$.*

33. Structure of Λ. Another simple $\mathcal{G}(\Lambda)$

All Dedekind-like rings Λ with finite normalization — in particular, all examples explicitly considered in this memoir — can be constructed from a direct sum of Dedekind domains by the following procedure.

CONSTRUCTION 33.1. Let $\mathcal{H}, \mathcal{U}, \mathcal{S}$ be finite index sets, and let data be given, as specified in (i)–(iii) and the paragraph below it.

(i) $\Gamma = \oplus_{h \in \mathcal{H}} \Gamma_h$, where each Γ_h is a Dedekind domain.
(ii) $(\forall u \in \mathcal{U})$ a surjective ring homomorphism $\rho_u\colon \Gamma_{h_u} \twoheadrightarrow F_u$, where F_u is a 2-dimensional separable field extension of a subfield k_u, and Γ_{h_u} is a coordinate ring of Γ.
(iii) $(\forall s \in \mathcal{S})$ a pair of surjective ring homomorphisms $\rho_{js}\colon \Gamma_{h_{js}} \twoheadrightarrow k_s$ $(j = 1, 2)$, where k_s is a field and $\Gamma_{h_{1s}}$ and $\Gamma_{h_{2s}}$ are (not necessarily distinct) coordinate rings of Γ.

We also assume: *no maximal ideal of any Γ_h appears more than once in the combined list of maximal ideals* $\ker(\rho_u)$ *and* $\ker(\rho_{js})$ $(u \in \mathcal{U},\ s \in \mathcal{S},\ j = 1, 2)$.

Extend each ρ_u and ρ_{js} to a homomorphism $\tilde{\rho}_u\colon \Gamma \twoheadrightarrow F_u$ and $\tilde{\rho}_{js}\colon \Gamma \twoheadrightarrow k_s$ by letting it equal zero on every Γ_h on which it is not already defined, and let

$$\Lambda = \{\gamma \in \Gamma \mid \tilde{\rho}_u(\gamma) \in k_u \ (\forall u \in \mathcal{U}) \text{ and } \tilde{\rho}_{1s}(\gamma) = \tilde{\rho}_{2s}(\gamma) \ (\forall s \in \mathcal{S})\} \tag{33.1.1}$$

PROPOSITION 33.2. *The ring Λ defined in Construction 33.1 is a Dedekind-like ring with finite normalization Γ. The singular maximal ideals of Λ are*

unsplit:	$\mathfrak{m}_u = \ker(\tilde{\rho}_u) \cap \Lambda$	$(u \in \mathcal{U})$
split:	$\mathfrak{m}_s = \ker(\tilde{\rho}_{1s}) \cap \ker(\tilde{\rho}_{2s}) \cap \Lambda$	$(s \in \mathcal{S})$

with $\mathfrak{m}_s$ strictly split if and only if $h_{1s} \neq h_{2s}$. Each $\mathfrak{m}_u$-residue inclusion of Λ is $k_u \subset F_u$ and each $\mathfrak{m}_s$-residue inclusion is the diagonal imbedding $k_s \subset k_s \times k_s$.

PROOF. Let $\bar{\Gamma}$ be the direct sum of the rings F_u and $k_s \times k_s$, and let $\mathfrak{c}$ be the intersection of the ideals $\mathfrak{m}_u$ and $\mathfrak{m}_s$ of Γ. Then the ring homomorphisms $\tilde{\rho}_u\colon \Gamma \twoheadrightarrow F_u$ and $[\tilde{\rho}_{1s}, \tilde{\rho}_{2s}]\colon \Gamma \twoheadrightarrow k_s \oplus k_s$ define a ring homomorphism $\rho\colon \Gamma \to \bar{\Gamma}$ whose kernel is $\mathfrak{c}$; and ρ is surjective by the Chinese Remainder theorem and the hypothesis that the kernels of the various maps ρ_u and ρ_{js} are all distinct maximal ideals. Therefore, Λ is the pullback of the following conductor square:

$$\begin{array}{ccc} \Lambda & \subset & \Gamma \\ \Big\downarrow{\scriptstyle\rho} & & \Big\downarrow{\scriptstyle\rho} \\ \bar{\Lambda} & \subset & \bar{\Gamma} \end{array} \tag{33.2.1}$$

where $\bar{\Lambda}$ is the direct sum of every k_u and *one* copy of every k_s (because of the conditions $\tilde{\rho}_{1s}(\gamma) = \tilde{\rho}_{2s}(\gamma)$ in (33.1.1)). The bottom row of this square is the direct sum of the residue inclusions given in the statement of the proposition. In particular, the diagonal imbeddings result from the conditions $\tilde{\rho}_{1s}(\gamma) = \tilde{\rho}_{2s}(\gamma)$.

Since $\Gamma/\Lambda \cong \bar{\Gamma}/\bar{\Lambda}$ as Λ-modules, ${}_{\Lambda}\Gamma$ is finitely generated. The ring Λ is noetherian by [**KL2**, Lemma 4.2], and Γ is the normalization of Λ because it is module-finite over Λ and integrally closed in their common total quotient ring.

To complete the proof that Λ is Dedekind-like and its singular maximal ideals are as specified, we must show that, for each maximal ideal $\mathfrak{m}$ of Λ, the localization $\Lambda_{\mathfrak{m}}$ is the corresponding type of local Dedekind-like ring [Theorem and Definitions 11.3]. This is a somewhat long but straightforward computation, based on the fact that tensoring (33.2.1) with the flat Λ-module $\Lambda_{\mathfrak{m}}$ yields another conductor square defining $\Lambda_{\mathfrak{m}}$ as a pullback. The details are best broken into three cases: (i) $\mathfrak{m} = \ker(\mathfrak{m}_u)$ for some u; (ii) $\mathfrak{m} = \ker(\mathfrak{m}_s)$ for some s; and (iii) any other maximal ideal. In cases (i) and (ii) the localized bottom row consists of the corresponding residue inclusion mentioned in the statement of the proposition. In case (iii) the localized bottom row becomes zero, yielding $\Lambda_{\mathfrak{m}} = \Gamma_{\mathfrak{m}}$, that is, yielding the nonsingular maximal ideals of Λ. □

The following instance of the above construction is one of the simplest Dedekind-like rings with nontrivial $\mathcal{G}(\Lambda)$. In a couple of places, we shall use the (known) formula for $\mathcal{G}(\Lambda)$ that it contains.

EXAMPLE 33.3 ($\mathcal{G}(\Lambda) = \mathrm{r}\mathcal{G}(\Lambda) = k^{\times}$). Let $\Gamma = k[x] \oplus k[x]$, the direct sum of two polynomial rings in one variable over a field k. Fix distinct elements $a, b \in k$, and let ρ_a and ρ_b denote the natural homomorphisms $k[x] \twoheadrightarrow k$ determined by $x \to a$ and $x \to b$. Then set

$$\Lambda = \{(\gamma_1, \gamma_2) \in \Gamma \mid \rho_a(\gamma_1) = \rho_a(\gamma_2) \quad \text{and} \quad \rho_b(\gamma_1) = \rho_b(\gamma_2)\} \tag{33.3.1}$$

By Proposition 33.2, Λ is a Dedekind-like ring with normalization Γ and with exactly two singular maximal ideals:

$$\mathfrak{m}_a = \big(k[x]\cdot(x-a) \times k[x]\cdot(x-a)\big) \cap \Lambda \quad \text{and}$$
$$\mathfrak{m}_b = \big(k[x]\cdot(x-b) \times k[x]\cdot(x-b)\big) \cap \Lambda$$

Both are strictly split.

In this example, $s\mathcal{G} = \mathcal{G}(\Lambda)$ [Corollary 27.11]. Moreover, since Γ is a principal ideal ring, we have $\mathcal{G}(\Lambda) = r\mathcal{G}(\Lambda)$, and hence Lemma 31.1 yields $\mathcal{G}(\Lambda) \cong k^\times$.

EXAMPLE 33.4. *Any finite configuration of* singspec(Λ) *can be realized inside the product of finitely many copies of any quadratic number field.* In more detail, let R be the ring of algebraic integers in any quadratic number field, $\mathcal{H}$ a finite nonempty index set, and $\Gamma = \oplus_{h\in\mathcal{H}} \Gamma_h$ where each $\Gamma_h = R$. Then there exists a Dedekind-like ring Λ with normalization Γ such that, for each index $h \in \mathcal{H}$ and each pair of distinct indices $h_1 \neq h_2$ in $\mathcal{H}$, Λ has exactly:

(i) a_h unsplit maximal ideals supporting Γ_h,
(ii) b_h nonstrictly split maximal ideals supporting Γ_h, and
(iii) c_{h_1,h_2} strictly split maximal ideals supporting Γ_{h_1} and Γ_{h_2},

where a_h, b_h, c_{h_1,h_2} are arbitrarily pre-assigned nonnegative integers.

Moreover, *the ring* Λ *is indecomposable if and only if its connections graph* $\mathcal{K}(\Lambda)$ *is connected.* [See Definition 16.5 and Theorem 16.8]

To prove (i)–(iii), recall that, by the Tchebotarev Density Theorem [**Ja**, Theorem 10.4], there are infinitely many primes in $\mathbb{Z}$ which split in R, and there are infinitely many primes in $\mathbb{Z}$ which remain inert in R. Thus, R contains infinitely many maximal ideals each of which is a 2-dimensional separable extension of some subfield, and R contains infinitely many pairs of maximal ideals with isomorphic residue fields. Therefore Λ exists by Construction 33.1 and Proposition 33.2.

34. Example: $s\mathcal{G} = \mathcal{G}(M) \neq \mathcal{G}(\Lambda)$

For simple examples in which $s\mathcal{G} = \mathcal{G}(\Lambda) \neq 0$, see Examples 31.2 and 33.3. Recall that when $s\mathcal{G} = \mathcal{G}(M) \neq \mathcal{G}(\Lambda)$, M can be chosen to have all of its ranks equal to 2 and, in this situation, cannot be torsionfree [Theorem 27.12]. The next example shows that this situation can actually occur.

EXAMPLE 34.1 ($s\mathcal{G} = \mathcal{G}(M) \neq \mathcal{G}(\Lambda) \neq 0$). A Dedekind-like ring (with finite normalization) and $M \in \text{fingen}_\infty(\Lambda)$ such that the map $\xi^{s\mathcal{G},M}$ is an isomorphism (informally, $s\mathcal{G} = \mathcal{G}(M)$) but the surjective map $\xi^{M,\Lambda}$ is neither zero nor an isomorphism (informally, $s\mathcal{G} \neq \mathcal{G}(\Lambda) \neq 0$).

Let $F \supset k$ be a separable field extension of degree 2, and $N = \text{im}(N_k^F)$ the multiplicative group of norms of nonzero elements of F. Let $\Gamma = F[x]$, the polynomial ring in one variable over F. Fix distinct elements $a, b \in F$, and let ρ_a and ρ_b denote the substitution homomorphisms $\Gamma \twoheadrightarrow F$ determined by $x \to a$ and $x \to b$ respectively, so that $\ker(\rho_a) = \Gamma(x-a)$ and $\ker(\rho_b) = \Gamma(x-b)$. Set:

$$\Lambda = \big\{\gamma \in \Gamma \mid \rho_a(\gamma) \in k \quad \text{and} \quad \rho_b(\gamma) \in k\big\}$$

Note that, in the notation of Construction 33.1, $\rho_a = \tilde{\rho}_a$ and $\rho_b = \tilde{\rho}_b$ because Γ has only one coordinate ring. By Proposition 33.2, Λ is a Dedekind-like domain with normalization Γ and exactly two singular maximal ideals $\mathfrak{m}_a = \Gamma(x-a) \cap \Lambda$ and $\mathfrak{m}_b = \Gamma(x-b) \cap \Lambda$, both unsplit.

Let $M \in \operatorname{fingen}(\Lambda)$ (of rank 2) be such that $\mathrm{s}\mathcal{G} = \mathcal{G}(M)$ and $\operatorname{eqstab}(M)(\mathfrak{m}) = \operatorname{im}(N_{k(\mathfrak{m})}^{F(\mathfrak{m})}) = N$ for $\mathfrak{m} = \mathfrak{m}_a$ and $\mathfrak{m} = \mathfrak{m}_b$ [Theorem 27.10]. We claim that:

$$\mathcal{G}(M) \cong \frac{F^\times \times F^\times}{(N \times N)(\operatorname{diag}(F^\times))} \cong F^\times/N \tag{34.1.1}$$

and

$$\mathcal{G}(\Lambda) \cong \frac{F^\times \times F^\times}{(k^\times \times k^\times)(\operatorname{diag}(F^\times))} \cong F^\times/k^\times \tag{34.1.2}$$

The group $\mathcal{G}_\Gamma(\Gamma)$ is trivial, because Γ is a principal ideal domain. Therefore, the Mayer-Vietoris sequence for M yields an isomorphism $\mu^M\colon \mathcal{I}/\operatorname{istab}(M) \cong \mathcal{G}(M)$ [Theorem 25.14]. We have $\mathcal{I} = F^\times \times F^\times$ [Lemma 31.1(i)], and $\operatorname{istab}(M) = \operatorname{eqstab}(M) \cdot \operatorname{im}(\Gamma^\times)$ [Theorem 23.5]. Moreover, $\operatorname{eqstab}(M) = \operatorname{eqstab}(M)(\mathfrak{m}_a) \times \operatorname{eqstab}(M)(\mathfrak{m}_b)$ [Definition 23.1] and this equals $N \times N$, by our choice of M. Thus we obtain the first isomorphism in (34.1.1); the second is clear.

The nontrivial first isomorphism in (34.1.2) follows easily from Lemma 31.1, and the fact that $\mathrm{r}\mathcal{G}(\Lambda) = \mathcal{G}(\Lambda)$ because Γ is a principal ideal ring.

By Mayer-Vietoris consistency [Lemma 27.2], the map $\xi^{M,\Lambda} : \mathcal{G}(M) \to \mathcal{G}(\Lambda)$ can be identified with the natural surjection of the middle side of (34.1.1) to that of (34.1.2), and hence with the natural surjection $F^\times/N \twoheadrightarrow F^\times/k^\times$. We conclude that:

$$\ker(\xi^{M,\Lambda}) = 0 \iff N = k^\times \tag{34.1.3}$$

We single out two special cases of interest.

(i) Let $k = \mathbb{R}$ and $F = \mathbb{C}$, the real and complex numbers, respectively. Then $N = \operatorname{im}(N_{\mathbb{R}}^{\mathbb{C}}) = \mathbb{R}^+$, the group of positive real numbers. Therefore, (34.1.3) shows that $\ker(\xi^{M,\Lambda}) \neq 0$; that is, the surjective group homomorphism $\xi^{M,\Lambda}\colon \mathcal{G}(M) \twoheadrightarrow \mathcal{G}(\Lambda)$ is not an isomorphism in this case. Less formally, $\mathrm{s}\mathcal{G} = \mathcal{G}(M) \neq \mathcal{G}(\Lambda)$ (even though $\mathcal{G}(M) \cong \mathcal{G}(\Lambda)$ as groups, in this case!).

(ii) Let k and F be finite fields. Then N_k^F is a surjection; that is, $N = F^\times$ [**LN**, Theorem 2.28(ii)]. Therefore, the surjective homomorphism $\xi^{M,\Lambda}\colon \mathcal{G}(M) \twoheadrightarrow \mathcal{G}(\Lambda)$ is an isomorphism in this case. Less formally, $\mathrm{s}\mathcal{G} = \mathcal{G}(M) = \mathcal{G}(\Lambda)$.

35. Power Isomorphism, Cancellation, Power Cancellation

Jacobinski [**J**, Proposition 2.8] showed that, for finitely generated torsionfree modules over the orders that arise in algebraic number theory, $\operatorname{genus}(A) = \operatorname{genus}(B)$ implies the existence of an integer e such that $A^{(e)} \cong B^{(e)}$. Guralnick [**G1**, Theorem B] extended this and many other of Jacobinski's results to all finitely generated modules over these rings. Levy and Wiegand [**LW**, Theorem 5.6] determined that — for torsionfree modules over Bass rings — the existence and value of e is determined by the Picard group $\mathcal{G}(\Lambda)$, and Levy [**L2**, Theorem 14.4] removed the torsionfree restriction if Λ is Dedekind-like with finite normalization and (in our present terminology) no unsplit maximal ideals. We now extend this last result to arbitrary Dedekind-like rings with finite normalization, by replacing the Picard group by the super genus class group

THEOREM 35.1 (Power isomorphism in genus). *Suppose that ${}_\Lambda\Gamma$ is finitely generated, and let $M \in \operatorname{fingen}(\Lambda)$ be such that $\mathrm{s}\mathcal{G} = \mathcal{G}(M)$. Then the implication*

$$\operatorname{genus}(A) = \operatorname{genus}(B) \implies (\exists e > 0) \quad A^{(e)} \cong B^{(e)} \tag{35.1.1}$$

holds in fingen(Λ) *if and only if* $\mathcal{G}(M)$ *is a torsion group. The integer* e *can be chosen independently of* A *and* B *if and only if* $\mathcal{G}(M)$ *has finite exponent, in which case the exponent of* $\mathcal{G}(M)$ *is the smallest* e *that works for all* A *and* B.

Moreover, there are examples of Dedekind-like rings for which each of these three possibilities occurs: $\mathcal{G}(M)$ *has finite exponent,* $\mathcal{G}(M)$ *is torsion but not of finite exponent, and* $\mathcal{G}(M)$ *is nontorsion.*

PROOF. As usual, we may work in $\text{fingen}_\infty(\Lambda)$ [see §7]. The isomorphism $A^{(e)} \cong B^{(e)}$, with $B \in \text{genus}(A)$ is equivalent to the equation $e{\cdot}[B] = 0$ in $\mathcal{G}(A)$. Thus, the implication in (35.1.1) holds if and only if every genus class group $\mathcal{G}(A)$ is a torsion group; and it holds for some fixed e if and only if every $\mathcal{G}(A)$ has exponent dividing e. But the super genus class group $\text{s}\mathcal{G} = \text{r}\mathcal{G}(M)$ maps onto all genus class groups [Theorem 28.3] and is itself a genus class group in the present situation. Therefore, every genus class group $\mathcal{G}(A)$ is a torsion group if and only if $\mathcal{G}(M)$ is a torsion group, and every genus class group $\mathcal{G}(A)$ has exponent dividing a fixed integer e if an only if the same is true of $\mathcal{G}(M)$. This concludes the proof of all but the statement in the final paragraph of the theorem.

For the supplementary statement, note that all of the possibilities listed already occur for Dedekind domains Λ, because every abelian group can occur as the class group of some Dedekind domain [**C**, Theorem 7]. For more subtle examples of these possibilities, where Γ is a principal ideal ring, use Example 33.3, with the field k respectively finite, the algebraic closure of a finite field, and the rational numbers. □

Another of Jacobinski's results is that, for torsionfree modules over number-theoretic orders, Γ is a test-module for direct-sum cancellation, (i.e., (iii) $\Longleftrightarrow$ (iv) in Theorem 35.2 below) [**J**, Proposition 2.10]. Again, Guralnick extended this result to all finitely generated modules over these rings [**G3**, Corollary 6.5]. For torsionfree modules, Wiegand extended the commutative case of this result to arbitrary reduced commutative noetherian rings of Krull dimension one with finite normalization [**W1**, Theorem 2.3], and Levy [**L2**, Proposition 13.6] again removed the torsionfree restriction if Λ is Dedekind-like with finite normalization and no unsplit maximal ideals. We further extend this last result to arbitrary Dedekind-like rings with finite normalization.

THEOREM 35.2 (${}_\Lambda\Gamma$ as test module for cancellation). *If* ${}_\Lambda\Gamma$ *is finitely generated, then the following conditions are equivalent for* $A, B \in$ fingen(Λ).

(i) $B \in \text{rgen}(A)$.
(ii) $A \oplus \Gamma_{\mathcal{C}(A)} \cong B \oplus \Gamma_{\mathcal{C}(A)}$.
(iii) $A \oplus \Gamma \cong B \oplus \Gamma$.
(iv) $A \oplus X \cong B \oplus X$ *for some* $X \in$ fingen(Λ).

PROOF. We may assume that $A, B, X \in \text{fingen}_\infty(\Lambda)$ since modules of finite length cancel from direct sums [see §7].

(i) $\Longleftrightarrow$ (ii). By Proposition 32.3, and Theorem 25.14, condition (i) is equivalent to the equation $\xi^{A,\Gamma_{\mathcal{C}(A)}}[B] = [\Gamma_{\mathcal{C}(A)}]$, because the right-hand side is the zero element of $\mathcal{G}(\Gamma_{\mathcal{C}(A)})$. Condition (ii) is merely a restatement of this equation.

(ii) $\Longrightarrow$ (iii) $\Longrightarrow$ (iv) are obvious.

(iv) $\Longrightarrow$ (iii). Given (iv), we get that $[B] \in \ker(\xi^{A,A\oplus X})$ [Theorem 26.2]. Now the map $\xi^{A\oplus X,\Gamma}$ is defined, by Proposition 32.3, so by transitivity of ξ [Theorem 27.1(iv)], $[B] \in \ker(\xi^{A,\Gamma})$. That is, $\xi^{A,\Gamma}[B] = [\Gamma]$, a restatement of condition (iii).

(iii) $\implies$ (ii). We have $\Gamma = \Gamma_{\mathcal{C}(A)} \oplus \Gamma_{\mathcal{D}}$, where $\Gamma_{\mathcal{D}}$ is the direct sum of all the Dedekind domain direct summands of Γ not contained in $\Gamma_{\mathcal{C}(A)}$. Thus, given condition (iii), we can cancel $\Gamma_{\mathcal{D}}$ from both sides of the isomorphism to obtain condition (ii) [Corollary 26.6]. □

Following Goodearl, we say that *power cancellation* holds for a class of R-modules if the implication (35.3.1) below holds for all R-modules A, B, and X in the class. It follows from [**J**, Proposition 2.8], for example, that power cancellation holds for finitely generated torsionfree modules over arithmetic orders. By considering a stable range condition on the endomorphism rings of modules, Goodearl was able to extend Jacobinski's result to prove power cancellation for all finitely generated modules over a class of rings that includes all commutative, torsionfree, finite rank $\mathbb{Z}$-algebras [**Go**, Theorem 3.10]. Guralnick extended the class of rings for which power cancellation of finitely generated modules is known to hold [**G2**] and proved a uniform bound for the cancellation exponent for (separable) orders over Dedekind domains all of whose residue fields are finite [**G3**, Theorem 5.5]. Levy and Wiegand determined when power cancellation holds for finitely generated torsionfree modules over Bass rings and, in this situation, determined the smallest exponent e which always works [**LW**, Theorem 6.6]. Levy removed the torsionfree restriction for Dedekind-like rings with finite normalization and no unsplit maximal ideals [**L2**, Theorem 13.7]. In both of these last two situations, it is the exponent of $\mathrm{r}\mathcal{G}(\Lambda)$ that is critical. We now extend this last result to arbitrary Dedekind-like rings with finite normalization, this time replacing the restricted Picard group $\mathrm{r}\mathcal{G}(\Lambda)$ by the restricted super genus class group $\mathrm{rs}\mathcal{G}$.

THEOREM 35.3 (Power cancellation). *Suppose that ${}_\Lambda\Gamma$ is finitely generated, and let $M \in \mathrm{fingen}(\Lambda)$ be such that $\mathrm{s}\mathcal{G} = \mathcal{G}(M)$. Then the implication*

$$(35.3.1) \qquad A \oplus X \cong B \oplus X \implies (\exists e > 0) \quad A^{(e)} \cong B^{(e)}$$

holds in $\mathrm{fingen}(\Lambda)$ *if and only if* $\mathrm{r}\mathcal{G}(M)$ *is a torsion group. The integer e can be chosen independently of A, B, and X if and only if* $\mathrm{r}\mathcal{G}(M)$ *has finite exponent, in which case the exponent of* $\mathrm{r}\mathcal{G}(M)$ *is the smallest e that works for all A, B, and X.*

Moreover, there are examples of Dedekind-like rings for which each of these three possibilities occurs: $\mathrm{r}\mathcal{G}(M)$ *has finite exponent,* $\mathrm{r}\mathcal{G}(M)$ *is torsion but not of finite exponent, and* $\mathrm{r}\mathcal{G}(M)$ *is nontorsion.*

PROOF. Without loss of generality we may work in $\mathrm{fingen}_\infty(\Lambda)$ [see §7]. The first isomorphism in (35.3.1) implies that $B \in \mathrm{rgen}(A)$, by Theorem 35.2. Moreover, for $B \in \mathrm{rgen}(A)$, the second isomorphsim in (35.3.1) is equivalent to $e{\cdot}[B] = 0$ in $\mathrm{r}\mathcal{G}(A)$. Thus, if every $\mathrm{r}\mathcal{G}(A)$ is a torsion group, then implication (35.3.1) holds. Moreover, if every $\mathrm{r}\mathcal{G}(A)$ has finite exponent dividing e, then implication (35.3.1) holds with exponent e.

Conversely, suppose that implication (35.3.1) holds. Take $B \in \mathrm{rgen}(A)$ for some A and B. Then $A \oplus \Gamma \cong B \oplus \Gamma$, by Theorem 35.2, so that, by implication (35.3.1), we have $A^{(e)} \cong B^{(e)}$ for some e. Therefore $e{\cdot}[B] = 0$ in $\mathrm{r}\mathcal{G}(A)$; that is, every $\mathrm{r}\mathcal{G}(A)$ is a torsion group. Similarly, if implication (35.3.1) holds with fixed e, then every $\mathrm{r}\mathcal{G}(A)$ has finite exponent dividing e.

Now consider $\mathrm{s}\mathcal{G} = \mathcal{G}(M)$. Recall that $\mathrm{rs}\mathcal{G} = \mathrm{r}\mathcal{G}(M)$, and $\mathrm{rs}\mathcal{G}$ maps onto $\mathrm{r}\mathcal{G}(A)$ for every $A \in \mathrm{fingen}(\Lambda)$ [Theorem 28.3]. Thus, $\mathrm{r}\mathcal{G}(M)$ is a torsion group if and only if $\mathrm{r}\mathcal{G}(A)$ is a torsion group for every $A \in \mathrm{fingen}(\Lambda)$. Similarly, $\mathrm{r}\mathcal{G}(M)$ has finite exponent dividing the integer e if and only the same is true for every $\mathrm{r}\mathcal{G}(A)$, and this explains the minimality property of the exponent of $\mathrm{r}\mathcal{G}(M)$ claimed in the theorem.

For the supplementary statement, let Λ be the Dedekind-like ring in Example 33.3. Since Λ has no unsplit maximal ideals, we have $\mathrm{s}\mathcal{G} = \mathcal{G}(\Lambda)$ [Theorem 27.10]. Moreover, since Γ is a principal ideal ring, we have $\mathrm{r}\mathcal{G}(\Lambda) = \mathcal{G}(\Lambda)$, where $\mathcal{G}(\Lambda) = k^\times$, as was also proved in Example 33.3. Thus, the three situations in the supplementary statement occur if we take k to be, respectively, a finite field, the algebraic closure of a finite field, and the rational numbers. □

We remark that there is a missing ingredient, with regard to the foregoing theorem: We do not have an example showing that $\mathrm{rs}\mathcal{G} = \mathrm{r}\mathcal{G}(M)$ cannot be replaced by $\mathrm{r}\mathcal{G}(\Lambda)$ in the statement of the theorem. [Problem 8 at the end of this memoir]

The next corollary relates the super genus class group to direct-sum cancellation, and sharpens Theorem 35.2.

COROLLARY 35.4. *Suppose that* ${}_\Lambda\Gamma$ *is finitely generated, and let* $M \in \mathrm{fingen}_\infty(\Lambda)$ *be such that* $\mathrm{s}\mathcal{G} = \mathcal{G}(M)$. *Then the following statements are equivalent.*

(i) *Direct-sum cancellation holds for finitely generated* Λ*-modules.*
(ii) $M \oplus \Gamma \cong M' \oplus \Gamma \implies M \cong M'$ *for all* $M' \in \mathrm{rgen}(M)$.
(iii) $\mathrm{r}\mathcal{G}(M) = 0$.

PROOF. (i) $\implies$ (ii) is obvious.

(ii) $\implies$ (iii) Take $M' \in \mathrm{rgen}(M)$. Then $M \oplus \Gamma \cong M' \oplus \Gamma$ by (i) $\implies$ (iii) of Theorem 35.2. But then (ii) of the present corollary implies that $M \cong M'$, that is, (iii) holds.

(iii) $\implies$ (i) follows from the case $e = 1$ of Theorem 35.3. □

The final application of our theory to direct-sum behavior displays an interesting difference between cancellation of torsionfree modules and more general cancellation. Levy and Wiegand proved the positive part of the next result for Bass rings [**LW**, Theorem 6.2], and Levy removed the torsionfree restriction if Λ is Dedekind-like with finite normalization and no unsplit maximal ideals [**L2**, Theorem 13.5].

THEOREM 35.5 (Cancellation of Λ). *If* Λ *is a Dedekind-like ring (with or without finite normalization) and* $A, B \in \mathrm{fingen}(\Lambda)$ *are torsionfree, then*

$$A \oplus \Lambda \cong B \oplus \Lambda \implies A \cong B \tag{35.5.1}$$

This implication can fail if A *and* B *are not torsionfree.*

PROOF. Suppose that $A \oplus \Lambda \cong B \oplus \Lambda$. Then $\mathrm{rgen}(A) = \mathrm{rgen}(B)$ [Theorem 35.2], so that $\mathcal{C}(A) = \mathcal{C}(B)$ and $B \cong A^{\boldsymbol{u}}$ for some $\boldsymbol{u} \in \mathcal{I}_{\mathcal{C}(A)}$ [Theorem 22.6]. Theorem 22.7 then shows that $A \oplus \Lambda \cong B \oplus \Lambda \cong A^{\boldsymbol{u}} \oplus \Lambda \cong (A \oplus \Lambda)^{\boldsymbol{u}}$; that is, $\boldsymbol{u} \in \mathrm{istab}(A \oplus \Lambda)$.

Consider the case that A is faithful. Since $\mathrm{genus}(A) = \mathrm{genus}(B)$, B is also faithful, and hence $\mathcal{C}(A) = \mathcal{C}(B) = \mathcal{C}(\Lambda)$. Also, every completion $\hat{A}_{\mathfrak{m}}$ and $\hat{B}_{\mathfrak{m}}$ is a faithful $\hat{\Lambda}_{\mathfrak{m}}$-module. By Theorem 24.5, we have that $\mathrm{eqstab}(\hat{A}_{\mathfrak{m}}) \supseteq k^\times = \mathrm{eqstab}(\hat{\Lambda}_{\mathfrak{m}})$ for every $\mathfrak{m}$. Recall that $\mathrm{eqstab}(M)(\mathfrak{m}) = \mathrm{eqstab}(\hat{M}_{\mathfrak{m}})$ for every M and $\mathfrak{m}$ [Lemma 24.3]. Therefore $\mathrm{eqstab}(A)\,\mathrm{eqstab}(\Lambda) = \mathrm{eqstab}(\Lambda)$. Thus, returning to the previous paragraph and using Corollary 24.7 and Theorem 23.5, we see that

$$\begin{aligned}\boldsymbol{u} \in \operatorname{istab}(A \oplus \Lambda) &= \operatorname{eqstab}(A)\operatorname{eqstab}(\Lambda)\operatorname{im}(\Gamma^{\times})\\ &= \operatorname{eqstab}(A)\operatorname{im}(\Gamma^{\times}) = \operatorname{istab}(A)\end{aligned}$$

Therefore $B \cong A^{\boldsymbol{u}} \cong A$, as desired.

If A is not faithful, write $\Gamma = \oplus_h \Gamma_h$, as usual. Let Λ_h be the projection of Λ in Γ_h for each index h; let $\Lambda' = \oplus\{\Lambda_h \mid \Gamma_h \notin \mathcal{C}(A)\}$; and let $A' = A \oplus \Lambda'$ and $B' = B \oplus \Lambda'$. Then the first isomorphism in (35.5.1) implies that $A' \oplus \Lambda \cong B' \oplus \Lambda$, so that $A' \cong B'$ by the "faithful" case of the theorem, and hence $A \cong B$ by Corollary 26.6.

For the supplementary statement, choose Λ with finite normalization and M such that $\mathrm{s}\mathcal{G} = \mathcal{G}(M)$; and assume that the surjective map $\xi^{M,\Lambda}$ is not injective. (For example, Λ and M can be as in Example 34.1(i).) Then we can choose a nonzero element $[B] \in \ker(\xi^{M,\Lambda})$. Thus, $M \not\cong B$, but $M \oplus \Lambda \cong B \oplus \Lambda$, by definition of $\xi^{M,\Lambda}$. (Note that, by implication (35.5.1), M cannot be not torsionfree!) □

Remark. Theorem 35.5 is related to the following more general question: If direct-sum cancellation holds in the category of finitely generated torsionfree modules over a reduced commutative noetherian ring Ω of Krull dimension 1, does direct-sum cancellation hold in fingen(Ω)? Very recently, Hassler and Wiegand have shown the answer to be "no", even if Ω is Dedekind-like [**HW**].

36. Unsplit Mayer-Vietoris; Example: $\mathcal{G}(\Lambda) \neq \mathrm{r}\mathcal{G}(\Lambda) \neq 0$; Cancellation in $\mathbb{Z}[\sqrt{n}\,]$

Our interest in the next example was motivated by the fact that the theory of direct-sum and local-global behavior in $\mathrm{fingen}_{\infty}(\Lambda)$ would be vastly simpler than what is in the memoir, if all Mayer-Vietoris sequences over Dedekind-like rings were split. A byproduct is our only explicit computation of groups $\mathcal{G}(\Lambda) \neq \mathrm{r}\mathcal{G}(\Lambda)$ (finite, in this case).

EXAMPLE 36.1 (Unsplit Mayer-Vietoris Sequence). Let $R = \mathbb{Z}[\sqrt{-5}\,]$, which is known to be a Dedekind domain with class number 2 (see, for example, Table 4 on page 425 of [**BS**]), and whose unit group is just $R^{\times} = \{\pm 1\}$. Clearly the rational prime 5 is ramified in R, since $(5) = (\sqrt{-5}\,)^2$, and $\mathfrak{n} = (\sqrt{-5}\,)$ is a maximal ideal of R. Let $\Gamma = R \oplus R$, and let ρ_1 and ρ_2 be the natural map of the first and second R, respectively, onto the residue field $k = R/\mathfrak{n}$. Also — as in Construction 33.1 — let $\tilde{\rho}_i$ $(i = 1, 2)$ be the extension of ρ_i to a map $\Gamma \twoheadrightarrow k$ obtained by letting ρ_i be zero on the coordinate ring of Γ on which it is not already defined.

The ring $\Lambda = \{\gamma \in \Gamma \mid \tilde{\rho}_1(\gamma) = \tilde{\rho}_2(\gamma)\}$ is a Dedekind-like ring with finite normalization Γ and exactly one singular maximal ideal $\mathfrak{m} = \ker(\tilde{\rho}_1) \cap \ker(\tilde{\rho}_2)$. Moreover, $\mathfrak{m}$ is strictly split. We can view Λ as being defined by the following conductor square. [See Proposition 33.2, but note that we can omit the "$\cap\,\Lambda$" part of the formula given there because, by (36.1.1), we have $\ker(\tilde{\rho}_1) \cap \ker(\tilde{\rho}_2) = \ker(\rho) \subseteq \Lambda$.]

(36.1.1)
$$\begin{array}{ccc} \Lambda & \subset & \Gamma = R \oplus R \\ \big\downarrow{\scriptstyle \rho} & & \big\downarrow{\scriptstyle \rho = (\rho_1, \rho_2)} \\ k & \subset & \bar{\Gamma} = k \oplus k \end{array}$$

The inclusion in the bottom row identifies the field k in the lower left with the diagonal of $\bar{\Gamma}$.

We claim that *the Mayer-Vietoris sequence* (25.14.1) *for* $\mathcal{G}(\Lambda)$ *is nonsplit.* The class group $\mathcal{G}_\Gamma(\Gamma) \cong \mathcal{G}_R(R) \times \mathcal{G}_R(R)$ is noncyclic of order 4 because, as already noted, the class group $\mathcal{G}_R(R)$ is cyclic of order 2. Using Lemma 31.1, one easily computes that

$$\mathrm{r}\mathcal{G}(\Lambda) \cong \frac{k^\times \times k^\times}{\big(\mathrm{diag}(k^\times)\big)\big(\{\pm 1\} \times \{\pm 1\}\big)} \cong \frac{k^\times}{\{\pm 1\}}$$

is cyclic of order 2 (since $k = R/\mathfrak{n}$ is isomorphic to the ring of integers modulo 5). Therefore, to prove that the Mayer-Vietoris sequence (25.14.1) for $\mathcal{G}(\Lambda)$ is nonsplit, it suffices to find a module $M \in \mathrm{genus}(\Lambda)$ such that $2 \cdot [M] \neq 0$ in $\mathcal{G}(\Lambda)$.

One easily checks that $I = (3, 1 + 2\sqrt{-5}\,)$ is a nonprincipal ideal of R, by computing that $(3, 1 + 2\sqrt{-5}\,)\cdot(3, 1 - 2\sqrt{-5}\,) = (3)$, so that the norm of I is 3, while no element of R has norm 3. It is also a straightforward computation that $I^2 = (2 + \sqrt{-5}\,)$ is principal.

Clearly I is relatively prime to $\mathfrak{n}$ in R, so the restriction to I of the natural surjection $R \twoheadrightarrow k = R/\mathfrak{n}$ is also a surjection $I \twoheadrightarrow k$. Therefore $\rho = (\rho_1, \rho_2)$ is a surjection from $R \oplus I$ onto $\bar{\Gamma}$. Let M (a Λ-module) be the pullback of the following conductor square. M is an ideal of Λ because the maps ρ_1 and ρ_2 defining it are restrictions of the maps in (36.1.1) defining Λ.

(36.1.2)
$$\begin{array}{ccc} M & \subset & R \oplus I \\ \big\downarrow \rho & & \big\downarrow \rho = (\rho_1, \rho_2) \\ \bar{\Lambda} & \subset & \bar{\Gamma} \end{array}$$

Note that M is in genus(Λ) because square (36.1.1) becomes equal to square (36.1.2) when localized at the singular maximal ideal $\mathfrak{m}$, while M becomes isomorphic to Γ at any maximal ideal outside singspec(Λ).

It now suffices to prove that $2\cdot[M] \neq 0$ in $\mathcal{G}(\Lambda)$. We have $[M] + [M] = [M^2]$ in $\mathcal{G}(\Lambda)$, because $M \oplus M \cong M^2 \oplus \Lambda$ [**R**, Theorem 38.13], where M^2 denotes the square of M in the ring Λ. Therefore, it suffices to show that $M^2 \not\cong \Lambda$.

Let P be the pullback of the following square.

(36.1.3)
$$\begin{array}{ccc} P & \subset & R \oplus I^2 \\ \big\downarrow \rho & & \big\downarrow \rho = (\rho_1, \rho_2) \\ \bar{\Lambda} & \subset & \bar{\Gamma} \end{array}$$

We claim that

(36.1.4) $$\Gamma M^2 = \Gamma P \qquad \text{and} \qquad M^2 \subseteq P \subseteq \Lambda$$

From (36.1.2) we have $M = \{(a, b) \in R \oplus I \mid \rho(a, b) \in \bar{\Lambda}\}$. Therefore M^2 is generated by the pairs $(ac, bd) = (a, b)(c, d)$ with $(a, b), (c, d) \in M$. Since the projection maps $M \to R$ and $M \to I$ are surjections, it follows that the projection maps $M^2 \to R$ and $M^2 \to I^2$ are surjections. In other words, $\Gamma M = R \oplus I^2 = \Gamma P$ as claimed. Moreover, since ρ is a ring homomorphism, we have $\rho(ac, bd) = \rho(a, b)\cdot\rho(c, d) \in \bar{\Lambda}$, which shows that $M^2 \subseteq P$. We have $P \subseteq \Lambda$ for the same reason that M is an ideal of Λ, completing the proof of (36.1.4).

Next, we claim that

$$M^2, P \in \text{genus}(\Lambda) \tag{36.1.5}$$

The proof that $P \in \text{genus}(\Lambda)$ is the same as the proof that $M \in \text{genus}(\Lambda)$. Since $M \in \text{genus}(\Lambda)$, M is an invertible ideal of Λ, and hence so is M^2. Therefore $M^2 \in \text{genus}(\Lambda)$.

Finally, we claim that

$$M^2 = P \tag{36.1.6}$$

Since $M^2 \subseteq P$, the claim can be verified locally (in P). Therefore, we may assume that Λ is local. From (36.1.5), we get $M^2 \cong P \cong \Lambda$. This, together with the inclusions (36.1.4), shows that $P = \Lambda p$ and $M^2 = \Lambda px$ for regular elements $p, x \in \Lambda$. Then $\Gamma M^2 = \Gamma P$ shows that $\Gamma px = \Gamma p$, and regularity of p allows it to be canceled, so that $x \in \Gamma^\times$. Hence $x \in \Gamma^\times \cap \Lambda$. But $\Gamma^\times \cap \Lambda = \Lambda^\times$, as can be verified from pullback diagram (36.1.1) defining Λ (or from the fact that Γ is integral over Λ). Therefore $M^2 = \Lambda px = \Lambda p = P$, proving (36.1.6). Thus, the pullback of square (36.1.3) is M^2.

Now we can show that $M^2 \not\cong \Lambda$. Suppose, to the contrary, that $M^2 \cong \Lambda$, so that $M^2 = (x)$ for some $x \in \Lambda$. Then as M^2 is the pullback of square (36.1.3), we can write $x = (x_1, x_2)$, where $R = (x_1)$ and $I^2 = (x_2)$, because M^2 projects onto both coordinates. Now the only units of R are ± 1, and $I^2 = (2 + \sqrt{-5}\,)$, so it follows that $x_1 = \pm 1$ and $x_2 = \pm(2 + \sqrt{-5}\,)$. But $x \in M^2$ implies $\rho(x) \in \bar{\Lambda}$ in (36.1.3); that is, $\rho_1(x_1) = \rho_2(x_2)$ in $k = R/\mathfrak{n} = \mathbb{Z}/(5)$. So in the residue field $\mathbb{Z}/(5)$, we get that $\pm\bar{1} = \rho_1(x_1) = \rho_2(x_2) = \pm\bar{2}$, contradiction. Thus, we conclude that $M^2 \not\cong \Lambda$, as claimed.

EXAMPLE 36.2 ($\mathcal{G}(\Lambda) \neq \mathrm{r}\mathcal{G}(\Lambda) \neq 0$). Let Λ be as in Example 36.1. Then:

$$\mathcal{G}(\Lambda) \cong C_2 \times C_4 \quad \text{and} \quad \mathrm{r}\mathcal{G}(\Lambda) \cong C_2 \tag{36.2.1}$$

where C_n denotes a cyclic group of order n.

This holds because the Mayer-Vietoris sequence $\mathrm{r}\mathcal{G}(\Lambda) \hookrightarrow \mathcal{G}(\Lambda) \twoheadrightarrow \mathcal{G}_\Gamma(\Gamma)$ is short exact and nonsplit, with $\mathrm{r}\mathcal{G}(\Lambda) \cong C_2$ and $\mathcal{G}_\Gamma(\Gamma) \cong C_2 \times C_2$.

Additional observation. If the above construction is carried out with $\Gamma = \mathbb{Z} \oplus R$ instead of $R \oplus R$, then one gets $\mathcal{G}(\Lambda) \cong C_4$.

For then both $\mathcal{G}(\Gamma)$ and $\mathrm{r}\mathcal{G}(\Lambda)$ are cyclic of order 2, and the module M above, suitably modified, still yields an element of order 4 in $\mathcal{G}(\Lambda)$.

EXAMPLE 36.3 (Cancellation in $\mathbb{Z}[\sqrt{n}\,]$). *The quadratic order $\Lambda = \mathbb{Z}[\sqrt{n}\,]$ is Dedekind-like for every squarefree integer n and is a Dedekind domain if and only if $n \equiv 2, 3 \mod 8$.* In more detail, the localization Λ_p is a semi-local PID for every odd prime p. On the other hand, the localization Λ_2 is a semi-local PID if $n \equiv 2, 3 \mod 4$. But Λ_2 is a (nonstrictly) split or an unsplit Dedekind-like ring if $n \equiv 1 \mod 8$ or $n \equiv 5 \mod 8$, respectively. (The details of the conductor square structure of $\hat{\mathbb{Z}}_2[(1+\sqrt{n}\,)/2]$ are stated and proved in [**KL2**, Example 12.4], and the corresponding details for and $\mathbb{Z}_2[(1+\sqrt{n}\,)/2]$ are interspersed in the proofs below.) Since Λ is noetherian, it is therefore Dedekind-like.

Therefore, we consider the two cases $n \equiv 1, 5 \mod 8$, either determining $\mathrm{r}\mathcal{G}(\Lambda)$ and when direct-sum cancellation holds, or reducing these question to properties of units in the normalization $\Gamma = \mathbb{Z}[(1 + \sqrt{n}\,)/2]$ of Λ. If $n \equiv 1, 5 \mod 8$, then $\mathrm{singspec}(\mathbb{Z}[\sqrt{n}\,])$ consists of a single maximal ideal $\mathfrak{m}$ (lying over 2), nonstrictly split if $n \equiv 1 \mod 8$, and unsplit if $n \equiv 5 \mod 8$. By Lemma 31.1, we have $\mathrm{r}\mathcal{G}(\Lambda) \cong \bar{\Gamma}^\times/(k^\times \cdot \operatorname{im} \Gamma^\times)$, where $\bar{\Gamma} = \Gamma/\mathfrak{m}$ and $k = \Lambda/\mathfrak{m}$. Before considering these two cases separately, we note that *direct-sum cancellation holds in* fingen(Λ) *if and only if* $\mathrm{r}\mathcal{G}(\Lambda)$ *is trivial.* (Since the residue fields of Λ are finite, we have $\mathrm{s}\mathcal{G} = \mathcal{G}(\Lambda)$ [Corollary 27.11], so the statement is part of Corollary 35.4.)

First suppose that $n \equiv 1 \mod 8$, so that the singular maximal ideal is split. Then $\bar{\Gamma} = k \oplus k$, where $k = \mathbb{Z}/2\mathbb{Z}$ is a field of order 2, and hence $\bar{\Gamma}^\times$ is the trivial group. We conclude that $\mathrm{r}\mathcal{G}(\Lambda) = 0$ *and cancellation holds.*

Suppose instead that $n \equiv 5 \mod 8$, so that the singular maximal ideal is unsplit. Then $\bar{\Gamma}$ is a quadratic field extension of the field $k = \mathbb{Z}/2\mathbb{Z}$, and hence $\bar{\Gamma}^\times$ is a group of order 3. Therefore, $\mathrm{r}\mathcal{G}(\Lambda)$ is either trivial or cyclic of order 3, depending on the existence of enough units in Γ. In more detail, there are three cases:

(i) If $n < -3$, then Γ has only the units ± 1, in which case $\Gamma^\times$ does not map onto $\bar{\Gamma}^\times$. Thus in this case, $\mathrm{r}\mathcal{G}(\Lambda)$ *is cyclic of order 3, and cancellation does not hold.*

(ii) If $n = -3$, then Γ contains the unit $(-1+\sqrt{-3}\,)/2$, which does map to a generator of $\bar{\Gamma}^\times$. Thus in this case, $\mathrm{r}\mathcal{G}(\Lambda) = 0$, *and cancellation holds.*

(iii) If $n > 0$, then Γ has infinitely many units, and in some cases $\Gamma^\times$ maps onto $\bar{\Gamma}^\times$, while in others it does not. (See, for example, Table 1 on page 422 of [**BS**].) *In the first situation,* $\mathrm{r}\mathcal{G}(\Lambda) = 0$, *and cancellation holds. In the second situation,* $\mathrm{r}\mathcal{G}(\Lambda)$ *is cyclic of order 3, and cancellation fails.*

APPENDIX A

37. Open Problems

1. Are the possible characteristic 2 exceptions to our tameness results [see Additional Hypothesis 10.2] genuine exceptions?

2. Is the tame-wild dichotomy in [**KL1, KL2**] really a dichotomy, in the sense that no complete local Λ can be both tame and wild?

Many of our results are incomplete when Λ has infinitely many unsplit maximal ideals with nonsurjective norm maps. [See the next two problems.]

3. Suppose that Λ has infinitely many unsplit maximal ideals with nonsurjective norm map. Is there an identification $\mathrm{s}\mathcal{G} = \mathcal{G}(M)$ (via $\xi^{\mathrm{s}\mathcal{G},M}$) for some M? [Compare with the situations in Theorems 27.10–27.12]. If so, Λ must not have too many units [Theorem 23.5 and Theorem 25.14].

4. Consider the super Mayer-Vietoris sequence (27.6.1). If Λ has infinitely many unsplit maximal ideals with nonsurjective norm maps, what is the relationship between between the group $\mathcal{I}$ of residue unit idèles and the restricted super genus class group $\mathrm{rs}\mathcal{G}$? If $\mathrm{singspec}(\Lambda)$ has only countably many such unsplit maximal ideals, we know that $\mathcal{I}$ maps naturally onto $\mathrm{rs}\mathcal{G}$ [Theorem 27.9]. In the uncountable case, we do not know even this.

5. Does the inverse-directed structure for genus class groups [§27] hold for rings more general than Dedekind-like rings?

6. Is it possible to choose a base point in every genus in such a way that the set of base points is closed under direct sums (up to isomorphism)? (If so, then the "correction term" $[M_0]$ in (29.1.1) can always be chosen to be 0.)

7. What happens to results in Chapter 8, "Finite Normalization", when Λ does not have finite normalization? For example: Is $\xi^{M,\Gamma}$ still defined for all faithful M (and equal to $\xi^{M,\Gamma}_{\Lambda,\Gamma}$) [Proposition 32.3]? Is ${}_{\Lambda}\Gamma$ still a test module for cancellation in $\mathrm{fingen}(\Lambda)$ [Theorem 35.2]? What about our results on power cancellation? [Unfortunately we know too few examples of Dedekind-like rings without finite normalization.]

8. Suppose that ${}_{\Lambda}\Gamma$ is finitely generated and hence $\mathrm{s}\mathcal{G} = \mathcal{G}(M)$ for some M. Then power cancellation holds in $\mathrm{fingen}(\Lambda)$ if and only if $\mathrm{r}\mathcal{G}(M)$ has finite exponent [Theorem 35.3]. Is this equivalent to requiring that $\mathrm{r}\mathcal{G}(\Lambda)$ has finite exponent?

9. See the "local versus global summands" problem following the proof of Theorem 30.12.

APPENDIX B

38. Terminology Index

39. Notation Index

Item	In/near Section
Λ (Dedekind-like ring)	3.1, 10.1
Γ (normalization of Λ or Ω)	3.1
Γ_h (coordinate ring of Γ)	(3.1.1)
$\hookrightarrow$, $\twoheadrightarrow$ (injection, surjection)	3.1
$\mathcal{H}$ (index set, decomposition of Γ)	(3.1.1)
fingen	3.1
finlen	3.1
fingen_∞	3.1
maxspec	3.1
minspec	3.1
singspec	3.1
maxsupp_Λ	3.1
$\bar{X}(\mathfrak{m})$ $(= X/\mathfrak{m}X)$	3.1
$\rho^{\mathfrak{m}}$ (nat. hom. $X \twoheadrightarrow \bar{X}(\mathfrak{m})$)	3.1
genus(M)	(2.2.1)
$M_{\mathfrak{m}}$ ($\mathfrak{m}$-localization)	3.1
$\hat{M}_{\mathfrak{m}}$, $(\ldots)\hat{}_{\mathfrak{m}}$ ($\mathfrak{m}$-adic completion)	3.1
(a$\forall\mathfrak{m}$) (for almost all $\mathfrak{m}$)	3.1
$(\ldots)^\times$ units group of $(\ldots)$	3.1
M_Q (Q-localization)	4.2
$P(\mathfrak{m})$ (primary component)	(6.1.1)
rad$(\ldots)$ (Jacobson radical)	10.1
μ_Λ (min. number of generators)	10.8
$k(\mathfrak{m})$ (residue field)	(11.2.2)
$F(\mathfrak{m})$ (residue field)	(11.5.1)
$N(\Omega)$ (nilradical)	13.1
$\mathcal{K}(M)$ (connections graph)	16.5
$(\Lambda \subseteq \Gamma)$-inclusion	17.1
$\Omega(\mathcal{F})$ (module-finite overring)	18.1
rgen (restricted genus)	19.1
$t(X)$ (torsion submodule)	19.2
X_∞	(21.2.1)
$\det(\psi_\infty)$	(21.2.2)
$\psi = \psi_\infty \oplus \psi_t$ (endo-id‘ele)	(21.2.3)
S^ψ (residue endo-idèle action)	21.4
$\mathcal{C}(M)$ (torsionfree support set)	(22.1.1)
$\Gamma_{\mathcal{C}(M)}$	22.1
$\mathcal{I}$ (group of residue unit idèles)	(22.1.2)
$\mathcal{I}_{\mathcal{C}(M)}$	(22.1.3)
M^u (residue unit idèle action)	22.4
eqstab (equality stabilizer)	23.1
istab (isomorphism stabilizer)	23.1
N_k^F (norm map)	24.4
$\mathcal{G}(M)$ (genus class group)	25.1
r$\mathcal{G}(M)$ (restricted genus cl. gp.)	25.1
$\xi^{M,N}$ (loss-of-structure map)	(25.4.1)
$\xi^{M,X}_{\Lambda,\Gamma}$ (loss-of-structure map)	(25.4.2)
s$\mathcal{G}$ (super genus)	27.5
rs$\mathcal{G}$ (restricted super gen. cl. gp)	27.5
s$\mathcal{G} = \mathcal{G}(M)$ (identification)	(27.5.1)
$\xi^{\text{s}\mathcal{G},N}$ (loss-of-structure map)	27.8
$\psi^{\text{s}\mathcal{G},N}$	28.2
$e(\mathcal{D})$ (idempotent)	30.3

Bibliography

[AM] M. F. Atiyah and I. G. MacDonald, *Introduction to Commutative Algebra,* Addison-Wesley, Reading 1969.

[BS] Borevitch and Shafarevich, *Number Theory,* Academic Press, New York 1966.

[C] L. Claborn, "Every abelian group is a class group," *Pac. J. Math.*, **18** (1966), 219–222.

[EE] D. Eisenbud and E. G. Evans, "Generating modules efficiently: theorems from algebraic K-theory," *J. Algebra* **27** (1973), pp. 278–305.

[E] E. G. Evans, Jr., "Krull-Schmidt and cancellation over local rings," *Pac. J. Math.*, **46** (1973), 115–121.

[F] A. Facchini, *Module Theory: Endomorphism Rings and Direct Sum Decompositions in some classes of Modules,* Birkhaüser, Boston, 1975.

[G] O. Goldman, "Determinants in projective modules," *Nagoya Math. J.*, **18** (1961), 27–36.

[Go] K. R. Goodearl, "Power cancellation of groups and modules," *Pacific J. Math.*, **64** (1976), 387–411.

[G1] R. M. Guralnick, "The genus of a module," *J. Number Theory,* **18** (1984), 169–177.

[G2] R. M. Guralnick, "Power cancellation of modules," *Pacific J. Math.*, **124** (1986), 131–144.

[G3] R. M. Guralnick, "The genus of a module II. Roiter's Theorem, power cancellation and extension of scalars," *J. Number Theory,* **26** (1987), 149–165.

[GL1] R. M. Guralnick and L. S. Levy, "Presentations of modules when ideals need not beprincipal," *Ill. J. Math.*, **32** (1988), 593–653.

[GL2] R. M. Guralnick, and L. S. Levy,"Cancellation and direct summands in dimension 1," *J. Alg.*, **142** (1991), 310–347.

[H] T. W. Hungerford, "On the structure of principal ideal rings," *Pacific J. Math.* **25** (1968), 543–547.

[HL] W. J. Heinzer and L. S. Levy, "Domains of dimension 1 with infinitely many singular maximal ideals," *Rocky Mountain J. Math.* (to appear).

[HW] W. Hassler and R. Wiegand, "Direct Sum Cancellation for Modules over One-Dimensional Rings," *J. Algebra* (to appear).

[J] H. Jacobinski, "Genera and decompositions of lattices over orders," *Acta Math.*, **121** (1968), 1–29.

[Ja] G. J. Janusz, *Algebraic Number Fields,* Academic Press, New York 1973.

[KL0] L. Klingler and L. S. Levy, "Direct-sum cancellation: modules versus lattices," *Comm. Alg.*, **18** (1990), 1857–1868.

[KL1] L. Klingler and L. S. Levy, "Representation type of commutative noetherian rings I: local wildness," *Pacific J. Math.*, **200** (2001), 345–386.

[KL2] L. Klingler and L. S. Levy, "Representation type of commutative noetherian rings II: local tameness," *Pacific J. Math.*, **200** (2001), 387–483.

[L] L. S. Levy, "Modules over pullbacks and subdirect sums," *J. Alg.* **71** (1981), 50–61.

[L1] L. S. Levy, "Krull-Schmidt uniqueness fails dramatically over subrings of $\mathbf{Z} \oplus \cdots \oplus \mathbf{Z}$," *Rocky Mountain J. Math.* **13** (1983), 659–678.

[L2] L. S. Levy, "Modules over Dedekind-like rings," *J. Alg.*, **93** (1985), 1–116.

[L3] L. S. Levy, "$\mathbb{Z}G_n$-modules, G_n cyclic of square-free order n," *J. Alg.*, **93** (1985), 354–375.

[LO] L. S. Levy and C. J. Odenthal, "Package deal theorems and splitting orders, in dimension 1," *Trans. Amer. Math. Soc.*, **348** (1996), 3457–3503.

[LW] L. S. Levy and R. Wiegand, "Dedekind-like behavior of rings with 2-generated ideals," *J. Pure Appl. Alg.*, **37** (1985), 41–58.

[LT] L. S. Levy and Jan Trlifaj, "Γ-Separated Covers" (in preparation).

[LN] R. Lidl and H. Niederreiter, *Finite Fields,* 2nd Ed (1997), *Encyclopedia of Mathematics and its Applications* **20** Cambridge University Press, Cambridge (1997).

[M] H. Matsumura, *Commutative Ring Theory,* Cambridge Studies in Advanced Mathematics, Vol. 8, Cambridge University Press, Cambridge 1980.

[N] M. Nagata, *Local Rings,* Interscience Publishers, New York, 1962.

[NR] L. A. Nazarova and A. V. Roiter, "Finitely generated modules over a dyad of two local rings and finite groups with an abelian normal divisor of Index p," *Zap. Nauch. Sem. Leningrad Odtel. Mat. Inst. Steklov (LOM1) Izv. Akad. Nauk SSSR,* Ser Mat. **33**, No. 1 (1969). English transl: *Math. USSR Izvestija* **3**, No. 1 (1969).

[NRSB] L. A. Nazarova, A. V. Roiter, V. V. Sergeichuk, and V. M. Bondarenko, "Application of modules over a dyad for the classification of finite p-groups that have an abelian subgroup of index p and of pairs of mutually annihilating operators" (Russian), *Zap. Nauch. Sem. Leningrad Odtel. Mat. Inst. Steklov (LOM1)* **28** (1972), 69–92. English transl: *J. Soviet Math.* **3** (1975), 636–653.

[R] I. Reiner, *Maximal Orders,* Academic Press, New York 1975.

[Ri] K. M. Ringel, "The representation type of local algebras," *Springer Lecture Notes in Mathematics* **488** (1975), 282–305.

[S] E. Steinitz, "Rechteckige Systeme und Moduln in Algebraischer Zahlkörper I,II," *Math. Ann.,* **71** (1911), 328–354, **72** (1912), 297–345.

[We] E. Weiss, *Algebraic Number Theory,* Chelsea Publishing Company, New York 1963.

[W1] R. Wiegand, "Cancellation over commutative rings of dimension one and two," *J. Alg.,* **88** (1984), 438–459.

[W2] R. Wiegand, "Failure of Krull-Schmidt for direct sums of copies of a module," *Advances in commutative ring theory (Fez, 1997)*, 541–547, Lecture Notes in Pure and Appl. Math. **205**, Dekker, New York, 1999.

[WW] R. Wiegand and S. Wiegand, "Stable isomorphism of modules over one-dimensional rings," *J. Alg.,* **107** (1987), 425–435.

Editorial Information

To be published in the *Memoirs*, a paper must be correct, new, nontrivial, and significant. Further, it must be well written and of interest to a substantial number of mathematicians. Piecemeal results, such as an inconclusive step toward an unproved major theorem or a minor variation on a known result, are in general not acceptable for publication. Papers appearing in *Memoirs* are generally longer than those appearing in *Transactions*, which shares the same editorial committee.

As of March 31, 2005, the backlog for this journal was approximately 7 volumes. This estimate is the result of dividing the number of manuscripts for this journal in the Providence office that have not yet gone to the printer on the above date by the average number of monographs per volume over the previous twelve months, reduced by the number of volumes published in four months (the time necessary for preparing a volume for the printer). (There are 6 volumes per year, each containing at least 4 numbers.)

A Consent to Publish and Copyright Agreement is required before a paper will be published in the *Memoirs*. After a paper is accepted for publication, the Providence office will send a Consent to Publish and Copyright Agreement to all authors of the paper. By submitting a paper to the *Memoirs*, authors certify that the results have not been submitted to nor are they under consideration for publication by another journal, conference proceedings, or similar publication.

Information for Authors

Memoirs are printed from camera copy fully prepared by the author. This means that the finished book will look exactly like the copy submitted.

The paper must contain a *descriptive title* and an *abstract* that summarizes the article in language suitable for workers in the general field (algebra, analysis, etc.). The *descriptive title* should be short, but informative; useless or vague phrases such as "some remarks about" or "concerning" should be avoided. The *abstract* should be at least one complete sentence, and at most 300 words. Included with the footnotes to the paper should be the 2000 *Mathematics Subject Classification* representing the primary and secondary subjects of the article. The classifications are accessible from `www.ams.org/msc/`. The list of classifications is also available in print starting with the 1999 annual index of *Mathematical Reviews*. The Mathematics Subject Classification footnote may be followed by a list of *key words and phrases* describing the subject matter of the article and taken from it. Journal abbreviations used in bibliographies are listed in the latest *Mathematical Reviews* annual index. The series abbreviations are also accessible from `www.ams.org/publications/`. To help in preparing and verifying references, the AMS offers MR Lookup, a Reference Tool for Linking, at `www.ams.org/mrlookup/`. When the manuscript is submitted, authors should supply the editor with electronic addresses if available. These will be printed after the postal address at the end of the article.

Electronically prepared manuscripts. The AMS encourages electronically prepared manuscripts, with a strong preference for $\mathcal{A}\mathcal{M}\mathcal{S}$-LaTeX. To this end, the Society has prepared $\mathcal{A}\mathcal{M}\mathcal{S}$-LaTeX author packages for each AMS publication. Author packages include instructions for preparing electronic manuscripts, the *AMS Author Handbook*, samples, and a style file that generates the particular design specifications of that publication series. Though $\mathcal{A}\mathcal{M}\mathcal{S}$-LaTeX is the highly preferred format of TeX, author packages are also available in $\mathcal{A}\mathcal{M}\mathcal{S}$-TeX.

Authors may retrieve an author package from e-MATH starting from `www.ams.org/tex/` or via FTP to `ftp.ams.org` (login as `anonymous`, enter username as password, and type `cd pub/author-info`). The *AMS Author Handbook* and the *Instruction Manual* are available in PDF format following the author packages link from `www.ams.org/tex/`. The author package can be obtained free of charge by sending email to `pub@ams.org` (Internet) or from the Publication Division, American Mathematical Society, 201 Charles St., Providence, RI 02904, USA. When requesting an author package, please specify AMS-LaTeX or AMS-TeX, Macintosh or IBM (3.5) format, and the publication in which your paper will appear. Please be sure to include your complete mailing address.

Sending electronic files. After acceptance, the source file(s) should be sent to the Providence office (this includes any TeX source file, any graphics files, and the DVI or PostScript file).

Before sending the source file, be sure you have proofread your paper carefully. The files you send must be the EXACT files used to generate the proof copy that was accepted for publication. For all publications, authors are required to send a printed copy of their paper, which exactly matches the copy approved for publication, along with any graphics that will appear in the paper.

TeX files may be submitted by email, FTP, or on diskette. The DVI file(s) and PostScript files should be submitted only by FTP or on diskette unless they are encoded properly to submit through email. (DVI files are binary and PostScript files tend to be very large.)

Electronically prepared manuscripts can be sent via email to `pub-submit@ams.org` (Internet). The subject line of the message should include the publication code to identify it as a Memoir. TeX source files, DVI files, and PostScript files can be transferred over the Internet by FTP to the Internet node `e-math.ams.org` (130.44.1.100).

Electronic graphics. Comprehensive instructions on preparing graphics are available at `www.ams.org/jourhtml/graphics.html`. A few of the major requirements are given here.

Submit files for graphics as EPS (Encapsulated PostScript) files. This includes graphics originated via a graphics application as well as scanned photographs or other computer-generated images. If this is not possible, TIFF files are acceptable as long as they can be opened in Adobe Photoshop or Illustrator. No matter what method was used to produce the graphic, it is necessary to provide a paper copy to the AMS.

Authors using graphics packages for the creation of electronic art should also avoid the use of any lines thinner than 0.5 points in width. Many graphics packages allow the user to specify a "hairline" for a very thin line. Hairlines often look acceptable when proofed on a typical laser printer. However, when produced on a high-resolution laser imagesetter, hairlines become nearly invisible and will be lost entirely in the final printing process.

Screens should be set to values between 15% and 85%. Screens which fall outside of this range are too light or too dark to print correctly. Variations of screens within a graphic should be no less than 10%.

Inquiries. Any inquiries concerning a paper that has been accepted for publication should be sent directly to the Electronic Prepress Department, American Mathematical Society, 201 Charles St., Providence, RI 02904, USA.

Editors

This journal is designed particularly for long research papers, normally at least 80 pages in length, and groups of cognate papers in pure and applied mathematics. Papers intended for publication in the *Memoirs* should be addressed to one of the following editors. In principle the Memoirs welcomes electronic submissions, and some of the editors, those whose names appear below with an asterisk (*), have indicated that they prefer them. However, editors reserve the right to request hard copies after papers have been submitted electronically. Authors are advised to make preliminary email inquiries to editors about whether they are likely to be able to handle submissions in a particular electronic form.

***Algebra** to ALEXANDER KLESHCHEV, Department of Mathematics, University of Oregon, Eugene, OR 97403-1222; email: ams@noether.uoregon.edu

Algebraic geometry to DAN ABRAMOVICH, Department of Mathematics, Brown University, Box 1917, Providence, RI 02912; email: amsedit@math.brown.edu

***Algebraic number theory** to V. KUMAR MURTY, Department of Mathematics, University of Toronto, 100 St. George Street, Toronto, ON M5S 1A1, Canada; email: murty@math.toronto.edu

***Algebraic topology** to ALEJANDRO ADEM, Department of Mathematics, University of British Columbia, Room 121, 1984 Mathematics Road, Vancouver, British Columbia, Canada V6T 1Z2; email: adem@math.ubc.ca

Combinatorics and Lie theory to SERGEY FOMIN, Department of Mathematics, University of Michigan, Ann Arbor, Michigan 48109-1109; email: fomin@umich.edu

Complex analysis and harmonic analysis to ALEXANDER NAGEL, Department of Mathematics, University of Wisconsin, 480 Lincoln Drive, Madison, WI 53706-1313; email: nagel@math.wisc.edu

***Differential geometry and global analysis** to LISA C. JEFFREY, Department of Mathematics, University of Toronto, 100 St. George St., Toronto, ON Canada M5S 3G3; email: jeffrey@math.toronto.edu

Dynamical systems and ergodic theory to ROBERT F. WILLIAMS, Department of Mathematics, University of Texas, Austin, Texas 78712-1082; email: bob@math.utexas.edu

***Functional analysis and operator algebras** to MARIUS DADARLAT, Department of Mathematics, Purdue University, 150 N. University St., West Lafayette, IN 47907-2067; email: mdd@math.purdue.edu

***Geometric analysis** to TOBIAS COLDING, Courant Institute, New York University, 251 Mercer St., New York, NY 10012; email: traneditor@cims.nyu.edu

***Geometric analysis** to MLADEN BESTVINA, Department of Mathematics, University of Utah, 155 South 1400 East, JWB 233, Salt Lake City, Utah 84112-0090; email: bestvina@math.utah.edu

Harmonic analysis, representation theory, and Lie theory to ROBERT J. STANTON, Department of Mathematics, The Ohio State University, 231 West 18th Avenue, Columbus, OH 43210-1174; email: stanton@math.ohio-state.edu

***Logic** to STEFFEN LEMPP, Department of Mathematics, University of Wisconsin, 480 Lincoln Drive, Madison, Wisconsin 53706-1388; email: lempp@math.wisc.edu

Number theory to HAROLD G. DIAMOND, Department of Mathematics, University of Illinois, 1409 W. Green St., Urbana, IL 61801-2917; email: diamond@math.uiuc.edu

***Ordinary differential equations, and applied mathematics** to PETER W. BATES, Department of Mathematics, Michigan State University, East Lansing, MI 48824-1027; email: bates@math.msu.edu

***Partial differential equations** to PATRICIA E. BAUMAN, Department of Mathematics, Purdue University, West Lafayette, IN 47907-1395; email: bauman@math.purdue.edu

***Probability and statistics** to KRZYSZTOF BURDZY, Department of Mathematics, University of Washington, Box 354350, Seattle, Washington 98195-4350; email: burdzy@math.washington.edu

***Real analysis and partial differential equations** to DANIEL TATARU, Department of Mathematics, University of California, Berkeley, Berkeley, CA 94720; email: tataru@math.berkeley.edu

All other communications to the editors should be addressed to the Managing Editor, ROBERT GURALNICK, Department of Mathematics, University of Southern California, Los Angeles, CA 90089-1113; email: guralnic@math.usc.edu.

Titles in This Series

832 **Lee Klingler and Lawrence S. Levy,** Representation type of commutative Noetherian rings III: Global wildness and tameness, 2005

831 **K. R. Goodearl and F. Wehrung,** The complete dimension theory of partially ordered systems with equivalence and orthogonality, 2005

830 **Jason Fulman, Peter M. Neumann, and Cheryl E. Praeger,** A generating function approach to the enumeration of matrices in classical groups over finite fields, 2005

829 **S. G. Bobkov and B. Zegarlinski,** Entropy bounds and isoperimetry, 2005

828 **Joel Berman and Paweł M. Idziak,** Generative complexity in algebra, 2005

827 **Trevor A. Welsh,** Fermionic expressions for minimal model Virasoro characters, 2005

826 **Guy Métivier and Kevin Zumbrun,** Large viscous boundary layers for noncharacteristic nonlinear hyperbolic problems, 2005

825 **Yaozhong Hu,** Integral transformations and anticipative calculus for fractional Brownian motions, 2005

824 **Luen-Chau Li and Serge Parmentier,** On dynamical Poisson groupoids I, 2005

823 **Claus Mokler,** An analogue of a reductive algebraic monoid whose unit group is a Kac-Moody group, 2005

822 **Stefano Pigola, Marco Rigoli, and Alberto G. Setti,** Maximum principles on Riemannian manifolds and applications, 2005

821 **Nicole Bopp and Hubert Rubenthaler,** Local zeta functions attached to the minimal spherical series for a class of symmetric spaces, 2005

820 **Vadim A. Kaimanovich and Mikhail Lyubich,** Conformal and harmonic measures on laminations associated with rational maps, 2005

819 **F. Andreatta and E. Z. Goren,** Hilbert modular forms: Mod p and p-adic aspects, 2005

818 **Tom De Medts,** An algebraic structure for Moufang quadrangles, 2005

817 **Javier Fernández de Bobadilla,** Moduli spaces of polynomials in two variables, 2005

816 **Francis Clarke,** Necessary conditions in dynamic optimization, 2005

815 **Martin Bendersky and Donald M. Davis,** V_1-periodic homotopy groups of $SO(n)$, 2004

814 **Johannes Huebschmann,** Kähler spaces, nilpotent orbits, and singular reduction, 2004

813 **Jeff Groah and Blake Temple,** Shock-wave solutions of the Einstein equations with perfect fluid sources: Existence and consistency by a locally inertial Glimm scheme, 2004

812 **Richard D. Canary and Darryl McCullough,** Homotopy equivalences of 3-manifolds and deformation theory of Kleinian groups, 2004

811 **Ottmar Loos and Erhard Neher,** Locally finite root systems, 2004

810 **W. N. Everitt and L. Markus,** Infinite dimensional complex symplectic spaces, 2004

809 **J. T. Cox, D. A. Dawson, and A. Greven,** Mutually catalytic super branching random walks: Large finite systems and renormalization analysis, 2004

808 **Hagen Meltzer,** Exceptional vector bundles, tilting sheaves and tilting complexes for weighted projective lines, 2004

807 **Carlos A. Cabrelli, Christopher Heil, and Ursula M. Molter,** Self-similarity and multiwavelets in higher dimensions, 2004

806 **Spiros A. Argyros and Andreas Tolias,** Methods in the theory of hereditarily indecomposable Banach spaces, 2004

805 **Philip L. Bowers and Kenneth Stephenson,** Uniformizing dessins and Belyĭ maps via circle packing, 2004

804 **A. Yu Ol'shanskii and M. V. Sapir,** The conjugacy problem and Higman embeddings, 2004

803 **Michael Field and Matthew Nicol,** Ergodic theory of equivariant diffeomorphisms: Markov partitions and stable ergodicity, 2004